The Chemistry and Application of Dyes: An Introduction and Exploration

THE CHEMISTRY AND APPLICATION OF DYES: AN INTRODUCTION AND EXPLORATION

Dr. Rose Marie O. Mendoza

ARCLER PRESS

www.arclerpress.com

The Chemistry and Application of Dyes: An Introduction and Exploration

Dr. Rose Marie O. Mendoza

Arcler Press
224 Shoreacres Road
Burlington, ON L7L 2H2
Canada
www.arclerpress.com
Email: orders@arclereducation.com

ISBN: 978-1-77469-825-9 (Hardcover)

ABOUT THE AUTHOR

Dr. Rose obtained her PhD in Chemical Engineering from the University of the Philippines-Diliman in 2013, while her Masters in Chemical Engineering and BS Chemical Engineering degree was obtained from Adamson University. She is also a Professor in the Graduate School Department under the Master of Engineering Program at Adamson University since 2006. In 2017, she obtained her Post Doctorate Degree in Green Power: Hydrogen Generation and Fuel Cell Development from the University of California Merced, USA (in collaboration with the Energy Storage and Conversion Materials Laboratory of the University of the Philippines, Diliman). She is a visiting scientist to several universities abroad such as the University of California Merced in USA, Kindai University in Japan, National Taiwan University and Chia Nan University of Pharmacy and Science in Taiwan. She has published more than 20 research papers in water treatment and remediation, green technology, fuel cells, electrodialysis, nanomaterials, high voltage spatial electricity, batteries in peer-reviewed journals, and have more than 15 textbooks published in chemical engineering and processes. As an educator, she is very active in publishing her research to local peer reviewed journals. An advocate of Climate Action, a Climate Reality Leader Corp Member and a budding Sustainability and Circularity talent, Dr. Mendoza is also presently actively engaged in several local and international Green Federation as Board Member and Officer such as the Green Party of the Philippines, and Treasurer of the Asian Pacific Federation of Green Women's Network. She is also one of the Board of Directors of the Philippine Society of Engineering Educators, Mark Energy Revolution Corporation and one of the pioneering members of the Philippine Institute of Chemical Engineers, CaMaNaVa Chapter, and an Associate Member of the National Research Council of the Philippines (NRCP).

TABLE OF CONTENTS

List of Figures .... *xi*
List of Tables .... *xv*
List of Abbreviations .... *xvii*
Preface .... *xix*

Chapter 1 Introduction to Dye Chemistry 1
Unit Introduction 2
1.1. Historical Development of Synthetic Dyes 3
1.2. Development of Color and Constitution Theory 18
1.3. Summary 22
References 23

Chapter 2 Classification of Dyes by Chemical Structure 29
Unit Introduction 30
2.1. Azo Dyes 31
2.2. Anthraquinone Dyes 42
2.3. Benzodifuranone (BDF) Dyes 44
2.4. Polycyclic Aromatic Carbonyl Dyes 45
2.5. Indigoid Dyes 45
2.6. Polymethine and Associated Dyes 46
2.7. Styryl Dyes 48
2.8. Di and Tri-Aryl Carbonium and Related Dyes 49
2.9. Phthalocyanines 50
2.10. Quinophthalones 51
2.11. Sulfur Dyes 53
2.12. Nitro and Nitroso Dyes 53
2.13. Miscellaneous Dyes 54
2.14. Summary 55
References 56

Chapter 3	**Basic Dyes**	**61**
	Unit Introduction	62
	3.1. Properties	62
	3.2. Uses of Basic Dyes	65
	3.3. Application of Basic Dyes	65
	3.4. Vastness Properties	68
	3.5. Triphenylmethane Dyes	69
	3.6. Other Basic Dyes	87
	3.7. Organic Pigments from Basic Dyes	91
	3.8. Manifold Record Systems	94
	3.9. Summary	96
	References	97
Chapter 4	**Cationic Dyes for Synthetic Fibers**	**105**
	Unit Introduction	106
	4.1. Classification of Cationic Dyes for Synthetic Fibers	107
	4.2. Pendant Cationic Dyes	108
	4.3. Delocalized Cationic Dyes	119
	4.4. Amine Salts	134
	4.5. Trends in Cationic Dyes for Synthetic Fibers	135
	4.6. Summary	136
	References	137
Chapter 5	**Photochemistry of Dyes**	**145**
	Unit Introduction	146
	5.1. Photochemical Reactions of Dyes in Solution	146
	5.2. Photochemistry of Dyes on Substrates	176
	5.3. Summary	190
	References	191
Chapter 6	**Dyeing of Textile Fiber Blends**	**205**
	Unit Introduction	206
	6.1. Key Fiber Blends	206
	6.2. Dyeing of Textile Blends	211
	6.3. Main Dyeing Methods	218

6.4. Future Trends 221
6.5. Summary 226
References 227

Chapter 7 **Dyes for the Medical Industry 235**
Unit Introduction 236
7.1. Dyes for Indication 238
7.2. Dyes as Colorants in Medicines 240
7.3. Dyes for Therapy 241
7.4. Medical Photosensitisers 248
7.5. Potential Adverse Effects of Dyes 253
7.6. Summary 255
References 256

Chapter 8 **Non-Textile Applications of Dyes 261**
Unit Introduction 262
8.1. Dyes for Displays 263
8.2. Laser Dyes 268
8.3. Dyes For Optical Data Storage 272
8.4. Organic Photoconductors (OPCS) 277
8.5. Nonlinear Optics 282
8.6. Summary 285
References 286

Index 293

LIST OF FIGURES

Figure 1.1. Illustration of (1) alizarin; (2) indigo; (3) indoxyl; and (4) murexide.

Figure 1.2. Illustration of two dyes making up aniline purple.

Figure 1.3. Illustration of (1) Rosaniline; (2) Hofmann's violet; and (3) Girard's violet imperial.

Figure 1.4. Structure of azo dye.

Figure 1.5. Structure of Congo red dye.

Figure 1.6. Illustration of alkali fusion.

Figure 2.1. Demonstration of simple structure of azo dyes.

Figure 2.2. Illustration of patterns of azo dyes with their structures.

Figure 2.3. Representation of synthesis of azo dyes.

Figure 2.4. Representation of the process of azo/hydrazine tautomerism.

Figure 2.5. Representation of effect of solvent on 4-phenylazo-1-naphthol.

Figure 2.6. Illustration of electronic effect of substituents in 4-phenylazo-1-naphthol.

Figure 2.7. Spectra of I-phenylazo-2-naphthol and 2-phenylazo-l-naphthol.

Figure 2.8. Illustration of disazo dyes.

Figure 2.9. Illustration of structure of various carbocyclic azo dyes.

Figure 2.10. Representation of structure of various heterocyclic azo structures.

Figure 2.11. Illustration of structures of various anthraquinone dyes.

Figure 2.12. Illustration of polycyclic aromatic carbonyl dyes structures.

Figure 2.13. Representation of structure of hemicyanines.

Figure 2.14. Representation of structure of diazaheinicyanines.

Figure 2.15. Representation of structural interrelationships of di and tri-aryl carbonium and related dyes.

Figure 2.16. Representation of condensation of quinaldine derivatives with phthalic anhydride to form quinophthalones.

Figure 3.1. Illustration of the structures of pararosaniline and carbinal bases.

Figure 3.2. Representation of the preparation reaction of malachite green dye.

Figure 3.3. Preparation of diaminotriphenylmethane derivative from the corresponding dimethoxy carbinol by condensation with an appropriate aromatic amine sulfonic acid.

Figure 3.4. Illustration of the structures of crystal violet and triphenylmethane dyes.

Figure 3.5. Illustration of the structures of bright greenish blue dye for acid-modified acrylic fiber and a similar dye prepared from benzaldehyde and N-ethyl-N - (2-cyanoethy l)-m-toluidine.

Figure 3.6. Illustration of the structures of blue alcohol-soluble dyes and dinitro crystal violet azide.

Figure 3.7. Illustration of the structures of tris-4 [bis (4-dimethylaminopheny l) methylamino phenylmethane.

Figure 3.8. Illustration of the structures of ferrocenes analogs of malachite green.

Figure 3.9. Illustration of the preparation of triarylmethane dye, m.p. 141°.

Figure 3.10. Illustration of the reaction of Michler's ketone with metalated 1, 4-cis-polybutadiene gave a carbinol, which changed to a deep red powder with iodine.

Figure 3.11. Illustration of the structures of biphenyl homologs of malachite green and crystal violet dyes.

Figure 3.12. Illustration of the structures of basic dyes containing oleophilic anions and pararosaniline antimonyl tartrate.

Figure 3.13. Illustration of the structure of ions obtained by the radiolytic reduction of laugh's violet.

Figure 4.1. Illustration of the structures of Janus red B, Alcian yellow GXS, and basic red 18 dyes.

Figure 4.2. Representation of synthesis reaction leading to lightfast pendant cationic azo dyes.

Figure 4.3. Illustration of the structures of various azo dyes.

Figure 4.4. Formation of the cationic group.

Figure 4.5. Illustration of examples of alternate cationic groups.

Figure 4.6. Illustration of the sampling tools.

Figure 4.7. Representation of the formation of red-blue and green-blue dyes.

Figure 4.8. Illustration of the structures of commercial blue and basic blue 49 dyes.

Figure 4.9. Structure of the fluorescent methane dye.

Figure 4.10. Structure of the lightfast quinophthalone yellows dye.

Figure 4.11. Structure of nitrodiphenylamine dyes.

Figure 4.12. Structure of malachite green and fuchsin dyes.

Figure 4.13. Illustration of bright greenish blue dyes prepared from o-chlorobensaldehyde.

Figure 4.14. Representation of reaction between 4-chlorobenzophenone with 1-methyl-2-phenylindole and p-anisidine.

Figure 4.15. Representation of structure of yellow chromophores.

Figure 4.16. Representation of reaction between N-ethylnaphthostyril with 1-methylindole in the presence of $POCl_3$ to produce red dyes.

Figure 5.1. Illustration of cis-trans isomerization of 11-cis-retinal (A); to all-£rans-retinal (B).

Figure 5.2. Representation of possible photochemical reaction of the photoperiodism chromophore phytochrome.

Figure 5.3. The present study aims to depict photochromic transitions through a simple state diagram.

Figure 5.4. Illustration of (A) structures that give dyes a photooxidizing reactivity against other compounds (corresponding to Class II). And (B) structures which give dyes a photoreducing reactivity, and which can lead to a photooxidation of the dye in the presence of oxygen with subsequent sensitized photooxidation of other compounds.

Figure 5.5. Illustration of n–π* excitation process: (A) simple atomic orbital representation; (B) usual valence bond notation; and (C) notation according to Zimmerman (λ-electrons are indicated as solid dots, s electrons as circular dots, and the p_y electrons as small y's).

Figure 5.6. Dependence of the position of 3(n, π *) and 3(π, π *) in anthraquinone on electron-donating substituents X: (A) high photoreactivity towards H abstraction; and (B) no photoreactivity towards H abstraction.

Figure 5.7. Lightfastness of azo dyes: (A) and (B) high lightfastness; and (C) small lightfastness. [Note: (B) corresponds to cis-azobenzol].

Figure 5.8. Illustration of types of fading-rate curves.

Figure 5.9. Illustration of explaining the abnormal fading behavior of a dyed material.

Figure 5.10. Illustration of (A) example of CFG curves. (B) Scheme of particle-growth possibility with increasing dye concentration: 1 – the number of particles increased (see curve 1 of (A)); 2 – the range of sizes of particles increased (see curve 2 of (A)).

Figure 5.11. Scheme of CF curves: 1 – low slope (fastness of lb higher than that of la); and 2 – high slope (fastness higher than dye-substrate system 1).

Figure 5.12. Illustration of comparison method: 1 – equivalence line; no difference in effect of fastness; 2 – dyes of fiber I have higher fastness than on II; and 3 – difference in chemical nature of fading.

Figure 6.1. Mechanism of modification of cellulose component and dyeing profile of cotton/polyester blend.

Figure 6.2. Mediated enzymatic indigo dyeing and soaping processes.

Figure 7.1. Rationale for dye use in healthcare.

Figure 7.2. Illustration of clofazimine (quinoneimine) and mauveine.

Figure 7.3. Illustration of phenanthridine-based trypanocidal agents.

Figure 7.4. Illustration of drugs based on the reduction of azoic dyes.

Figure 7.5. Representation of antimicrobial textile dyes.

Figure 7.6. Representation of antibacterial/antifungal bis(azoic) dyes.

Figure 7.7. Phthalocyanines for use in photodynamic therapy.

Figure 8.1. Representation of a dyed liquid crystal display.

Figure 8.2. Illustration of azo dyes tried in liquid crystal mixtures.

Figure 8.3. Representation of fluorescent dyes.

Figure 8.4. Representation of electrochemical reduction of the colorless bispyridinium salt to the highly colored radical cation.

Figure 8.5. Graph of relative positions for dye absorption and lasing wavelength.

Figure 8.6. Structure of coumarin dye.

Figure 8.7. Configuration of a simple optical disk.

Figure 8.8. Illustration of photoconductive drum.

Figure 8.9. Representation of latent image formation.

Figure 8.10. Development of latent image.

LIST OF TABLES

Table 3.1. Fastness to washing and light of some basic dyes on wool and cotton.

Table 3.2. Spectra of malachite green has p-methoxy and ferrocenyl derivatives.

Table 4.1. Shade and fastness of azines and related structures.

Table 6.1. Major textile blends and their suitable uses.

Table 6.2. Recommended dye combinations for proper dyeing of most important fiber.

Table 6.3. Some useful ternary blends and their proper dye combination.

Table 8.1. Emission of common lasers.

LIST OF ABBREVIATIONS

BDF	Benzodifuranone
BDFs	Birefringent Dichroic Filters
BS	British Standard
CCA	Charge Control Agent
CD	Compact Disc
CF	Characteristic Fading
CFG	Characteristic Fastness-Grade
CGL	Charge Generation Layer
CRT	Cathode Ray Tube
GS	Gray Scale
HMO	Hückel Molecular Orbital
MC540	Merocyanine 540
MCT-ß-CD	Monochlorotriazine ß-Cyclodextrin
MO	Molecular Orbital
MRSA	Methicillin-Resistant *Staphylococcus aureus*
$NO+HSO_4$	Nitrosyl Sulfuric Acid
NOT	Nitronium Ion
OPC	Optical Proximity Correction
OPC	Organic Photoconductor
OPCs	Operational Process Conditions
OPCs	Organic Photovoltaic Cells
PDT	Photodynamic Therapy
PMA	Phosphomolybdic Acid
PPP	Pariser-Parr-Pople
PTA	Phosphotungstic Acid

PTMA	Phosphotungstomolybdic Acid
SFP	Standard Fading Period
UV	Ultraviolet
YAG	Yttrium-Aluminum-Garnet

PREFACE

Dyes have been an integral part of human civilization for millennia. From the earliest natural dyes extracted from plants and minerals to the modern synthetic dyes synthesized in laboratories, the quest for color has always captivated our imagination. The vibrant hues that adorn our world, from the colors of a sunrise to the vivid tones of a painted masterpiece, owe their existence to the remarkable chemistry of dyes.

The development of the dye industry has been inexorably linked not only with the development of the chemical industry but also with organic chemistry itself since the middle of the last century. The position of dye chemistry at the forefront of chemical advance has declined somewhat since 1945 and more markedly so during the last 15 years, with pharmaceutical and medicinal chemistry assuming an increasingly prominent position. Nevertheless, dye production still accounts for a significant portion of the business of most major chemical companies.

This book comprises eight chapters; the first chapter provides the introduction of dye chemistry with emphasis on the historical development of dyes and development of color theory. The second chapter deals with the classification of different dyes by their chemical structure.

The third chapter provides an insight into the properties, uses, and applications of basic dyes. This chapter also discusses the organic pigments obtained from basic dyes. The fourth chapter discusses the classification of cationic dyes for synthetic fibers.

The fifth chapter describes the photochemistry of dyes and discusses the photochemical reactions of dyes in solution. The sixth chapter deals with the dyeing of fiber blends and highlights the key fiber blends along with the main dyeing methods. The seventh chapter discusses the application of dyes in the medical industry and provides an insight about the uses of dye for medical indication, as a colorant along with its use in therapy. The eighth chapter provides an insight into the non-textile applications of dyes such as in displays, optical data storage, and photoconductors.

The chemistry and application of dyes is intended as a comprehensive resource for students, researchers, and professionals in the field of chemistry, chemical and materials science. It provides a solid exploration of the chemistry underlying dyes and their multifaceted applications.

CHAPTER 1

Introduction to Dye Chemistry

Contents

Unit Introduction 2

1.1. Historical Development of Synthetic Dyes 3

1.2. Development of Color and Constitution Theory 18

1.3. Summary 22

References 23

UNIT INTRODUCTION

The intricate and extensive nature of dye chemistry presents numerous challenges for individuals needing specialized knowledge in the field. An effective approach to enhancing comprehension of the topic involves analyzing the progression of the chemistry and technology of commercial synthetic dyes, which can be evaluated from various perspectives. Certain advancements have been primarily focused on the utilization of colorants on different substrates, such as the implementation of azoic dyeing or prereduced vat dyes. An alternative methodology involves examining the exploration and establishment of dye chromophores, which is intricately linked to the historical progression of color and constitution theory (Laurent et al., 2014). The historical account of synthetic dyes is characterized by a multifaceted interplay of distinct factors, intimately connected with the concurrent advancement of theoretical organic chemistry, and subject to the influence of social and economic forces in the emerging industrialized countries of the latter part of the 19th century. This brief review will analyze the progression of new dyes and dyeing techniques in a chronological manner. Additionally, the contemporary comprehension of the molecular basis of color will be explored independently, while also being linked to the practical advancements being achieved in the industrial sector.

The study reveals a significant observation, which is the substantial contribution of dye chemistry towards the advancement of contemporary structural organic chemistry. The motivations of dye chemists may not have been entirely altruistic, however, their contributions to practical organic chemistry provided impetus for academic research. Despite a lag in structural organic chemistry during the initial period of new dye introductions from 1850–1880, theoretical advancements rapidly progressed and were able to direct the course of dye research by the early 1900s. The historical account of dye chemistry serves as a captivating preface to contemporary dye chemistry and imparts valuable insights into the fundamental principles regarding structural organic chemistry which are presently considered commonplace.

Did You Know?

Mauveine, the first synthetic dye, was named after the color mauve, which comes from the French word "malva" meaning mallow flower.

1.1. HISTORICAL DEVELOPMENT OF SYNTHETIC DYES

1.1.1. Early History up to 1865

Until the mid-1800s, the vast majority of colorants were derived from vegetable, animal, or mineral sources. These colorants often had to be transported over long distances and were susceptible to the unpredictable forces of nature. The economic constraints associated with the utilization of these colorants became more evident during the latter part of the 18th century, coinciding with the onset of the industrial revolution in Europe. In the 1850s the textile industry was primed for the adoption of cost-effective synthetic dyes that offered a higher level of dependability in terms of supply. Madder and Indigo were two natural dyes of significant value that posed a considerable challenge to experts in the field of dye chemistry (Ardila-Leal et al., 2021). The replacement of natural dyes with their synthetic counterparts represented two significant milestones in the chronicles of synthetic dyes.

The Rubia tinctorum plant root was the source of madder, with alizarin (1,2-dihydroxyanthraquinone) being the primary coloring agent. The process of mordant dyeing, which typically involves the use of alum, was utilized to dye cellulosic fabrics with a particular material (Barnett, 2007). This resulted in the formation of an insoluble red metal complex of alizarin within the fiber. The widely adopted Turkey Red method was a complex and laborious process that entailed a production timeline of up to a month to achieve the final dyed fabric (Meshnick & Dobson, 2001). Despite yielding bright red dyes with exceptional rapidity properties, even by contemporary standards, the economic viability of the process was less than optimal. The production of madder was primarily concentrated in Turkey, however, a significant portion had to be procured from distant regions, particularly the British East Indies, which posed a notable economic drawback.

Indigo, being a vegetable-based natural blue colorant, held significant importance, much like Madder. The primary origin of this pigment was derived from the Indigofera tinctoria plant indigenous to India, which possesses indoxyl in the form of a glycoside derivative within its leaves (Saxena & Raja, 2014). The process of leaf fermentation results in the release of indoxyl, which subsequently undergoes swift oxidation upon exposure to air, ultimately yielding the intended indigo that is not soluble in water (Pickering, 2005). The woad plant had been conventionally utilized in Europe to obtain indigo for dyeing purposes. However, the importation of the Indian indigo commenced in the twelfth century, eventually leading to the exclusive use of the Indian dye for cotton dyeing by the 1850s (Figure 1.1).

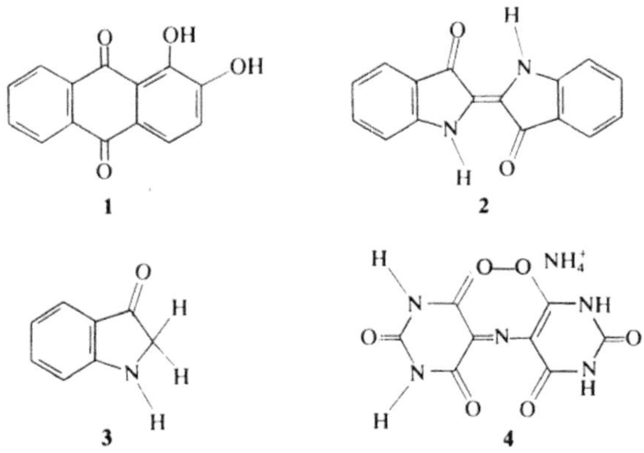

Figure 1.1. *Illustration of (1) alizarin; (2) indigo; (3) indoxyl; and (4) murexide.*

Source: J. Griffiths, Creative Commons License.

Prior to the occurrence of significant attacks on natural dyes, numerous advancements in dye chemistry were made, dating back to 1771 when Woulfe initiated his research on indigo and identified picric acid, which can be considered the first synthetic colorant, through the reaction of natural dye with nitric acid (Vettese Forster & Christie, 2013). The discovery that silk could be dyed to a greenish yellow hue using picric acid was made earlier, however, it was not in 1845 that a cost-effective method for synthesizing picric acid from phenol was developed, thereby enabling the commercialization of this "dye" (Hisano, 2016). Despite its limited usefulness in shading indigo dyeings to produce bright greens, the product was deemed unsuccessful due to its poor lightfastness. Murexide, an additional dye, was identified prematurely for industrial application. The initial synthesis of this compound was accomplished by Prout in 1818 through the reaction of uric acid with concentrated nitric acid, subsequently treated with ammonia (Ferreira et al., 2001; Streb et al., 2007). However, its utilization as a mordant dye on a commercial scale was not feasible until 1851, when a cost-effective method was discovered. Despite its lack of success, the fastness properties exhibited by the material were notably deficient.

Remember

The discovery of synthetic dyes had a significant impact on the fashion industry as these dyes made colorful fabrics accessible to a wider population.

The chemical compound indigo garnered the interest of multiple chemists, ultimately leading to the discovery of aniline, a crucial intermediate in the production of dyes. In 1826, Unverdorben (1826) discovered that indigo could be subjected to "distillation" resulting in the formation of aniline (Desrochers, 2008). Consequently, it is not unexpected that the initial practitioners of dye chemistry employed aniline as a fundamental component to synthesize pigmented substances. Nevertheless, aniline remained an infrequent chemical phenomenon until a significant series of occurrences had transpired. In 1825, Faraday made the discovery of benzene (Brenner & Murmann, 2016). Subsequently, in 1845, Zinin (1880); and Hofmann (1892) were able to determine the process of nitrating this chemical and reducing the resulting product to aniline. Mansfield's work in the last years of the 1840s demonstrated that coal tar could serve as a plentiful source of the necessary benzene. The outcome of these distinct inquiries served to demonstrate a significant element in the triumph narrative of Perkin's Aniline Purple (Van den Belt & Rip, 1987).

The historical context surrounding Perkin's discovery has been thoroughly extensively documented and will be succinctly summarized in this text. During Easter of 1856, William Henry Perkin conducted studies on the dichromate oxidation of aniline in his home laboratory while on vacation from the Royal College of Chemistry (Yeh & Lim, 2007). Perkin was eighteen years old at the time. He succeeded in extracting a minute quantity of a purple pigment from the unremarkable black substance, which seemed to have captured his interest. The dye produced vivid hues of purple on silk, prompting the individual to seek the professional assessment of a commercial dyer named Pullar, who provided a positive evaluation. The process was expeditiously patented in August of the same year, prompting Perkin to resign from his post at the Royal College to establish Perkin and Sons with his father, G. F. Perkin, and brother, T. D. Perkin, for the purpose of large-scale production of the dye (Gürses et al., 2016). It is noteworthy that undertaking such a venture during the period prior to the enactment of the Limited Liability Act of 1862 carried a significantly elevated level of risk, thereby rendering the prospect of incurring personal financial ruin a distinct possibility (Liu et al., 2016; Stepien et al., 2017). The fact that Perkin senior invested a significant portion of his personal savings into the enterprise is noteworthy.

The construction of the Greenford Green plant commenced in June 1857. However, the process was not without its challenges, as several technical obstacles had to be surmounted before the cost-effective production of the dye could be achieved. The pioneering of massive amounts of nitration of benzene and the reduction of nitrobenzene was necessary. The optimization of the oxidation step and dye isolation procedures was necessary due to the formation of Aniline Purple in significantly low yields (Baker, 2011). By the year 1864, advancements had been made in the process to such an extent that Aniline Purple could be presented as a solid in crystalline form, thereby making it commercially available. Despite the financial success that his dye brought him, Perkin continued to actively engage in experimental chemistry. The author proceeded to demonstrate that Aniline Purple is composed of two dyes, as depicted in Figure 1.2, with the former dye being the predominant constituent, albeit in varying proportions (Park & Shore, 2007). The structures were not inferred until a considerable amount of time had elapsed.

Although Aniline Purple demonstrated a favorable outcome in Britain soon after its inception, its reception in France was notably more remarkable. Regrettably, this did not yield significant monetary compensation for Perkin. The nullification of Perkin's French patent facilitated the prompt entry of French manufacturers into the market, enabling them to produce the dye with notable competitiveness (Burton, 2019). So successful were the French that Britain began importing the dye and the alternative name Mauveine, or Mauve, proved more popular for the dye than Aniline Purple.

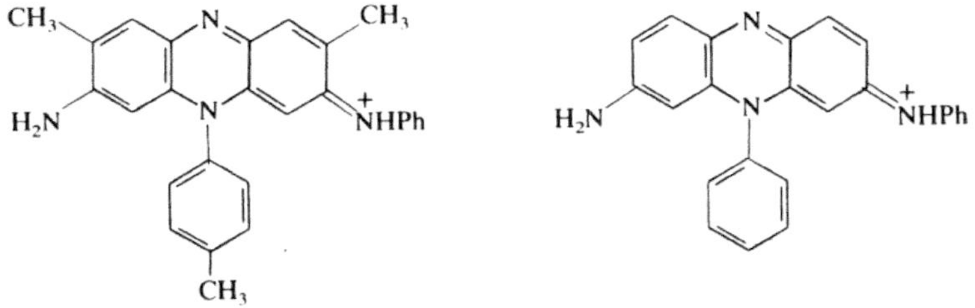

Figure 1.2. *Illustration of two dyes making up aniline purple.*

Source: D. R. Waring, Creative Commons License.

The achievement of Perkins in developing Aniline Purple had a significant impact on the quest for novel synthetic dyes in both Britain and France, serving as a catalyst for further research in this area. Chemists focused on aniline as a primary precursor and, utilizing a purely empirical methodology, due to the nascent state of structural chemistry and limited knowledge regarding the molecular structure of the majority of organic compounds, they expediently identified numerous novel primary dyes that exhibit distinctively vivid hues. The initial triphenylmethane dye, known as Magenta, was first identified in 1856 by Natanson (Pirok et al., 2019; Sandström et al., 2023). Subsequently, in 1859, Verguin demonstrated that it could be produced on a commercial scale through the economical oxidation of crude aniline with stannic chloride. Homo rosaniline, in conjunction with rosaniline, constitutes a fundamental constituent of the dye. Hofmann demonstrated that the presence of toluidine is a crucial component in the reaction. Magenta demonstrated a higher degree of commercial viability in comparison to Aniline Purple, leading to its widespread production in the United Kingdom by notable firms such as Perkin and Sons and Simpson, Maule, as well as Nicholson. (Van Bommel et al., 2007).

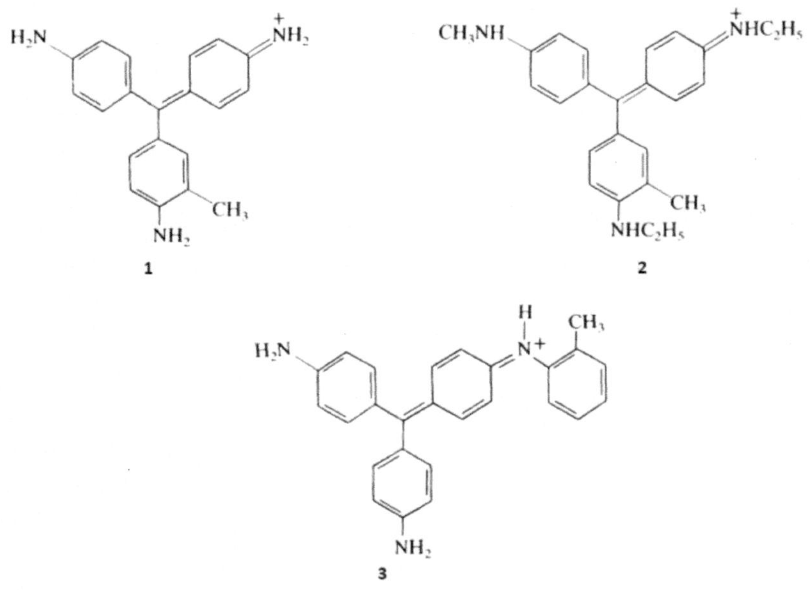

Figure 1.3. *Illustration of (1) Rosaniline; (2) Hofmann's violet; and (3) Girard's violet imperial.*

Source: G. Hallas, Creative Commons License.

During the late 1850s and 1860s, numerous discoveries were made in the triphenylmethane category, including Hofmann's Violet (1863) and Girard's Violet Imperial (1860). However, the initial indications of a methodical approach to dye synthesis were introduced by Nicholson and Gilbee in 1862. They intentionally sulfonated N-phenyl derivatives of Magenta to produce dyes with improved water solubility (Waring & Hallas, 2013). These particular products were demonstrated to be the initial acid dyes utilized for wool, thereby initiating the utilization of sulfonation as a conventional methodology for producing dyes that possess favorable attributes such as a high solubility in water and an affinity for protein fibers (Brooks, 2002) (Figure 1.3).

It is worth noting that the discoveries made during this era were not limited to the triphenylmethane basic dye category. In fact, a significant advancement was achieved when Martius successfully synthesized the azo dye Bismarck Brown in 1863 (Hagan & Poulin, 2021) (Figure 1.4).

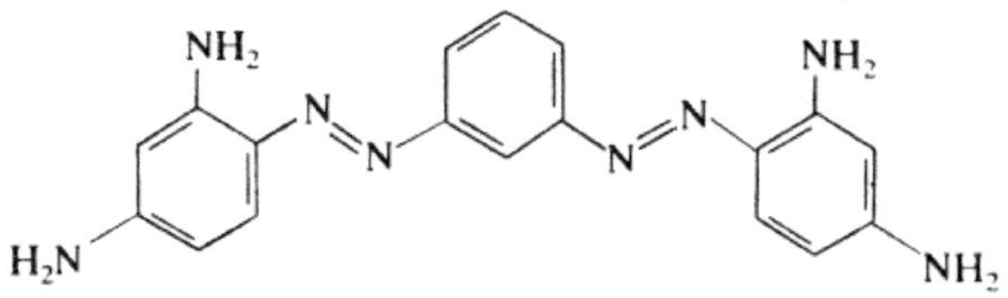

Figure 1.4. *Structure of azo dye.*

Source: J. Griffiths, Creative Commons License.

The comprehension of the configuration of this particular dye was limited until Griess explicated the azo coupling reaction that takes place among diazonium ions and amines in a prompt manner. The discovery of the diazotization reaction was made by Griess in 1858 (Rasmussen, 2018). Subsequently, Martius chanced upon his dye while pursuing research in this area. The establishment of the necessary circumstances for a self-coupling reaction was achieved through the reaction of m-phenylenediamine with nitrous acid. The initial trials conducted during that period eventually resulted in the emergence of the azo dyes, which currently represent the most significant chemical category of dyes employed in various applications.

During the mid-1860s, chemists began to focus more on the unresolved structural issues in organic chemistry. This led to the emergence of a more scientific methodology for dye chemistry. The approach of "trial and error" was no longer deemed a viable economic strategy (Peters & Freeman, 1991). Nowhere was the appreciation for systematic research more acute than in

Germany. Throughout the era of Aniline Purple and Magenta, Germany maintained a relatively inconspicuous presence. The lack of substantial patent legislation in the aforementioned nation facilitated the utilization of British and French innovations by German enterprises, thereby augmenting their financial gains. However, this did not contribute to the promotion of industrial research. Conversely, during that period, scientific instruction in Germany surpassed that of any other European nation, as demonstrated by the abundance of exceptional German chemists who were recruited by foreign dye corporations. By 1865, the German dye industry had started to make its presence felt, prompting several chemists to return to their homeland. The adoption of a new scientific approach led to a swift progress of the industry in Germany, while the industries in Britain and France experienced a corresponding decline.

1.1.2. Developments in Dye Chemistry

The transition from empirical dye research to a more deductive approach is not precisely defined, but the year 1865 is often considered a significant turning point. This is due to Kekule's proposal of the "cyclohexatriene" formulation of benzene in that year, which subsequently led to his explanation of the aromatic nature of benzene through the concept of oscillating bonds (Khatri et al., 2015). The significant progress made in the field of structural organic chemistry was especially relevant to the discipline of dye chemistry. This is because all dyes that were known during that period comprised of benzenoid residues. Consequently, this development facilitated a deeper comprehension of the molecular makeup of dyes and the chemical reactions that led to their formation.

A noteworthy advancement during the post-1865 era was the laboratory production of alizarin, which serves as the primary constituent of Madder, as previously demonstrated. The aforementioned progress was achieved by Graebe and Liebermann in the year 1868, which was approximately 37 years subsequent to the isolation of pure alizarin from Madder through sublimation. The researchers successfully determined the fundamental anthraquinonoid structure of the dye, which enabled them to undertake a deliberate synthesis process from anthraquinone, incorporating bromination and alkali fusion (Gürses et al., 2016). The fortuitous success of the experiment was attributed to the unknown positioning of the two hydroxy groups. Nonetheless, this outcome served to demonstrate the efficacy of the scientific method. The exorbitant expense associated with bromine impeded the commercialization of this technique. However, this circumstance served as a catalyst for the emergence of alternative synthetic approaches that could viably rival the economic feasibility of natural Madder. The pivotal moment occurred in the subsequent year and was achieved autonomously by both Perkin and Caro. The novel methodology entails the sulfonation of anthraquinone at the 2-position, succeeded by an alkali fusion step. The two companies, Perkin and Sons in England and Badische Anilin und Soda-Fabrik in Germany, arrived at a mutually agreeable arrangement to divide the production of synthetic alizarin (Mahalingam et al., 2021).

Did You Know?

The color of a dye is determined by its molecular structure and the way it absorbs and reflects light.

The advent of the novel synthetic material resulted in significant profitability, ultimately leading to the obsolescence of the Madder industry. During the time frame spanning from 1869 to 1873, Perkin & Sons experienced a lack of significant competition within the British market, leading to a substantial increase in the production of the dye, ultimately reaching a peak of 435 tons in 1873 (Iftikhar et al., 2019). Nevertheless, this event marked a significant shift in the company's trajectory. Due in large part to intense competition from German counterparts, the company was divested in 1874 and manufacturing operations at the Greenford facility were terminated by 1876. The synthetic dye industry caused the displacement of madder, which was the initial significant natural dye to be affected.

Despite the establishment of diazotization and diazo- coupling reactions by 1860 and the introduction of Bismarck Brown to the market, the progress of commercial azo dyes was initially sluggish. Subsequent research of a more scholarly character was necessary to fully recognize the possible uses of this category of pigments. Martius and Griess demonstrated in 1866 that Aniline Yellow, a dye with an ambiguous structure, was actually 4-aminoazobenzene (Willkomm et al., 2016). The formation of this compound involved coupling diazotized aniline to aniline, followed by an acid-catalyzed modification of the diazoamino compound. Kekule made the discovery of the coupling reaction between diazonium ions and phenols in 1870. He also utilized chemical methods to establish the structure of the hydroxyazo dyes (Kokkonen et al., 2021). During the mid-1870s, dye chemists began to recognize the adaptability of the coupling reaction. In

1875, Caro synthesized the fundamental azo dye, Chrysoidine, while in the subsequent year, Roussin produced the initial acid azo dyes, Orange I and Orange II, specifically for wool (Yen et al., 2012). Subsequently, the utilization of azo dyes has been progressively introduced and currently, they hold a significant position as the most crucial and extensively employed dye category.

Despite the efficacy of the scientific methodology in the realm of dye research, there remained an opportunity for the experimental chemist to fortuitously uncover valuable discoveries. This was aptly demonstrated by Croissant's 1873 discovery of sulfur dyes. The individual conducted an experiment involving the utilization of sawdust and other cellulosic materials, which were subjected to heating with sodium polysulfide resulting in the production of products with a deeply pigmented appearance (Higashino & Imahori, 2015). The aforementioned reduced products exhibited a notable attraction towards cotton and subsequent oxidation led to the attainment of the ultimate hue. The resultant dyeings displayed commendable fastness characteristics. Although the class of chromophores was novel, the dyeing process bore a striking resemblance to the traditional vat dyeing technique and was effortlessly embraced by conventional dyers. In 1880, the commencement of azoic dyeing marked the advent of a novel method for dyeing.

The aforementioned procedure was patented by the Read Holliday and Sons firm, located in Huddersfield, England. It entailed impregnating the textile with an alkaline solution containing either a phenol or naphthol, followed by subjecting the fabric to a solution of a diazotized aromatic amine. The dyer was responsible for executing the diazotization process, which led to the creation of dyes commonly referred to as "ice colors" (Fleischmann et al., 2015). These dyes proved to be highly effective in calico printing. During the early 1880s, cotton remained the primary textile material. Its dyeing process involved several techniques, among which vat dyeing with indigo and mordant dyeing with alizarin were the most effective in terms of fastness properties. Nevertheless, these procedures posed difficulties and yielded limited spectrums of hues. The fastness outcomes of basic dyes on cotton that was mordanted with tannin were found to be suboptimal, despite being the most practical techniques employed. In 1883, Griess made significant progress by creating disazo sulfonated dyes that exhibited an inherent attraction to cellulose, eliminating the requirement for a mordant (Polo et al., 2004). The initial set of direct dyes for cotton were introduced as the foremost ones, followed by the introduction of significantly enhanced dyes that employed benzidine as tetrazo component by Bottiger in the subsequent year. Among these, Congo Red is widely recognized as the most prominent (Park & Shore, 2007) (Figure 1.5).

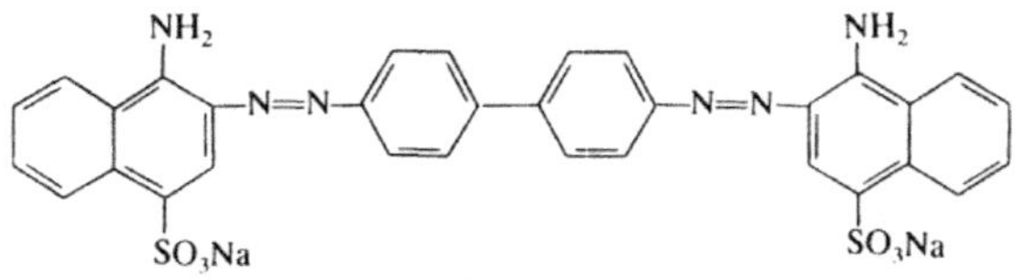

Figure 1.5. *Structure of Congo red dye.*

Source: G. Hallas, Creative Commons License.

The utilization of mordanting techniques, exemplified by the Turkey Red dyeing approach, has been a longstanding practice spanning millennia. However, the initial application of this methodology to azo dyes occurred in 1885, marking a significant advancement in the field. Meldola and Nietzki have successfully synthesized dyes through the coupling reaction among diazonium ions and salicylic acid (Mustroph et al., 2006). Their findings indicate that these dyes are capable of imparting good wash-fastness properties to chrome-treated wool. The aforementioned event marked the inception of metal-complex azo dye chemistry, which subsequently paved the way for numerous significant advancements in the field.

During the 1890s, the dyer was presented with a diverse selection of artificial dyes to utilize. However, despite this, indigo remained the dominant blue dye due to its exceptional fastness, thereby retaining its status as the most significant natural dye that persisted. During the late 19th century, there was a significant increase in research efforts towards discovering a feasible method for synthesizing indigo, as European dye manufacturers recognized the potential benefits that could be derived from such a breakthrough. The configuration of indigo had been firmly established by 1890 (Tan et al., 2015). Baeyer (1878) had successfully synthesized the dye in through the utilization of o- nitrophenyl acetic acid as a precursor in the laboratory. Both the aforementioned pathway and a subsequently identified substitute originating from o-nitrocinnamic acid were deemed uneconomical for commercial purposes. Consequently, industrial research was undertaken to explore the synthesis of indigo. The German company BASF was the most notable player in this domain. Through a remarkable infusion of financial resources and human labor, the company successfully introduced synthetic indigo to the market in 1897 as a viable substitute for the naturally occurring product. The manufacturing procedure was predicated on the transformation of anthranilic acid into o-carboxyphenylglycine, followed by the cyclization of the latter to indoxyl through alkali fusion, as illustrated in Figure 1.6 (Sanda et al., 2021).

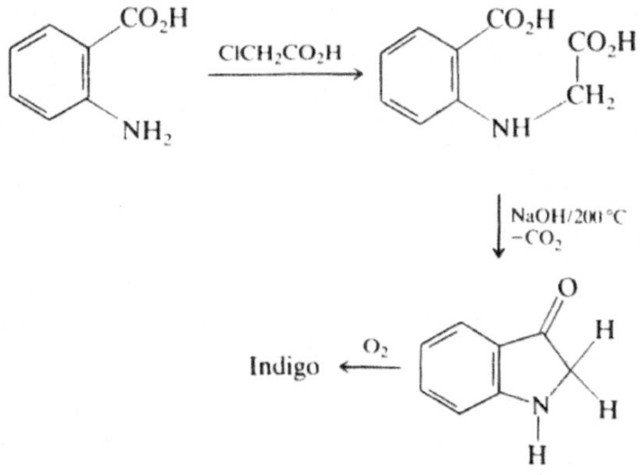

Figure 1.6. *Illustration of alkali fusion.*

Source: G. R. Waring, Creative Commons License.

Although the economic conditions were not optimal, the indigo that was produced fetched a bit more expensive than the natural material. The superior level of purity and the greater consistency in dyeing performance exhibited by the former option constituted a favorable attribute for which certain dyers were willing to incur costs. Roessler discovered a cost-effective pathway in 1902 whereby indoxyl could be produced by cyclizing phenyl glycine at moderate temperatures using sodamide as the base (Grätzel, 2000). The implementation of this new method resulted in a significant enhancement compared to the previous approach. Consequently, the sales of natural indigo experienced a substantial decline in a brief period, leading to an irreversible deterioration of the indigo plantations. Therefore, the final remaining exemplar of organic pigments had been vanquished.

By the beginning of the 20th century, the field of chemistry had undergone significant advancements and the discipline of dye chemistry had become progressively more intricate. Novel dye products were being introduced to the market at an unparalleled pace. During the early 1900s, there was a significant focus on synthesizing vat dyes, likely motivated by the notable achievements in producing synthetic indigo. This resulted in a substantial synthetic undertaking. In 1901, Bohn of BASF made a significant discovery of great importance. The individual endeavored to synthesize an indigo framework that comprises anthraquinone residues by employing a specific methodology. This involved the reaction of 2-amino anthraquinone with chloroacetic acid, followed by subjecting the resultant product to alkali fusion (Asim et al., 2014). Despite acquiring a blue dye, the resultant product did not conform to the anticipated indigo derivative. Subsequently, Scholl demonstrated that it possessed the quinonoid structure (1S), which is presently recognized as indanthrone. The novel blue chromophore demonstrated comparable application and vatting procedures to indigo and exhibited exceptional fastness properties when applied to cotton. Consequently, it was commercialized under the name Indanthren Blue R. The identification of quinones as prospective vat dye systems prompted extensive exploration of other associated structures, leading to the introduction of numerous novel vat colors during the 1900–1910 timeframe (Griffiths, 1990). Illustrative instances comprise of colorants derived from indanthrone, flavanthrone, pyranthrone, violanthrone, isoviolanthrone, 1,5-dibenzamidoanthraquinone, anthraquinone-acridones, and anthraquinone-carbazoles. It is noteworthy that the aforementioned dyes held significant commercial value, as cellulosics were the predominant textiles utilized during that era and solely the vat dyes were capable of delivering exceptional results. Presently, vats continue to hold a significant and seemingly steadfast portion of the overall dye industry.

It is important to note that research on vats in the dye industry did not completely overshadow advancements in other dye categories. Significant progress was made in other domains as well. The J-acid direct dyes were first introduced during the period of 1900–1902 (Nachiyar et al., 2023). Subsequently, in the year 1912, the introduction of Naphtol AS and its analogs led to a noteworthy enhancement in the color spectrum and fastness characteristics of azoic dyes. The initial application of cyanine dyes as photographic sensitizers was in the non-textile domain. Numerous breakthroughs were achieved in the realm of sulfur dyes, while in the anthraquinone domain, Herzberg made a significant discovery in 1913 regarding the blue acid dyes that are produced from the combination of bromamine acid and arylamines (Griffiths, 1988).

The onset of the First World War in 1914 resulted in a deceleration of advancements in dye chemistry, leading to a dearth of noteworthy discoveries in the subsequent four years. During this period, manufacturers prioritized the preservation of established dyes by allocating resources towards their production. Nevertheless, notable innovations were made, with the most noteworthy being the Neolan pigments developed by the Society of Chemical

Industry (Basle) in 1915 (Vittum & Weissberger, 1954). The aforementioned were preformed complexes of chromium that were utilized in the dyeing process of wool and silk with mordant azo dyes. During the period of war, there was a heightened focus on enhancing the state of dye intermediates. As an illustration, a cost-effective method for producing phthalic anhydride was discovered, which entailed the air-oxidation of naphthalene using vanadium pentoxide. The aforementioned finding had a significant impact on the economic aspects of indigo and anthraquinone synthesis. Additionally, it sparked a surge of interest in the advancement of novel anthraquinone dyes.

Following the war, the resurgence of the industry led to a continuation of advancements in dye chemistry. However, due to the absence of significant breakthroughs in textile polymers, this era was characterized more by the consolidation of existing knowledge rather than the pursuit of novel chromophore innovations. The color spectrums of azo, anthraquinone, and vat dyes were enhanced, and producers endeavored to discover novel techniques for dye application to simplify the process for dyers and to gain a competitive advantage. In 1920, ICI introduced the first green vat dye, known as Caledon Jade Green, which effectively addressed a significant gap in the vat dye spectrum. During the period of 1920–1930, advancements in stabilized diazonium salts and sulfuric esters of pre-reduced vat dyes led to enhanced flexibility and ease of use in the azoic and vat dyeing procedures (Chan et al., 2011). By the 1920s, cellulose acetate, a synthetic fiber, had become increasingly significant and posed challenges for the dyeing process. Consequently, a challenge arose that necessitated a completely novel approach in the field of dye application chemistry. The year 1922 marked the introduction of lonamine dyes, which facilitated the successful dyeing of cellulose acetate to desirable depths while exhibiting commendable fastness properties (Zollinger, 2003). The aforementioned compounds are derivatives of amino-dyes that have been modified with N-methylsulfonic acid. Despite their water solubility, these compounds underwent gradual hydrolysis in the dyebath, resulting in the formation of a water-insoluble dye. This dye, which is currently recognized as a disperse dye, was responsible for dyeing the hydrophobic acetate fiber (Kabir et al., 2019). Subsequently, it was recognized that the identical outcome could be attained through the utilization of a water-insoluble pigment that is finely dispersed in water. The aforementioned observation led to the possibility of utilizing a broader spectrum of dye structures compared to the lonamine method. Consequently, the initial disperse dyes were developed, namely the Duranols (British Dyestuffs) and the SRA Colors (British Celanese). It is noteworthy that the contemporary categorization of this particular group as "disperse dyes" was not embraced until the year 1953.

The chromophore that has been deemed the most significant of the current century was first made available for commercial use in the year 1934. The substance in question is copper phthalocyanine, which was commercialized by ICI under the trade name Monastral Fast Blue BS (Murakami et al., 2003). Similar to numerous significant discoveries, it is common to retrospectively identify instances where previous researchers had observed comparable or identical phenomena but were unable to fully utilize their observations. The production of metal-free phthalocyanine was reportedly achieved by Braun & Tcherniac in 1907 through the extended heating of o-cyanobenzamide above its melting point. However, it took several years before the blue product was accurately identified. The synthesis of copper phthalocyanine was first reported by de Dierbach and Van der Weid in 1927 (Ihmels & Otto, 2005). The compound was prepared through the heating of o-dibromobenzene and cuprous cyanide in pyridine under sealed conditions. Nevertheless, the compound was perceived as a composite of the constituents, and its architecture was not subjected to further investigation. One year later, Dendridge from the Scottish Dyes Corporation assessed the production of phthalimide from molten phthalic anhydride and ammonia. The reaction was conducted in an enamel-lined iron autoclave and resulted in the formation of a bright blue-green impurity (Costela et al., 2003). This observation marked a significant difference from the previous situation. Subsequent examinations revealed that the chromatic substance was a result of the interaction among the reactant substances and the iron substrate, which had fissures in its coating. In this instance, the material's commercial viability was postulated, prompting an endeavor to elucidate its structure. In the following years, as the structure was determined and additional metal derivatives were produced, it was discovered that copper phthalocyanine exhibited exceptional pigment characteristics. Consequently, Monastral Fast Blue BS was introduced into the market. Currently, copper phthalocyanine derivatives are extensively utilized as blue pigments, while appropriately adapted phthalocyanines have been broadly employed as textile dyes, including the acid, direct, ingrain, and reactive dye categories.

Following the 1930s, the introduction and subsequent commercial expansion of novel synthetic fibers emerged as a crucial factor influencing the trajectory of dye research. The year 1939 marked the introduction of both nylon 66 by DuPont and nylon 6 by IG to the market (Langhals, 1995). The military potential of these fibers during the war years provided a significant boost to their production. In contrast, poly(ethylene terephthalate) was first identified by Whinfield and Dickson in 1941; however, due to wartime limitations, its industrial production was delayed until the 1950s. In 1954, "Terylene" manufactured by ICI and "Dacron" produced by Du Pont were made available for purchase. Poly(acrylonitrile) fibers were first developed in the late 1940s, with the initial introduction of "Orlon" by Du Pont. Currently, synthetic fibers possess a substantial portion of the global market (Muthyala, 2006). The inherent hydrophobic nature of synthetic materials presented challenges for the dyeing process. To address this issue, disperse dyes originally intended for cellulose acetate were employed and yielded satisfactory results. Notwithstanding, there were deficiencies, specifically in relation to polyester, prompting manufacturers to focus on altering the dyeing characteristics and stability attributes of extant chromophoric systems for these novel polymers. During the period of the 1950s and 1960s, a greater number of novel dyes were introduced on a yearly basis compared to previous years. However, it is noteworthy that the majority of these dyes were fundamentally composed of azo compounds and anthraquinones, albeit with altered side chains and substitution patterns, particularly in relation to the disperse types. Novel cationic structures have emerged for the purpose of dyeing acrylic fibers.

Despite the rapid growth in importance of man-made fibers during the early periods, cotton remained the predominant textile fiber and has retained its significance to this day. The introduction of cotton reactive dyes in 1956 marked a significant advancement in the field of dye application chemistry and is considered to be one of the most noteworthy developments of the century. The feasibility of achieving covalent bonding among the dye and fiber under practical dyehouse conditions was explored by ICI in 1953 (Waring & Hallas, 2013). This work led to the examination of dichlorotriazinyl dyes by Stephen and Rattee, which ultimately resulted in the creation of the Procion reactive dyes and the establishment of effective application procedures. The significant potential of these dyes has incited further research by various companies, leading to the patenting and successful commercialization of numerous other reactive dye systems. Several chemical compounds that have demonstrated resilience in the commercial sector are chloropyrimidines, chloroquinoxalines, fluoropyrimidines, and vinylsulfones. Polyamide and wool fibers have been the subject of considerable interest in relation to reactive dyes, albeit with comparatively fewer notable achievements (Bao et al., 2021). Following 1970, dye research experienced a period of decline, primarily due to the dye manufacturer's fulfillment of the dyer's requirements and the industrial recession of the 1970s. The manufacturers shifted their focus towards optimizing the current dye ranges and enhancing the cost-effectiveness of dye manufacturing, while reducing their emphasis on introducing new dyes (Lewis, 2014). The aforementioned trend persists in contemporary times, with the majority of novel advancements in dye chemistry being concentrated in high-tech domains, including medicine, diagnostics, lasers, and electrooptics, as per the available patent literature. It is noteworthy that a significant number of chromophores utilized in these applications were initially synthesized over half a century ago and were subsequently abandoned as textile dyes due to the progression of dye chemistry. Consequently, numerous antiquated dye systems, previously deemed solely of historical significance, are experiencing a resurgence in contemporary technology.

1.2. DEVELOPMENT OF COLOR AND CONSTITUTION THEORY

The co-evolution of dye chemistry and organic chemistry during the 19th century has been observed. At the outset, the progress of dye advancements outpaced theoretical advancements, and these advancements played a significant role in promoting the progression of structural organic chemistry. An additional significant impetus to structural theory did not arise from the chemistry of dyes, but rather from the endeavors of researchers to comprehend the genesis of color in dyes. During the late 19th and early 20th centuries, there was a notable development in the electronic theory of molecular structure. Consequently, it is pertinent to provide a succinct overview of the evolution of color and constitution theory since the inception of synthetic dyes. This will enable us to comprehend how this progression has culminated in the current ability to anticipate the color of a dye using contemporary computers even before its synthesis.

During the period when Perkin made his discovery, the field of organic chemistry was still in its early stages of development. The comprehension of the structure of benzenoid compounds was not fully established until Kekule's theory was published in 1865 (Dähne, 1978). It is noteworthy that the electron was not detected until the year 1897, and it was not until 1916 that the notion of the electron-pair bond was introduced by Lewis in his renowned publication titled "The Atom and the Molecule" (Hunger, 2007). In 1868, Graebe & Liebermann presented the notion that color in dye molecules was somehow linked to unsaturation, based on the ample data that had been gathered on synthetic dyes up to that point. The aforementioned deduction was derived from the empirical observation that the diminution of dyes resulted in the annihilation of their chromaticity. In 1876, Witt made the initial endeavor to elucidate the molecular composition of dyes by analyzing their color (Shapiro, 1993). At the time in question, the field of structural chemistry had advanced significantly, leading to a comprehensive comprehension of functional groups. Witt postulated that a dye molecule ought to be deconstructed into three constituent parts. The chemical structure comprised of benzene or fused benzene rings, with simple unsaturated groups such as $–NO_2$, –N≡N–, –C=O attached to them, were identified as chromophores by the researcher. Ultimately, there existed fundamental clusters, such as $-NH_2$ and -OH, which were classified as auxochromes, and the co-occurrence of these three constituents resulted in a vivid chromatic manifestation (Vittum & Weissberger, 1954). The aforementioned simplistic illustration remains relevant in contemporary times. Furthermore, Witt's theory serves as a precursor to contemporary notions such as electron-donating (+ M) and electron-withdrawing (-M) groups, as well as conjugation.

Witt's theoretical framework was found to be easily adaptable to the azo and anthraquinone dyes, however, it posed challenges in reconciling with the triphenylmethane cationic dyes. Armstrong and Nietzki independently proposed an alternative interpretation of the origin of color in 1888 and 1889, respectively. The coloration in this instance was attributed to the existence of quinonoid structures, which are characterized by benzenoid rings that have been altered to a cyclohexadiene residue featuring two exocyclic double bonds (Dähne, 1978). The inclusion of said units was deemed necessary in the compositional framework of the cationic dyes. Multiple quinonoid structures can typically be delineated, leading to the logical inference that the absorption of light is attributable to a form of structural oscillation between these potentialities. The electromagnetic wave theory of light remained widely accepted for a considerable duration. The interaction between a vibrating wave and an oscillating bond system was deemed a compelling mechanism for the absorption process (Lubai et al., 1986). Nevertheless, basic theoretical analyses indicated that the frequency of atomic oscillations was significantly insufficient to correspond with the frequency of electromagnetic radiation. Following the discovery of the electron, it was observed that oscillations of said particle would entail the appropriate frequency for light absorption. However, this proposition was not accorded significant consideration by chemists during the turn of the century. The field of theoretical chemistry witnessed substantial advancements during the early 1900s. Following Lewis's seminal work on the electronic theory of molecular bonding, the correlation between electronic excitation and light absorption gained widespread acceptance (Dewar, 1950).

The theoretical concept of mesomerism has proven to be highly advantageous in the realm of color and constitution studies. Thiele's "conjugation and partial valence hypothesis" was published in 1899 to account for the unusual chemical characteristics of conjugated polyene systems (Abdou, 2013). The hypothesis proposed that bond equalization took place in such systems, resulting in partially unsatisfied valencies at the ends of the chains. In 1911, Kauffmann expanded upon Thiele's theories and proposed the notion of the "decentralization of chemical functions." This concept acknowledged the resonance transmission of substituent effects in conjugated molecules. Kauffmann's theory was extended to include the notion of limiting structures, which are used to represent the actual structure of a molecule. These structures are an average of several possible representations. The aforementioned concepts served as the basis for mesomerism, and Ismailsky denoted the "average structure" as the chromostate due to its perceived correlation with color. A triphenylmethane dye can be considered as an amalgamation of three constraining structures, resulting in a chromostate that exhibits a strong coloration (Olsen, 2012). Despite the widespread acceptance of these concepts, the mechanism linking the different constraining frameworks remained unclear. In 1922, Weitz & Konig established a definitive differentiation between structural tautomerism, which pertains to authentic equilibria, and valence tautomerism, also known as mesomerism.

The paramount progression in theoretical chemistry was the emergence of quantum mechanics, which was initially mathematically formulated in 1925 through the autonomous contributions of Schrodinger and Heisenberg. In the initial years of the 1930s, the principles of quantum mechanics were employed to address the issue of molecular bonding. A mathematical representation of the n-bond was established in 1930 as an outcome of this endeavor. At this juncture, the organic chemist's longstanding utilization of the mesomerism concept and the novel concept of resonance were to be demonstrated as being commensurate through the application of quantum mechanical computations (Fabian & Hartmann, 2013). The concept under consideration was derived from the approximation methodology of valence bond theory, which was subsequently employed for the computation of the absorption spectra of dyes. During the early 1930s, a substitute mathematical approximation for the calculation of molecular properties using quantum mechanics was developed, known as molecular orbital (MO) theory. By employing this technique, Hückel succeeded in quantifying numerous characteristics of intricate conjugated organic molecules. His methodology, commonly known as Hückel molecular orbital (HMO) theory, proved to be of enduring significance and established the foundation for contemporary MO techniques (Zollinger, 2003).

The refinement and application of valence bond and HMO methods in subsequent years were significantly influenced by dyes and endeavors to forecast color. In 1939, Mulliken effectively employed the HMO method to analyze the absorption spectra of polyenes and cyanine dyes (Dewar, 1950). Similarly, in the same year, Pauling utilized valence bond theory to explain the chromaticity of cationic and anionic dyes. Forster conducted a spectroscopic analysis in 1940 to determine the shifts in polyene dyes as a result of chain length. Two years later, Herzfeld and Sklar utilized valence bond theory to predict the convergent and nonconvergent vinylene shifts in unsymmetrical and symmetrical cyanine dyes, respectively. By the year 1940, the literal meaning of the structure of organic molecules and the methods for computing the light absorption characteristics of dyes had been satisfactorily established.

In 1942, three distinct researchers independently introduced a novel quantum mechanical methodology for computing the characteristics of conjugated organic molecules. The method employed was the "electron gas" or "free electron" approach, as proposed by Bayliss, Kuhn, and Simpson (Maccoll, 1947). Although this approach proved to be highly effective in computing the spectroscopic characteristics of uncomplicated dyes and underwent numerous enhancements in subsequent years, it did not attain widespread recognition, primarily due to the heightened prominence of the HMO technique. In the initial stages of spectroscopic calculations, electronic interaction effects were disregarded. Consequently, in order to advance color prediction, it was necessary to refine the theories further. The majority of the progress made in MO theory after 1940 was focused on addressing the intricate issue of accommodating these effects. Despite this, Dewar managed to compute the color of various cyanine-type dyes with remarkable accuracy in 1950 by utilizing a properly parameterized version of HMO theory. At present, Valence Bond Theory has lost its prominence due to its unwieldiness and mathematical intricacy in addressing complex molecules.

1.3. SUMMARY

Dye chemistry is the branch of chemistry that deals with the study of dyes, which are substances used to impart color to various materials. Dyes have been used by humans for thousands of years, and their development and application have a rich historical background. The earliest evidence of dye usage dates back to ancient civilizations such as the Egyptians, who employed natural dyes derived from plants and minerals. The development of synthetic dyes began in the late 19th century. Dye chemistry continued to evolve with advancements in dye synthesis, understanding of dyeing processes and the development of more sustainable and eco-friendly dyes. Dye chemistry plays a vital role in various industries, including textile, printing, cosmetics, food and plastics.

REFERENCES

1. Abdou, M. M., (2013). Thiophene-based azo dyes and their applications in dyes chemistry. American *Journal of Chemistry, 3*(5), 126–135.
2. Ardila-Leal, L. D., Poutou-Piñales, R. A., Pedroza-Rodríguez, A. M., & Quevedo-Hidalgo, B. E., (2021). A brief history of color, the environmental impact of synthetic dyes and removal by using laccases. *Molecules, 26*(13), 3811–3813.
3. Asim, N., Ahmadi, S., Alghoul, M. A., Hammadi, F. Y., Saeedfar, K., & Sopian, K., (2014). Research and development aspects on chemical preparation techniques of photoanodes for dye sensitized solar cells. *International Journal of Photoenergy, 1*, 1–21.
4. Baker, R. M., (2011). *19th Century Synthetic Textile Dyes* (Vol. 1, pp. 4–10). Their history and identification on fabric Doctoral dissertation, University of Southampton.
5. Bao, G., Wen, S., Lin, G., Yuan, J., Lin, J., Wong, K. L., & Jin, D., (2021). Learning from lanthanide complexes: The development of dye-lanthanide nanoparticles and their biomedical applications. *Coordination Chemistry Reviews, 429*(1), 213–642.
6. Barnett, J. C., (2007). Synthetic organic dyes, 1856–1901: An introductory literature review of their use and related issues in textile conservation. *Studies in Conservation, 52*(1), 67–77.
7. Brenner, T., & Murmann, J. P., (2016). Using simulation experiments to test historical explanations: The development of the German dye industry 1857–1913. *Journal of Evolutionary Economics, 26*(1), 907–932.
8. Brooks, N. M., (2002). Nikolai Zinin and synthetic dyes: The road not taken. *Bulletin for the History of Chemistry, 27*(1), 26–36.
9. Burton, R. J., (2019). The potential impact of synthetic animal protein on livestock production: The new "war against agriculture." *Journal of Rural Studies, 68*(1), 33–45.
10. Chan, S. H. S., Yeong Wu, T., Juan, J. C., & Teh, C. Y., (2011). Recent developments of metal oxide semiconductors as photocatalysts in advanced oxidation processes (AOPs) for treatment of dye wastewater. *Journal of Chemical Technology and Biotechnology, 86*(9), 1130–1158.
11. Costela, A., García-Moreno, I., & Sastre, R., (2003). Polymeric solid-state dye lasers: Recent developments. *Physical Chemistry Chemical Physics, 5*(21), 4745–4763.

12. Dähne, S., (1978). Color and constitution: 100 years of research. *Science, 199*(4334), 1163–1167.
13. Desrochers, P., (2008). Bringing interregional linkages back in: Industrial symbiosis, international trade and the emergence of the synthetic dyes industry in the late 19th century. *Progress in Industrial Ecology, an International Journal, 5*(5–6), 465–481.
14. Dewar, M. J. S., (1950). Color and constitution. Part I. Basic dyes. *Journal of the Chemical Society (Resumed), 1*, 2329–2334.
15. Fabian, J., & Hartmann, H., (2013). *Light Absorption of Organic Colorants: Theoretical Treatment and Empirical Rules* (Vol. 12, pp. 7, 8). Springer Science & Business Media.
16. Ferreira, E. S., Quye, A., McNab, H., Wouters, J., Boon, J. J., & Hulme, A. N., (2001). Development of analytical techniques for the study of natural yellow dyes in historic textiles. *Dyes in History and Archaeology, 16*(1), 7–10.
17. Fleischmann, C., Lievenbrück, M., & Ritter, H., (2015). Polymers and dyes: Developments and applications. *Polymers, 7*(4), 717–746.
18. Grätzel, M., (2000). Perspectives for dye□sensitized nanocrystalline solar cells. *Progress in Photovoltaics: Research and Applications, 8*(1), 171–185.
19. Griffiths, J., (1988). Specialty dyes□new trends in modern dye chemistry. *Journal of the Society of Dyers and Colorists, 104*(11), 416–424.
20. Griffiths, J., (1990). Introduction: The evolution of present-day dye technology. *The Chemistry and Application of Dyes, 1*, 1–16.
21. Gürses, A., Açıkyıldız, M., Güneş, K., Gürses, M. S., Gürses, A., Açıkyıldız, M., & Gürses, M. S., (2016). Dyes and pigments: Their structure and properties. *Dyes and Pigments, 1*, 13–29.
22. Hagan, E., & Poulin, J., (2021). Statistics of the early synthetic dye industry. *Heritage Science, 9*(1), 33.
23. Higashino, T., & Imahori, H., (2015). Porphyrins as excellent dyes for dye-sensitized solar cells: Recent developments and insights. *Dalton Transactions, 44*(2), 448–463.
24. Hisano, A., (2016). The rise of synthetic colors in the American food industry, 1870–1940. *Business History Review, 90*(3), 483–504.
25. Hunger, K. (Ed.)., (2007). *Industrial Dyes: Chemistry, Properties, Applications* (Vol. 1, pp. 5–10). John Wiley & Sons.

26. Iftikhar, H., Sonai, G. G., Hashmi, S. G., Nogueira, A. F., & Lund, P. D., (2019). Progress on electrolytes development in dye-sensitized solar cells. *Materials, 12*(12), 1998.
27. Ihmels, H., & Otto, D., (2005). Intercalation of organic dye molecules into double-stranded DNA – General principles and recent developments. *Supermolecular Dye Chemistry, 1*, 161–204.
28. Kabir, F., Bhuiyan, M. M. H., Manir, M. S., Rahaman, M. S., Khan, M. A., & Ikegami, T., (2019). Development of dye-sensitized solar cell based on combination of natural dyes extracted from Malabar spinach and red spinach. *Results in Physics, 14*(1), 102474.
29. Khatri, A., Peerzada, M. H., Mohsin, M., & White, M., (2015). A review on developments in dyeing cotton fabrics with reactive dyes for reducing effluent pollution. *Journal of Cleaner Production, 87*(1), 50–57.
30. Kokkonen, M., Talebi, P., Zhou, J., Asgari, S., Soomro, S. A., Elsehrawy, F., & Hashmi, S. G., (2021). Advanced research trends in dye-sensitized solar cells. *Journal of Materials Chemistry A, 9*(17), 10527–10545.
31. Langhals, H., (1995). Cyclic carboxylic imide structures as structure elements of high stability. Novel developments in perylene dye chemistry. *Heterocycles, 1*(40), 477–500.
32. Laurent, A. D., Adamo, C., & Jacquemin, D., (2014). Dye chemistry with time-dependent density functional theory. *Physical Chemistry Chemical Physics, 16*(28), 14334–14356.
33. Lewis, D. M., (2014). Developments in the chemistry of reactive dyes and their application processes. *Coloration Technology, 130*(6), 382–412.
34. Liu, J., Zhou, Y., Zhao, F., Peng, Z., & Wang, S., (2016). Identification of early synthetic dyes in historical Chinese textiles of the late nineteenth century by high-performance liquid chromatography coupled with diode array detection and mass spectrometry. *Coloration Technology, 132*(2), 177–185.
35. Lubai, C., Xing, C., Kunyu, G., Jiazhen, H., & Griffiths, J., (1986). Color and constitution of azo dyes derived from 2-thioalkyl-4, 6-diaminopyrimidines and 3-cyano-1, 4-dimethyl-6-hydroxy-2-pyridone as coupling component. *Dyes and Pigments, 7*(5), 373–388.
36. Maccoll, A., (1947). Color and constitution. *Quarterly Reviews,*

Chemical Society, 1(1), 16–58.

37. Mahalingam, S., Manap, A., Omar, A., Low, F. W., Afandi, N. F., Chia, C. H., & Abd Rahim, N., (2021). Functionalized graphene quantum dots for dye-sensitized solar cell: Key challenges, recent developments and future prospects. *Renewable and Sustainable Energy Reviews, 144*(1), 110–999.
38. Meshnick, S. R., & Dobson, M. J., (2001). The history of antimalarial drugs. *Antimalarial Chemotherapy: Mechanisms of Action, Resistance, and New Directions in Drug Discovery* (Vol. 1, 15–25).
39. Murakami, T. N., Kijitori, Y., Kawashima, N., & Miyasaka, T., (2003). UV light-assisted chemical vapor deposition of TiO_2 for efficiency development at dye-sensitized mesoporous layers on plastic film electrodes. *Chemistry Letters, 32*(11), 1076–1077.
40. Mustroph, H., Stollenwerk, M., & Bressau, V., (2006). Current developments in optical data storage with organic dyes. *Angewandte Chemie International Edition, 45*(13), 2016–2035.
41. Muthyala, R., (2006). *Chemistry and Applications of Leuco Dyes* (Vol. 1, pp. 4–7). Springer Science & Business Media.
42. Olsen, S., (2012). A quantitative quantum chemical model of the Dewar–Knott color rule for cationic diarylmethanes. *Chemical Physics Letters, 532*(1), 106–109.
43. Otto Unverdorben, (1826). Ueber das verhalten der organischen körper in höheren temperaturen. *Annalen der Physik., 8*(11), 397–410. Bibcode: 1826AnP.84.397U. doi: 10.1002/andp.18260841109.
44. Park, J., & Shore, J., (2007). Significance of dye research and development for practical dyers. *Coloration Technology, 123*(4), 209–216.
45. Peters, A. T., & Freeman, H. S., (1991). *Color Chemistry: The Design and Synthesis of Organic Dyes and Pigments* (Vol. 1, pp. 193). London, New York: Elsevier applied science.
46. Pickering, A., (2005). Decentering sociology: Synthetic dyes and social theory. *Perspectives on Science, 13*(3), 352–405.
47. Polo, A. S., Itokazu, M. K., & Iha, N. Y. M., (2004). Metal complex sensitizers in dye-sensitized solar cells. *Coordination Chemistry Reviews, 248*(13–14), 1343–1361.
48. Rasmussen, S. C., (2018). Revisiting the early history of synthetic

polymers: Critiques and new insights. *Ambix, 65*(4), 356–372.

49. Sanda, M. D., Badu, M., Awudza, J., & Boadi, N., (2021). Development of TiO_2-based dye-sensitized solar cells using natural dyes extracted from some plant-based materials. *Chemistry International, 7*(1), 9–20.
50. Sandström, E., Vettorazzo, C., Mackay, C. L., Troalen, L. G., & Hulme, A. N., (2023). Development and application of desorption electrospray ionization mass spectrometry for historical dye analysis. *Analytical Chemistry, 95*(11), 4846–4854.
51. Saxena, S., & Raja, A. S. M., (2014). Natural dyes: Sources, chemistry, application and sustainability issues. *Roadmap to Sustainable Textiles and Clothing: Eco-Friendly Raw Materials, Technologies, and Processing Methods, 2*(2), 37–80.
52. Shapiro, A. E., (1993). *Fits, Passions and Paroxysms: Physics, Method, and Chemistry and Newton's Theories of Colored Bodies and Fits of Easy Reflection*. Cambridge University Press.
53. Stepien, M., Gonka, E., Żyła, M., & Sprutta, N., (2017). Heterocyclic nanographenes and other polycyclic heteroaromatic compounds: Synthetic routes, properties, and applications. *Chemical Reviews, 117*(4), 3479–3716.
54. Streb, J., Wallusch, J., & Yin, S., (2007). Knowledge spill-over from new to old industries: The case of German synthetic dyes and textiles (1878–1913). *Explorations in Economic History, 44*(2), 203–223.
55. Tan, K. B., Vakili, M., Horri, B. A., Poh, P. E., Abdullah, A. Z., & Salamatinia, B., (2015). Adsorption of dyes by nanomaterials: Recent developments and adsorption mechanisms. *Separation and Purification Technology, 150*(1), 229–242.
56. van Bommel, M. R., Berghe, I. V., Wallert, A. M., Boitelle, R., & Wouters, J., (2007). High-performance liquid chromatography and non-destructive three-dimensional fluorescence analysis of early synthetic dyes. *Journal of Chromatography A, 1157*(1–2), 260–272.
57. Van den Belt, H., & Rip, A., (1987). The Nelson-Winter-Dosi model and synthetic dye chemistry. *The Social Construction of Technological Systems: New Directions in the Sociology and History of Technology, 1*, pp. 129–54.
58. Vettese Forster, S., & Christie, R. M., (2013). The significance of the introduction of synthetic dyes in the mid-19th century on the

democratization of western fashion. *JAIC–Journal of the International Color Association, 11*(1), 4–9.

59. Vittum, P. W., & Weissberger, A., (1954). The chemistry of dye-forming development. *The Journal of Photographic Science, 2*(3), 81–96.
60. Waring, D. R., & Hallas, G., (2013). *The Chemistry and Application of Dyes* (Vol. 1, pp. 6–9). Springer Science & Business Media.
61. Yeh, B. J., & Lim, W. A., (2007). Synthetic biology: Lessons from the history of synthetic organic chemistry. *Nature Chemical Biology, 3*(9), 521–525.
62. Yen, Y. S., Chou, H. H., Chen, Y. C., Hsu, C. Y., & Lin, J. T., (2012). Recent developments in molecule-based organic materials for dye-sensitized solar cells. *Journal of Materials Chemistry, 22*(18), 8734–8747.
63. Zollinger, H., (2003). *Color Chemistry: Syntheses, Properties, and Applications of Organic Dyes and Pigments* (Vol. 1, pp. 5–7). John Wiley & Sons.

CHAPTER 2

Classification of Dyes by Chemical Structure

Contents

Unit Introduction 30

2.1. Azo Dyes 31

2.2. Anthraquinone Dyes 42

2.3. Benzodifuranone (BDF) Dyes 44

2.4. Polycyclic Aromatic Carbonyl Dyes 45

2.5. Indigoid Dyes 45

2.6. Polymethine and Associated Dyes 46

2.7. Styryl Dyes 48

2.8. Di and Tri-Aryl Carbonium and Related Dyes 49

2.9. Phthalocyanines 50

2.10. Quinophthalones 51

2.11. Sulfur Dyes 53

2.12. Nitro and Nitroso Dyes 53

2.13. Miscellaneous Dyes 54

2.14. Summary 55

References 56

UNIT INTRODUCTION

There exist multiple methods for categorizing dyes. One possible classification scheme for dyes is based on the type of fiber they are intended for, such as nylon, cotton, polyester, and other fibers. Dyes can be categorized based on their application technique to the substrate. The categorization in question would encompass a range of dye types, such as, disperse dyes, reactive dyes, vat dyes, azoic dyes, direct dyes and various others. The aforementioned categorizations are better suited for a publication that focuses on the technological aspects of dye application onto a substrate, as opposed to a work that primarily addresses dyes' synthesis and chemical properties. Dyes have been classified by Griaths solely on a theoretical basis, considering the electronic beginnings of color. There are only four classes that have been defined specifically (El Harfi & El Harfi, 2017):

- Donor-acceptor chromogens;
- Cyanine-type chromogens;
- Polyene chromogens; and
- n chromogens.

Despite its apparent simplicity, this system presents significant drawbacks from the perspective of a practicing dye chemist. The initial category encompasses the majority of dye types that hold significant commercial value, whereas the final category does not include any such types, and the remaining two categories comprise only a limited number of dye types. Consequently, this method does not classify the significant dyes in any way. Moreover, the aforementioned nomenclature is unfamiliar not only to the majority of chemists involved in dye research but also to the dyehouse scientist. The optimal approach for categorizing dyes is based on their chemical composition. This system presents numerous benefits. The identification of dyes is facilitated by their classification into distinct groups based on their characteristic properties, such as the strong and cost-effective azo dyes and the weak and expensive anthraquinone dyes. Secondly, it generates a limited quantity of clusters, approximately twelve. Thirdly, it is noteworthy that this classification is extensively utilized by both synthetic dye chemists and dye technologists, rendering it the most prevalent categorization. Hence, the terminologies such as azo yellow, anthraquinone red, and phthalocyanine blue are familiar to both chemists and technologists. Thus, the categorization of dyes based on their chemical composition is the methodology employed in this literary works and technologists who utilize these dyes (Abuamer et al., 2014).

Did You Know?

In 1856, Willian Henry Perkin made the discovery of the first synthetic dye, known as Mauveine.

The Color Index already established the classification of dyes (and pigments) based on their chemical structure. Reiterating this classification in this context would be redundant for various reasons. Initially, the reader may refer to the Color Index classification. The classification of color index encompasses pigments and dyes, whereas this publication is limited to the discussion of dyes exclusively. Therefore, it is noteworthy that significant categories of pigments, such as quinacridones, have been excluded. The present chapter's classification system, while retaining the fundamental structure of the Color Index classification, aims to surpass it by means of streamlining. Demonstrating the structural interdependencies among dyes, which are categorized into different categories by the Color Index, is accomplished. Ultimately, the classification selection aims to emphasize the latest advancements in the field of dye chemistry (Benkhaya et al., 2020).

The focus will be on the commercial significance of various dyes and recent advancements in the field. Notably, the blue reactive triphen-dioxazine dyes, red disperse tricyanovinyl styryl dyes, red disperse benzodifuranone (BDF) dyes, and blue disperse styryl dyes will be highlighted. The larger and more significant classes of dyes, namely azo, anthraquinone, and polymethine, will be subjected to further subcategorization.

For several years, dyestuff research's primary focus has been enhancing cost efficiency rather than advancing technical proficiency. Typically, the substitution involves the replacement of dyes that exhibit weak tinctorial properties, such as anthraquinone, which is the second most abundant class of dyes after azo dyes, with dyes that possess stronger tinctorial properties, like as heterocyclic azos, triphendioxazines, and BDFs. This particular theme will be consistently explored and analyzed throughout the chapter (Singh et al., 2017).

2.1. AZO DYES

The class of Azo dyes holds significant importance, constituting more than 50% of the entire range of commercially available dyes. As a result, azo dyes have drawn more attention in research than any other category, leading to a significant amount of available data on the subject. It is evident that this section can only cover the significant facets of azo dyes (Chung & Cerniglia, 1992).

Did You Know?

Azo dyes are prevalent among synthetic dyes and constitute more than 60% of the total dyes employed globally.

2.1.1. Basic Structure

Azo dyes are characterized by the presence of a minimum of one azo group (–N=N–), although they may also possess two, three, or, in rare instances, four azo groups. The azo moiety is covalently bound to two radicals, typically both of which are aromatic, although at least one is commonly so. The molecules are present in a trend format and exhibit a bond angle of approximately 120°. The nitrogen atoms display hybridization (Pavlović et al., 2009) (Figure 2.1).

Figure 2.1. *Demonstration of simple structure of azo dyes.*

Source: J. Griffiths, Creative Commons License.

Monoazo dyes, which are the most significant type, feature an A radical that comprises electron-accepting groups, while the E radical comprises electron-donating groups, such as hydroxy and amino groups. In academic literature, Azo dyes that solely consist of aromatic radicals are referred to as carbocyclic azo dyes, or occasionally as homocyclic azo dyes. Dyes with one or more heterocyclic radicals are classified as heterocyclic azo dyes. The structures presented depict representative instances of diverse azo dyes. These examples demonstrate the vast array of potential permutations that can arise in azo dyes, especially in the case of polyazo dyes. It is important to remember that the predominant type of commercially available azo dyes are monoazos, while trisazo and tetrakisazo dyes are comparatively less common (Alsantali et al., 2022) (Figure 2.2).

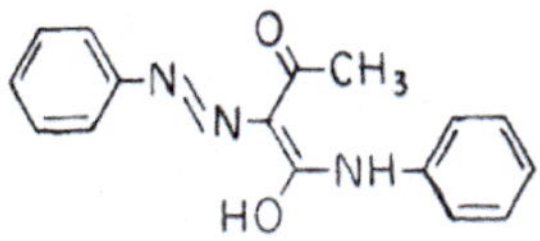

C.I. Disperse Yellow 16
Carbocyclic Monoazo
A→E

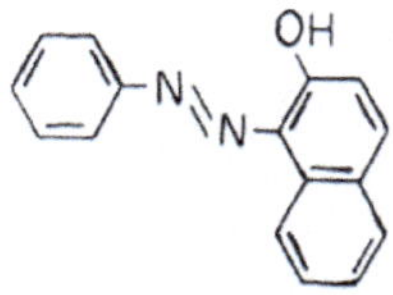

C.I. Solvent Yellow 14
Carbocyclic Monoazo
A→E

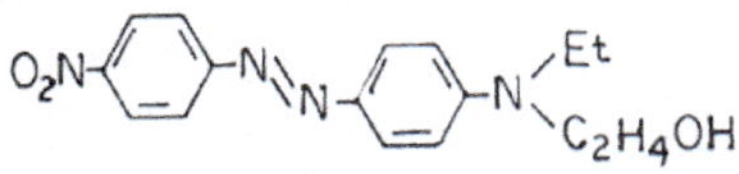

C.I. Disperse Red 1
Carbocyclic Monoazo

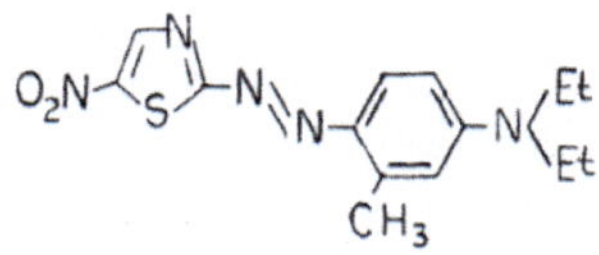

Disperse Blue
Heterocyclic Monoazo

Figure 2.2. *Illustration of patterns of azo dyes with their structures.*

Source: Geoffrey Hallas, Creative Commons License.

The correlation between the quantity of azo groups in each molecule and the potential permutations is apparent. The number of disazo dyes surpasses that of monoazo dyes, while the number of trisazo dyes exceeds that of disazo dyes, and so forth. The text highlights that tris- and tetrakis-azo dyes have limited commercial significance when compared to monoazo and disazo dyes. The examples provided in the text are few and far between (Ince & Tezcanli-Güyer, 2004).

Remember

The chromaticity of a dye is ascertained by its molecular configuration and its capacity to absorb and reflect light. Diverse chromatic hues are produced by distinct chemical compositions.

2.1.2. Synthesis

In the vast majority of cases, azo dyes are synthesized through the process of diazotization of an essential aromatic amine, which is subsequently coupled with an electron-rich nucleophile to form the desired diazonium salt. The contemporary dyestuffs sector is indebted to Peter Greiss, the chemist who identified the diazotization and coupling reaction in 1858.

The process of diazotization involves the treatment of a main aromatic amine with nitrous acid, which is typically produced in situ through the use of hydrochloric acid and sodium nitrite. The amine is nitrosated by nitrous acid to produce the N-nitroso compound, which subsequently undergoes tautomerization to form the diazo hydroxide. The resonance-stabilized diazonium salt can be generated through the process of protonation of the hydroxy group, which is subsequently followed by the elimination of water (Figure 2.3). Nitrosyl sulfuric acid ($NO+HSO_4$) is employed as the nitrosating agent for amines with weak basicity, characterized by the presence of multiple electron-withdrawing groups.

In contrast to a nitronium ion (NOT), a diazonium salt is considered a feeble electrophile. Consequently, it will exclusively interact with exceedingly electron-rich entities, including amino and hydroxy compounds (Shahab et al., 2016).

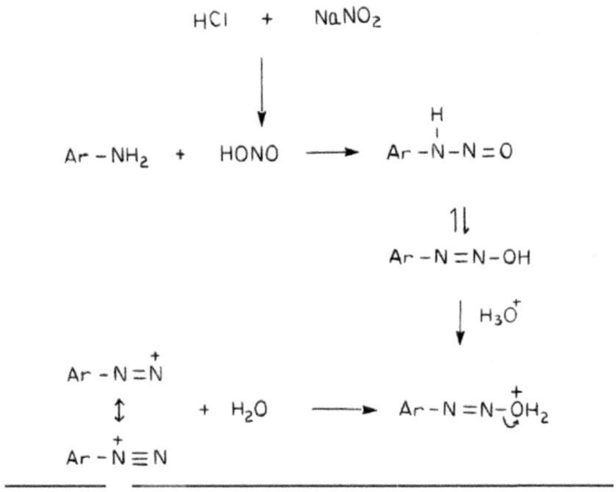

Figure 2.3. *Representation of synthesis of azo dyes.*

Source: David R. Waring, Creative Commons License.

Ionization of hydroxy compounds is a prerequisite for any chemical reaction to take place. As a result, alkaline conditions ($pH \geq pK_a$ of phenol or naphthol) are conducive to the coupling of hydroxy compounds, such as phenols and naphthols, while slightly acidic conditions are preferred for the coupling of aromatic amines, such as IN-dialkylanilines. The aforementioned approach ensures maximal stability of the diazonium salt, which exhibits stability under acidic conditions while simultaneously avoiding the deactivation of the nucleophile through amine protonation (Mallikarjuna & Keshavayya, 2020).

Stepwise coupling of coupling components that possess either amino or hydroxy groups, including H-acid, is feasible. The coupling process is initially conducted in an acidic environment to facilitate the formation of azo bonds within the ring containing amino groups. The pH is subsequently elevated to achieve ionization of the hydroxy group, typically to a pH value of ≥ 7, to facilitate coupling within the "naphtholate" ring, utilizing either the exact same or a distinct diazonium salt.

The extensive range of diazo parts and coupling components at disposal implies that an innumerable combination of azo dyes can be synthesized. The significant synthetic adaptability of azo dyes confers a notable benefit, as it enables the customization of molecules to meet specific application requirements. The rationale behind including hydroxy or amino groups in most azo dyes can be attributed to the synthetic pathway.

The production of azo dyes requires specific conditions, including the combination of strong acid and nitrous acid for diazotization, maintenance of low temperatures to stabilize the diazonium salt, and the presence of amino or hydroxy compounds with electron-rich properties for coupling. Consequently, azo dyes lack natural counterparts (Yildiz & Boztepe, 2002).

2.1.3. Tautomerism

Theoretically, it is possible for azo dyes to undergo tautomerism – or assume to have two or more structures that differ in the relative position of atoms attached, particularly that of hydrogen connectivity - wherein hydroxyazo dyes can exhibit azo/hydrazone tautomerism, amino azo dyes can exhibit azo/imino tautomerism, and protonated azo dyes can exhibit azonium/ammonium tautomerism (Lee et al., 2010).

2.1.3.1. Azo/Hydrazone Tautomerism

The aforementioned phenomenon was first identified by Zincke and Binderwald in 1884. The laborers procured identical orange-hued pigments by coupling benzene diazonium chloride with 1-naphthol and condensing phenylhydrazine with 1,4-naphthoquinone. The anticipated outcomes comprised the azo dye and the hydrazone dye. Zincke and Binderwald accurately postulated the existence of a dynamic equilibrium among the two molecular structures, known as tautomerism (Ball & Nicholls, 1982) (Figure 2.4).

The aforementioned finding has instigated comprehensive investigation into the phenomenon of azo/hydrazone tautomerism. This occurrence is not only intriguing but also holds significant importance in the context of commercial azo dyes. The reason for this phenomenon is that tautomers exhibit distinct characteristics such as varying hues, diverse attributes such as resistance to fading, divergent toxicological properties, and notably, different dyeing abilities. The cost-effectiveness of a dye is primarily determined by its tinctorial strength. Therefore, it is desirable for commercial azo dyes to exist in their strongest tautomeric form, which is the hydrazone form.

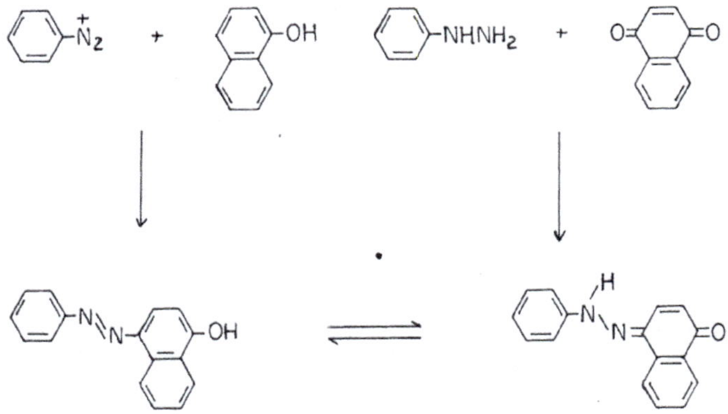

Figure 2.4. *Representation of the process of azo/hydrazine tautomerism.*

Source: F. Walker, Creative Commons License.

Hydroxyazo dyes can be classified into three main categories, with:

- The first being those that primarily exist as azo tautomers;
- Compounds that have the potential to exist as a blend of azo and hydrazone tautomers; and
- The entities that primarily exist as hydrazone tautomers (Özen et al., 2007).

The majority of azophenol dyes are classified under the first category. With the exception of a limited number of specific instances, they are entirely present in the azo configuration.

Did You Know?

Triphenylmethane dyes have been historically utilized as main indicators in acid and base titrations. An illustration can be observed in the case of phenolphthalein, which exhibits a pink hue upon encountering a basic substance.

Azonaphthol dyes exemplify the second category. Both the azo and hydrazone forms of 4-phenylazo-1-naphthol dyes coexist due to their comparable energies. The azo tautomers exhibit a yellow hue with a maximum wavelength absorption of approximately 410 nm and a maximum molar extinction coefficient of around 20,000. In contrast, the hydrazone tautomers display an orange-yellow coloration with a maximum wavelength absorption of approximately 480 nm and a maximum molar extinction coefficient of approximately 40,000. The tautomeric ratios are subject to the influence of both solvents (as depicted in Figure 2.5) and substitutions (as illustrated in Figure 2.6).

The hydrazone form of 2-phenylazo-1-naphthols and 1-phenylazo-2-naphthols is more prevalent than the azo form, as depicted in Figure 2.7.

Azopyrazolones, azopyridones, and azoacetoacetanilides are significant dye classes that exist entirely in the hydrazone form (Chen et al., 2012).

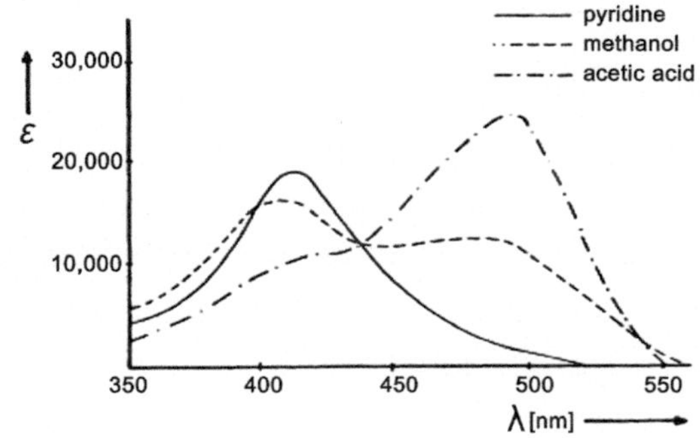

Figure 2.5. *Representation of effect of solvent on 4-phenylazo-1-naphthol.*

Source: P. Gregory, Creative Commons License.

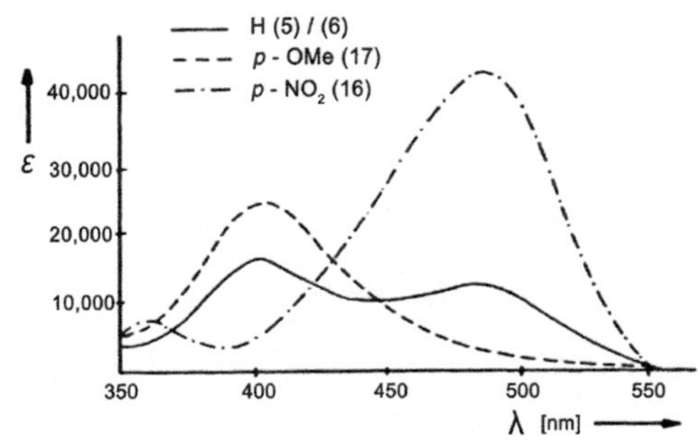

Figure 2.6. *Illustration of electronic effect of substituents in 4-phenylazo-1-naphthol.*

Source: P. F. Gordon, Creative Commons License.

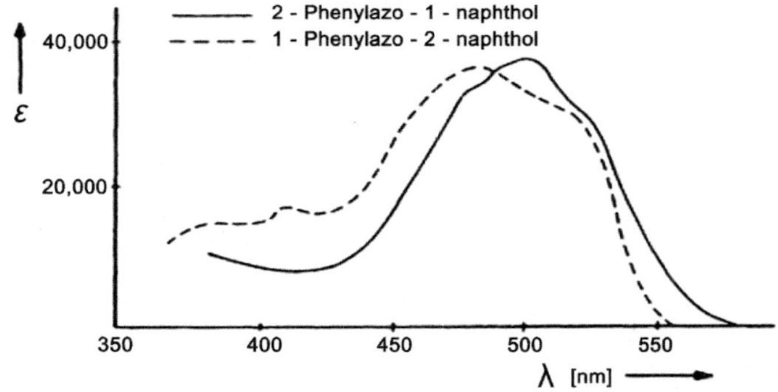

Figure 2.7. *Spectra of l-phenylazo-2-naphthol and 2-phenylazo-l-naphthol.*

Source: J. Griffiths, Creative Commons License.

2.1.3.2. Azo/Imino Tautomerism

The azo configuration is the sole form observed for aminoazo dyes, with no discernible indication of the imino configuration. The relative unpredictability of the imino categorizing will likely be a crucial factor.

Disazo dyes derived from aminonaphthols comprise two distinct groups, namely a genuine azo group and a hydrazo group (Gregory, 1990) (Figure 2.8).

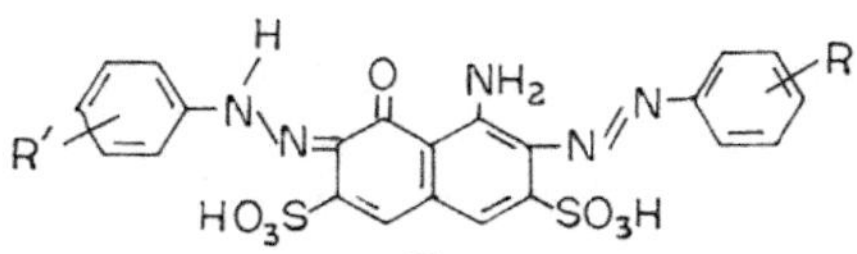

29

Figure 2.8. *Illustration of disazo dyes.*

Source: Geoffrey Hallas, Creative Commons License.

2.1.3.3. Azonium/Ammonium Tautomerism

The process of protonation in aminoazo dyes can occur at the terminal nitrogen atom, resulting in the formation of the ammonium tautomer, which is essentially colorless and has a maximum wavelength of absorption at approximately 325 nm. Alternatively, protonation can occur at the β-nitrogen atom of the azo group, leading to the formation of a resonance-stabilized azonium tautomer. The azonium tautomer exhibits enhanced brightness, greater bathochromic shift, and increased strength (with a maximum molar extinction coefficient of approximately (ε_{max} = 70,000) in comparison to the neutral azo dye. The azonium tautomers exhibit a correlation with the diazahemicyanine dyes that are employed for the purpose of imparting color to polyacrylonitrile (Venkatesh et al., 2011).

The protonation of azo dyes finds its most common application in indicator dyes, exemplified by Methyl Orange and Methyl Red. Further elaboration on the subject of tautomerism in azo dyes can be located in alternative sources (Stoyanov et al., 1996).

2.1.4. Metalized Azo Dyes

In this particular context, the significance lies solely in three specific metals, namely copper, chromium, and cobalt. The preeminent copper dyes are the copper(II) azo dye structures of formula 1:1 exhibit a planar configuration.

By way of comparison, the formation of 2:1 dye: metal complexes by chromium(III) and cobalt(III) results in nonplanar structures. Geometric isomerism is a phenomenon that exists. The Drew-Pfitzner or the mer type is formed by the o,o' dihydroxyazo dyes, whereas the Pfeiffer-Schetty or face type is formed by the o-hydroxy, o'-carboxyazo dyes (Devi & Reddy, 2010).

The process of metallization of dyes was initially implemented during the mordanting procedure to facilitate the fixation of the dye onto the substrate.

The utilization of premetallized dyes has become prevalent across diverse industries as a means to enhance the dye's characteristics, specifically its resistance to fading when exposed to light. Nevertheless, it should be noted that utilizing metallized azo dyes comes with a trade-off in luminosity, as they tend to exhibit a lower degree of brightness than their nonmetallized counterparts (Zhan et al., 1999).

Did You Know?

Historically, direct dyes were employed for the purpose of dyeing soldiers' uniforms throughout World War I, owing to their ease of application and extensive spectrum of colors.

2.1.5. Carbocyclic Azo Dyes

Carbocyclic azo dyes are widely used in commercial dye ranges due to their fundamental importance. These dyes offer a range of colors including yellow, red, blue, and green, for a variety of substrates such as polyester, cellulose, nylon, polyacrylonitrile, and leather. They are derived solely from benzene and naphthalene. Figure 2.9 displays common structures (Patil et al., 2023).

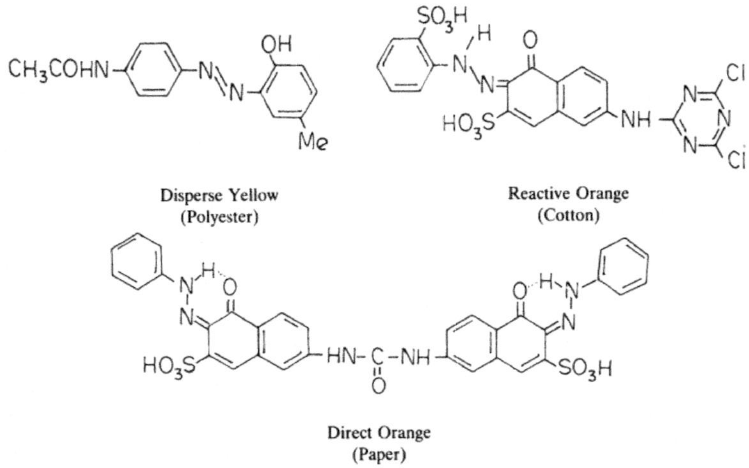

Figure 2.9. *Illustration of structure of various carbocyclic azo dyes.*

Source: David R. Waring, Creative Commons License.

The majority of azoic dyes are classified as carbocyclic azo compounds. The sole disparity lies in the formation of dyes within the internal pores throughout the process.

The carbocyclic azo category of dyes is known for its favorable cost-effectiveness and commendable overall fastness properties. Nevertheless, a characteristic that they do not possess is luminosity. As a result, their brightness cannot rival that of anthraquinone dyes. The deficiency exhibited by carbocyclic azo dyes has been discussed through the emergence of heterocyclic azo dyes (Abdou, 2013).

2.1.6. Heterocyclic Azo Dyes

The primary objectives of research on dyestuffs have been to enhance technical proficiency and achieve higher cost efficiency. A persistent objective was to amalgamate anthraquinone dyes' luminosity and superior permanence characteristics with the robustness and cost-effectiveness of azo dyes. The objective is currently being achieved through the utilization of heterocyclic azo dyes. Heterocyclic azo dyes can be classified into two primary categories: those obtained from heterocyclic coupling parts and those obtained from heterocyclic diazo components (Anderson & Nickless, 1967).

The heterocyclic coupling constituents that yield azo dyes of commercial significance solely comprise nitrogen as the heteroatom. Indoles, pyrazolones, and pyridones are classes of compounds that are known for their ability to produce yellow to orange dyes on a variety of substrates.

Azopyrazolone type dyes were commonly used as yellow dyes until a recent time period. Currently, the azo pyridone dyes have largely replaced these. Azopyridone yellow dyes exhibit superior brightness, strength, and fastness properties compared to azo pyrazolone dyes. As previously mentioned, they are present in the tautomeric form of hydrazone.

In contradistinction to the heterocyclic coupling constituents, the heterocyclic diazo constituents that furnish commercially significant azo dyes comprise sulfur, either in isolation or in conjunction with nitrogen. The S or S/N heterocyclic azo dyes exhibit vivid and intense hues spanning from red to blue to green, which effectively supplement the yellow/orange tones of the nitrogen heterocyclic azo dyes, resulting in a comprehensive spectrum of shades.

The thiadiazole red, nitrothiazole reddish blue, and thiophene greenish blue are examples of representative dyes. All of the dyes mentioned are classified as disperse dyes and are commonly used for coloring polyester, which is considered the second most significant substrate after cellulose (Mahmood et al., 2015) (Figure 2.10).

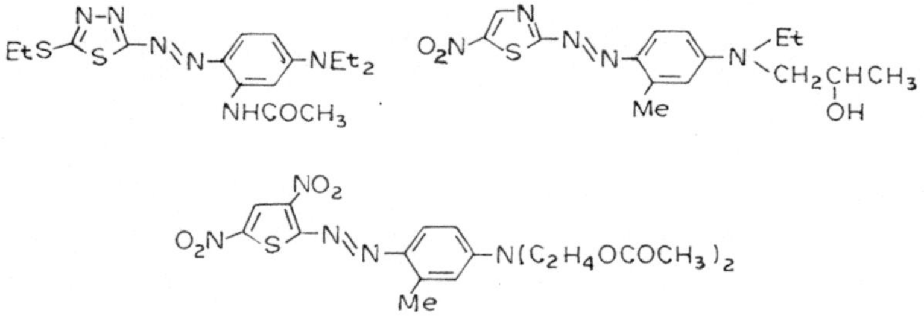

Figure 2.10. *Representation of structure of various heterocyclic azo structures.*

Source: C. W. Greenhalgh, Creative Commons License.

Azo dyes, as a collective group, are economically advantageous and exhibit a comprehensive range of hues, encompassing yellow, red, blue, and green, while also demonstrating favorable and well-rounded durability characteristics. The primary inadequacy of the heterocyclic azo dyes is their inferior luminosity compared to their primary competitors, the anthraquinone dyes. However, this shortcoming is currently being addressed (Vinodkumar et al., 2020).

2.2. ANTHRAQUINONE DYES

Following azo dyes, anthraquinone dyes constitute the second most significant category. Natural dyes have a long history and can be traced back to ancient times, as evidenced by their discovery in the coverings of mummies that date back more than four millennia. Anthraquinones were found to be the sole source of all significant natural red dyes, whereas azo dyes lacked natural counterparts (Routoula & Patwardhan, 2020).

2.2.1. Basic Structure

The basis for anthraquinone dyes is the organic compound 9,10 anthraquinone. Anthraquinone is predominantly devoid of color. In order to generate dyes that have commercial viability, electron-donor groups with significant potencies, including amino or hydroxy, are incorporated into any of the four alpha places (1, 4, 5, and 8). The prevalent substitution patterns observed are those of 1,4-, 1,2,4-, and 1,4,5,8-. In order to optimize the characteristics, both primary and secondary amino groups (excluding tertiary), as well as hydroxy groups, are utilized. These conditions optimize the degree of overlap between a-orbitals while minimizing steric hindrance and are further improved by intramolecular hydrogen bonding. The aforementioned characteristics are exemplified in the chemical compound known as C.I. Disperse Red 60 (Zhuo & Sun, 2013).

The ascending order of electron-donor group strength is hydroxyl < primary amine < secondary amine < aromatic amine. Anthraquinones that are tetra-substituted at positions 1, 4, 5, and 8 exhibit a greater degree of bathochromic shift than those that are di-substituted at positions 1 and 4 or tri-substituted at positions 1, 2, and 4. Hence, diverse hues can be attained through a judicious choice of donor groups and substitution patterns (Li et al., 2019). Figure 2.11 displays common dyes.

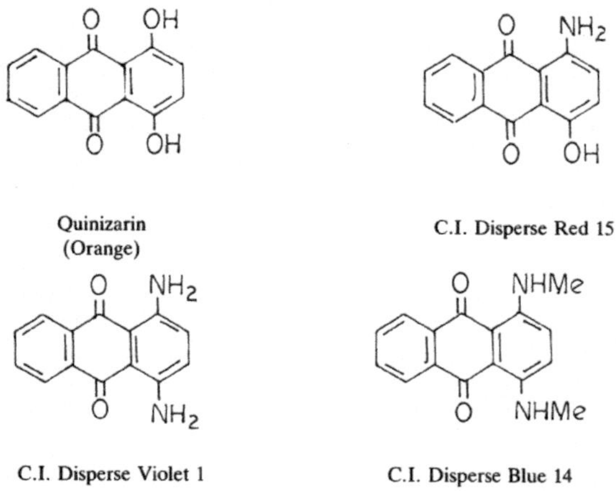

Figure 2.11. *Illustration of structures of various anthraquinone dyes.*

Source: B. Parton, Creative Commons License.

2.2.2. Synthesis

In stark contrast to azo dyes, anthraquinone dyes are synthesized through a gradual process of introducing substituents into the pre-existing anthraquinone framework, as depicted in the equation provided.

This underscores two crucial points of significance. Initially, the capacity to generate a diverse range of structures is limited. Additionally, the limited number of substitution centers, which amounts to only eight, presents an additional constraint on synthetic adaptability. Hence, the synthetic adaptability of anthraquinone dyes is considerably lower compared to azo dyes, which is a disadvantage of the former (Liu & Sun, 2008).

2.2.3. Tautomerism/Metal Complexes

While tautomerism has the potential to occur in amino and hydroxy anthraquinone dyes, empirical evidence of such a phenomenon has yet to be documented. Research conducted using carbon nuclear magnetic resonance spectroscopy has provided compelling evidence that amino and hydroxy derivatives of 9, 10-anthraquinone are present in their respective forms. Anthraquinone dyes are characterized by their notable brightness attributes and excellent fastness properties, encompassing lightfastness. The metallization process could negatively impact the former, and enhancing the latter is unnecessary. As a result, anthraquinone dyes that have been metallized do not hold any significant value (Waring & Hallas, 2013).

2.2.4. Properties

Anthraquinone dyes offer significant benefits in terms of their high brightness and excellent fastness properties. Nonetheless, it is important to note that these characteristics are counterbalanced by a significant drawback, namely their lack of cost-effectiveness due to their high cost and weak tinting strength. Anthraquinone dyes continue to be widely utilized, particularly in regions where red and blue hues are prominent, due to their unique combination of properties that cannot be replicated by alternative dyes, albeit at a higher cost. The current scenario indicates a shift in the trend as anthraquinone dyes are progressively facing competition from alternative dye varieties (Iwanaga et al., 2001).

2.3. BENZODIFURANONE (BDF) DYES

Benzodifuranone (BDF) dyes are a category of dyes that compete with anthraquinone dyes. The BDF chromogen represents a recent discovery among chromogens, with few others being identified in the current century. Similar to numerous significant dye breakthroughs, the BDF chromogen was fortuitously discovered. The validity of the pentacenequinone structure proposed by Junek for the highly red-hued substance produced from the reaction between p-benzoquinone and cyanoacetic acid was challenged by Greenhalgh et al. from ICI. The absence of conjugation in a compound renders it unsuitable for coloring. The BDF was accurately identified as the red compound.

Birefringent dichroic filters (BDFs) exhibit a unique characteristic of encompassing the entire range of colors in the visible spectrum, ranging from yellow to red and extending up to blue. The initial commercialization of a red disperse dye specifically designed for polyester fabrics has already generated a significant impact. The luminosity of this substance exceeds that of anthraquinone reds, and its superior tinctorial potency (approximately 3–4 times that of anthraquinones) renders it a cost-efficient option (Zeng et al., 2019).

2.4. POLYCYCLIC AROMATIC CARBONYL DYES

The dyes in question possess a structural composition characterized by the presence of one or more carbonyl groups that are interconnected via a quinonoid system. Typically composed of anthraquinones, these molecules are characterized by their relatively large size resulting from the assembly of smaller units. The polycyclic aromatic carbonyl dyes are commonly referred to as anthraquinonoid vat dyes due to their application onto the substrate, typically cellulose, through a matting process.

While polycyclic aromatic carbonyl dyes exhibit a wide range of colors, commercially significant hues are limited to blue dyes and tertiary shades such as browns, greens, and blacks. The commonly used polycyclic aromatic carbonyl dyes include indanthrone (blue), CI Vat Brown 3 (brown), CI Vat Black 27 (black), and CI Vat Green 1 (green), which is widely recognized as the most prominent among them (Figure 2.12).

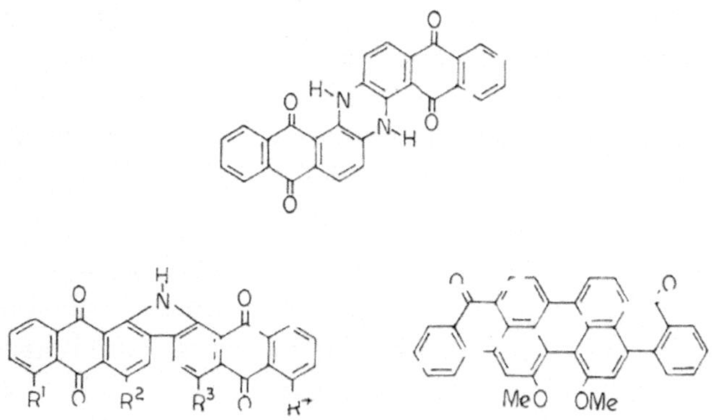

Figure 2.12. *Illustration of polycyclic aromatic carbonyl dyes structures.*

Source: W. Luttke, Creative Commons License.

The polycyclic aromatic carbonyl dyes, as a group, demonstrate the utmost degree of lightfastness and wet fastness. The lack of electron-donating and electron-withdrawing groups is a key factor in the high-lightfastness of the molecules. This limitation reduces the number of photochemical sites available. The direct cause of the high wet rapidity can be attributed to the specific application procedure for the dyes (Raval et al., 2017).

2.5. INDIGOID DYES

Similar to anthraquinone, BDF, and polycyclic aromatic carbonyl dyes, indigoid dyes are characterized by the presence of carbonyl groups. Additionally, they exhibit properties of vat dyes.

Indigoid dyes are considered to be among the most ancient types of dyes that have been discovered. Tyrian Purple, the renowned dye utilized by the Romans, is derived from 6,6' dibromoindigo. The dye known as Tyrian purple was highly priced, thereby limiting its accessibility to the affluent class who could afford clothing items colored with it. Undoubtedly, "born to the purple" is employed to connote affluence.

Despite the synthesis of numerous indigoid dyes, the significance of indigo remains unparalleled in contemporary times. Indigo is primarily utilized for the purpose of coloring denim jeans and jackets. The younger demographic holds it in high regard due to its distinctive blue hue and gradual tonal fade, resulting in increasingly lighter shades of blue (Stasiak et al., 2014).

The intriguing aspect of indigo lies in the fact that despite its diminutive size, it exhibits a blue hue. Typically, the production of blue dyes necessitates the utilization of extensive conjugation, such as in phthalocyanines (Section X), and/or the incorporation of multiple potent donor and acceptor groups, such as azo and anthraquinone blues. Numerous hypotheses have been put forth to elucidate the underlying cause of the blue hue of indigo. However, the dispute was not settled until recently, when Klessinger and Luttke's seminal research shed light on the matter. The researchers demonstrated with certainty that the chromophore, which is the molecular entity accountable for the coloration of indigo is the conjugated system that is cross-linked.

The aforementioned revelation led to the introduction of a novel nomenclature for this particular chromogen, which was denoted as an H-chromogen owing to its fundamental configuration that bore a striking resemblance to the uppercase letter H (Chowdhury et al., 2020).

2.6. POLYMETHINE AND ASSOCIATED DYES

Polymethine dyes that are most widely recognized are cyanine dyes. Currently, the utilization of exposing dyes for silver halide pictures is the primary commercial application. Cyanine dye derivatives are significant dyes for polyacrylonitrile.

2.6.1. Azacarbocyanines

A type of cyanine that possesses a three-carbon atom linkage between its heterocyclic nuclei is referred to as a carbocyanine. The substitution of carbon atoms with nitrogen atoms results in the formation of azacarbocyanines, diazacarbocyanines, and triazacarbocyanines, which correspond to one, two, and three nitrogen atoms, respectively. The yellow dyes belonging to these three categories hold significant importance as colorants for polyacrylonitrile (Gordon et al., 1987).

2.6.2. Hemicyanines

The dyes denoted as "half-cyanine" can be depicted by the structure presented herein. These compounds can be classified as cyanines that have undergone structural modification through the insertion of a benzene ring into the conjugated chain. Hemicyanines offer vivid red fluorescent dyes suitable for polyacrylonitrile (Mustroph, 2022) (Figure 2.13).

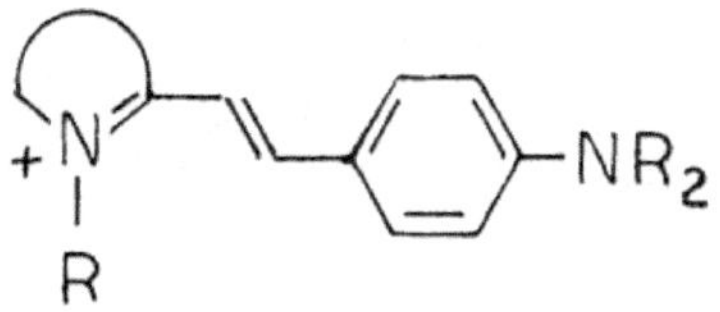

Figure 2.13. *Representation of structure of hemicyanines.*

Source: M. Klessinger, Creative Commons License.

2.6.3. Diazaheinicyanines

The Diazahemicyanine dyes represent a crucial category of polymethine dyes, possessing a fundamental structure (Figure 2.14).

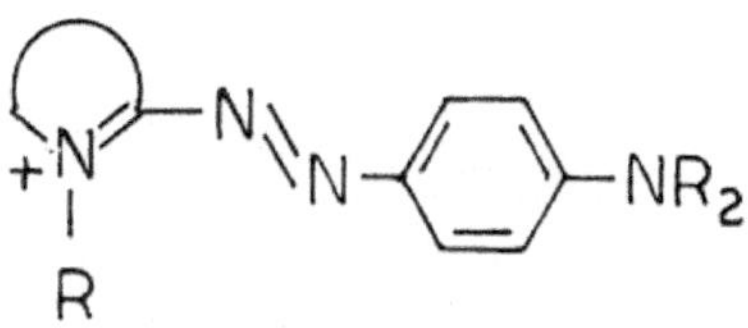

Figure 2.14. *Representation of structure of diazaheinicyanines.*

Source: M. Klessinger, Creative Commons License.

Typically, the heterocyclic ring is comprised of nitrogen atoms, ranging from one (e.g., pyridinium) to three (e.g., triazolium), or a combination of sulfur and nitrogen atoms, such as in (benzo)thiazolium and thiadiazolium.

Other examples of heterocyclic rings include pyrazolium and imidazolium, which contain two nitrogen atoms. The resulting red dyes for polyacrylonitrile in the market are provided by triazolium dyes, whereas the market-leading blue dye is a benzothiazolium dye (Matsubara et al., 1993).

2.7. STYRYL DYES

Styryl dyes have been categorized distinctly from polymethine dyes due to their non-polymethine nature. According to the Color Index, polymethine dyes are characterized as dyes with a conjugated carbon chain that an ammonium group caps and a nitrogen, sulfur, or oxygen atom. Alternatively, within the context of this particular section, they may refer to neutral molecules that feature a styryl group.

Styryl dyes have been historically significant as a collection of yellow dyes that have found diverse applications across various substrates. The compounds in question are produced through the process of condensation, wherein an active methylene compound, specifically malononitrile (where X and Y both represent CN), is combined with a carbonyl group, particularly an aldehyde. Styryl dyes possess compact molecular structures, rendering them highly suitable for the coloration of tightly packed hydrophobic materials, such as polyester. CI The compound Disperse Yellow 31, featuring R = Bu, $R^1 = C_2H_2Cl$, $R^2 = H$, X = CN, and Y = CO_2Et, represents a dye.

Although yellow styryl dyes were once commonly used, they have been replaced by more advanced dyes like azo pyridines. However, renewed attention has been given to red and blue styryl dyes (Jędrzejewska et al., 2013).

Remember

Styryl dyes can serve as vital dyes for staining and visualizing a diverse range of cellular components, including cell membranes, organelles, and lipid droplets. Live cell imaging examines can benefit significantly from their usage.

Incorporating a third cyano group to generate a tricyanovinyl moiety induces a significant bathochromic shift, thereby yielding dyes that exhibit a vivid red hue instead of the greenish yellow tint observed in dicyanovinyl dyes. Mitsubishi has obtained a patent for using tricyanovinyl dyes in the transfer printing process of polyester substrates. There are two methods for achieving cyanation: the first involves substituting a cyano group in tetracyanoethylene, while the second method entails oxidative cyanation of a dicyanovinyl dye using cyanide. The utilization of aforementioned hazardous substances may impede the process of bringing tricyanovinyl dyes to the market. Blue styryl dyes can be produced by utilizing an electron withdrawing group that possesses greater potency than tricyanovinyl. It was discovered by Sandoz that the condensation of a certain compound with an aldehyde result in the production of a vivid blue dye that is suitable for use with polyester. Furthermore, this particular dye exhibits a notable level of luminosity and a significant degree of tinctorial potency (with a maximum molar extinction coefficient of approximately 70,000). Nonetheless, the lightfastness of the aforementioned substance is merely moderate (Fedyunyayeva et al., 2011).

2.8. DI AND TRI-ARYL CARBONIUM AND RELATED DYES

Figure 2.15 illustrates the structural interrelationships among the diarylcarbonium dyes, triarylcarbonium dyes, and their heterocyclic derivatives.

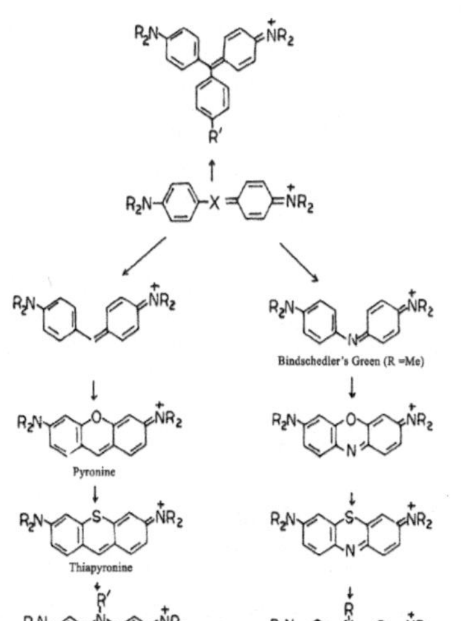

Figure 2.15. *Representation of structural interrelationships of di and tri-aryl carbonium and related dyes.*

Source: J. M. Robertson, Creative Commons License.

Collectively, the dyes exhibit vivid and potent coloration, yet they tend to exhibit inadequate resistance to fading when exposed to light. As a result, they find application in settings where luminosity and economic viability take precedence over durability, such as in the tinting of paper. Numerous dyes belonging to this category, particularly pyronine derivatives (xanthenes), are considered to be highly fluorescent dyes.

Commonly employed dyes include diphenylmethane, Auramine O (R = Me, X = CNH_2; CI Basic Yellow 2), triphenylmethane, Malachite Green (R = Me; R' = H; CI Basic Green 4), thiazine dye Methylene Blue (R = Me; CI Basic Blue 9), which is widely utilized in the gram staining test for bacteria, and oxazine dye, CI Basic Blue 3 (R = Et). The final dye is utilized to a certain degree in the process of dyeing acrylic fibers (Ganjoo et al., 2022).

The oxazine dyes merit further elaboration owing to the significant progress achieved in triphendioxazine dyes in recent times. The direct dyes of triphen dioxazine class have been a well-established research subject for several years, with CI Direct Blue 106 being a representative dye of this class. ICI and Parton's modification of the triphendioxazine chromogen resulted in the successful creation of blue reactive dyes for cotton, thereby renewing interest in these dyes, which are known for their inherent strength and brightness. The blue reactive dyes exhibit a combination of benefits attributed to both azo dyes and anthraquinone dyes. Therefore, these dyes exhibit high chromaticity, robustness, and excellent resistance to fading. The structural characteristics of these reactive dyes are conventional.

Similar to phthalocyanines, triphendioxazines are characterized by their large molecular size, which limits their applicability to coloring substrates with more open structures, which include paper and cotton (Gray, 1985).

2.9. PHTHALOCYANINES

Phthalocyanine represents the sole recently discovered chromogen of significant commercial significance subsequent to the nineteenth century. Similar to numerous significant dye discoveries, the discovery of phthalocyanines was fortuitous in nature. In 1928, while engaging in the customary production of phthalimide via the reaction between phthalic anhydride and ammonia (as represented by the equation below), it was discovered that the resultant substance was tainted with a blue impurity. An independent synthesis was conducted by chemists from Scottish Dyes Ltd, which is currently a subsidiary of ICI.

The blue substance is produced by introducing ammonia gas into hot phthalic anhydride containing iron fillings. The significance of the colorant was recognized due to its intense hue and remarkable stability, leading to the submission of a patent application in the corresponding year (Dahlen, 1939).

The elucidation of the structure of the blue material was accomplished in 1934 by Linstead, a researcher from the Imperial College of Science and Technology in London. It was determined that the material is the iron complex of. The novel substance was given the name phthalocyanine by Linstead, owing to its source (derived from phthalic anhydride) and its aesthetically pleasing blue hue (reminiscent of cyanine dyes). Robertson subsequently validated Linstead's structure using X-ray crystallography, one of this technique's earliest applications.

Phthalocyanines are chemical compounds that share structural similarities with the naturally occurring pigments chlorophyll and hemin. In contrast to natural pigments that exhibit notably low stability, phthalocyanines demonstrate remarkable stability and are arguably the most stable colorants currently utilized. The impact of substituents on the coloration of phthalocyanines is relatively insignificant, thereby limiting their chromatic range to shades of blue and green. Among the various metal complexes that were assessed, copper phthalocyanines exhibited the most favorable amalgamation of color and properties. As a result, almost all phthalocyanine dyes are founded on copper phthalocyanine, with CI Direct Blue 86 serving as a representative dye. In addition to their exceptional stability, phthalocyanines exhibit high brightness and tinting strength (with a maximum molar extinction coefficient of approximately 100,000), making them a cost-effective option. As a result, phthalocyanines find widespread application in the fields of inks for printing and paints. The widespread usage of phthalocyanine-based printing inks is evidenced by the prevalence of blue and green identification on consumer goods, while the majority of blue and green automobiles are the result of the application of phthalocyanine-based paints (Wöhrle et al., 2012).

2.10. QUINOPHTHALONES

Theoretically, quinophthalone dyes have the potential to display tautomerism, similar to hydroxy azo dyes. The synthesis of dyes involves the condensation of quinaldine products with phthalic anhydride, resulting in a structural representation as illustrated in Figure 2.16. The statement is erroneous as the presence of two single bonds hinders any possibility of conjugation among the two segments of the molecule. The dyes are present in a structural form where the pyrrole-type nitrogen atom acts as a donor and is conjugated to the two acceptor carbonyl groups through an ethylenic bridge. Furthermore, the enhanced conjugation is accompanied by additional structure stabilization by forming a six-membered intramolecular hydrogen bond that links the imino and carbonyl oxygen atoms.

Did You Know?

Quinophthalone dyes are frequently employed in the automotive sector to fabricate enduring and vivid coatings on automobiles and other modes of transportation.

Quinophthalones are significant dyes utilized for the pigmentation of plastics, such as CI Solvent Yellow 33 (where R = H), and for the pigmentation of polyester. The primary dye utilized for the image transfer distributing of polyester in the yellow hue is CI Spread Yellow 54 (R = OH) (Choi et al., 2014).

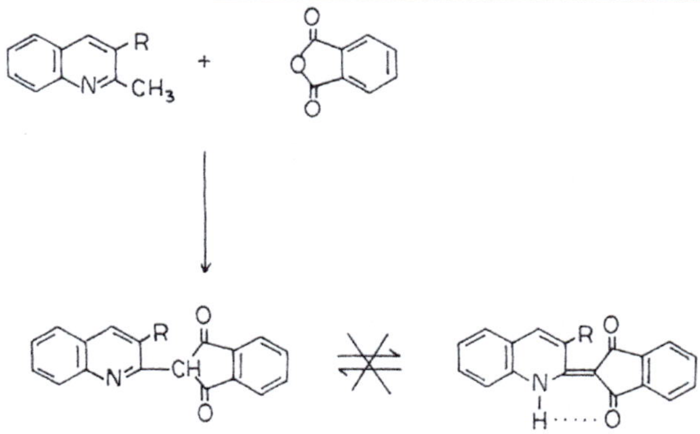

Figure 2.16. *Representation of condensation of quinaldine derivatives with phthalic anhydride to form quinophthalones.*

Source: B. Parton, Creative Commons License.

2.11. SULFUR DYES

Sulfur dyes are produced through a synthetic process that involves the application of heat to aromatic amines, phenols, or nitro compounds in the presence of sulfur or, more commonly, alkali polysulfides. Defining a chromogen for sulfur dyes is challenging, unlike other dye types. The probable composition of these entities comprises macromolecules featuring the phenothiazonethianthrone structure, wherein sulfur is present in the form of bridging links (sulfides) and thiazine groups.

Sulfur-based dyes are commonly employed in the dyeing process of cellulosic fibers. These compounds exhibit insolubility in aqueous solutions and require reduction to their leuco form, which is water-soluble, prior to application onto the substrate. This reduction process is achieved through the utilization of a sodium sulfide solution. The formation of the sulfur dye occurs inside the fiber pores through atmospheric oxidation.

Sulfur dyes are a significant category of dyes utilized in producing economical tertiary hues, particularly black, on cellulosic fibers. CI Sulfur Black 1 is a crucial dye that is synthesized by heating 2, 4-dinitrophenol with sodium polysulfide, as represented by the equation provided (Aspland, 1992).

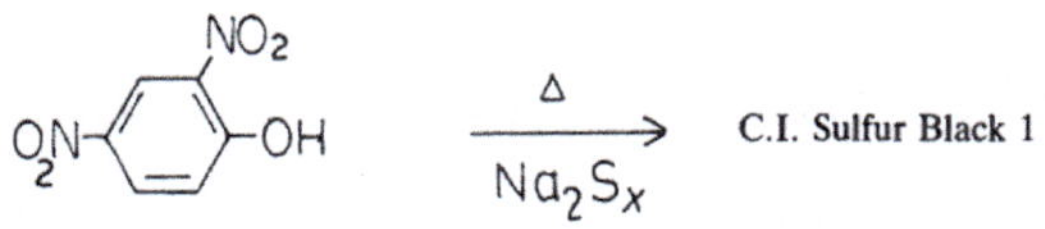

2.12. NITRO AND NITROSO DYES

Although these dyes currently hold little significance in the commercial realm, they remain intriguing due to their diminutive molecular structures.

Did You Know?

Nitro dyes represent one of the initial types of synthetic dyes that were created and subsequently employed in commercial applications.

The initial nitro dyes were categorized as acid dyes and were employed in the dyeing process of naturally occurring animal fibers, including wool and silk. The aforementioned compounds are examples of nitro derivatives of phenols, such as Picric Acid, CI 10305, or naphthols, such as CI Acid Yellow 1.

The nitrodiphenylamines are considered to be the most significant category of nitro dyes. The petite molecules are well-suited for infiltrating compact fibers, specifically polyester, and are consequently employed as dye dispersion molecules for polyester. Yellow is the predominant color of significant dyes. Despite their relatively low strength (with a maximum molar extinction coefficient of approximately 20,000), these dyes are considered cost-effective due to their facile synthesis from inexpensive intermediates, as illustrated in the equation below (Horobin, 2020).

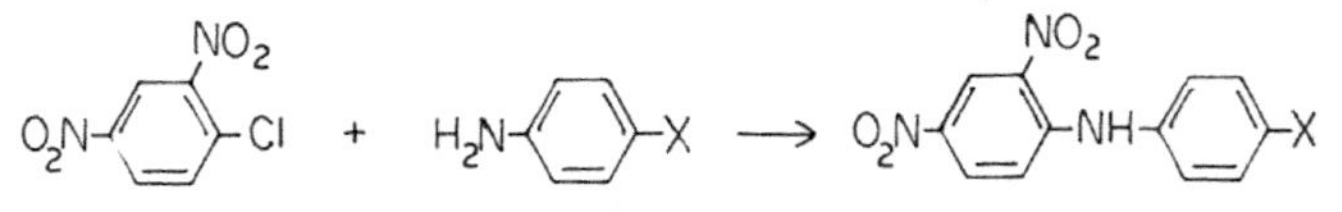

Nitroso dyes are compounds that result from the complexation of metals with o-nitrosophenols or naphthols. The metal-free precursor exhibits the potential for tautomerism between the nitrosohydroxy and quinoneoxime tautomeric forms.

The sulfonated 1-nitroso-2-naphthol iron complexes, such as CI Acid Green 1, are the sole commercially significant nitroso dyes. These cost-effective pigments are primarily utilized for paper coloration purposes (Taylor et al., 1982).

2.13. MISCELLANEOUS DYES

The stilbene dyes and formazan dyes remain significant classes of dyes in contemporary times. Stilbene dyes are commonly comprised of dye mixtures with uncertain composition, resulting from the sulfonated nitroaromatic compounds' condensation in aqueous caustic alkali. This process may occur independently or in conjunction with other aromatic compounds, such as arylamines. The sulfonated nitrostilbene holds a significant position among nitroaromatics, while the aminoazobenzenes are considered to be the most crucial arylamines. CI Direct Orange 34 is a representative example of a stilbene dye, which is derived from the condensation products of aminoazobenzene (Hammond et al., 1976).

Formazan dyes exhibit a formal similarity to azo dyes, owing to the presence of an azo group in their chemical structure. However, their distinct structural characteristics warrant their classification as a distinct category of dyes. Metal complexes, specifically copper complexes, of tetradentate formazans are considered to be the most significant formazan dyes. Reactive dyes are commonly utilized for cotton dyeing, with one of the representative examples being their use as such (Alsantali et al., 2022).

Did You Know?

The quest for environmentally sustainable dyes has resulted in the emergence of natural dyes sourced from flora, fauna, and microorganisms. The aforementioned dyes present a viable and eco-friendly substitute to their synthetic counterparts.

2.14. SUMMARY

From the perspective of both the dye chemist as the, the most effective approach to categorizing dyes is through their chemical composition. This results in a feasible quantity of approximately twelve categories. The class of Azo dye holds significant importance among all other dye classes. More than 50% of the present-day commercial dyes belong to the category of azo dyes. Anthraquinones are considered as the second most significant category of dyes, however, they are currently facing competition from heterocyclic azo dyes and more recent chromogens like BDFs. Acidic dyes refer to anionic dyes that are soluble in water and are predominantly utilized for the purpose of dyeing protein fibers, including but not limited to silk, wool, and nylon. Cationic dyes, commonly known as basic dyes, possess a positive charge and are predominantly utilized for dyeing acrylic, nylon, and protein fibers. Direct dyes are a type of water-soluble dye that can form a direct chemical bond with the fiber, eliminating the requirement for a mordant.

REFERENCES

1. Abdou, M. M., (2013). Thiophene-based azo dyes and their applications in dyes chemistry. *American Journal of Chemistry, 3*(5), 126–135.
2. Abuamer, K. M., Maihub, A. A., El-Ajaily, M. M., Etorki, A. M., Abou-Krisha, M. M., & Almagani, M. A., (2014). The role of aromatic Schiff bases in the dye's techniques. *International Journal of Organic Chemistry, 54*, 265–266.
3. Alsantali, R. I., Raja, Q. A., Alzahrani, A. Y., Sadiq, A., Naeem, N., Mughal, E. U., & Ahmed, S. A., (2022). Miscellaneous azo dyes: A comprehensive review on recent advancements in biological and industrial applications. *Dyes and Pigments, 199*, 110050.
4. Anderson, R. G., & Nickless, G., (1967). Heterocyclic azo dyestuffs in analytical chemistry. A review. *Analyst, 92*(1093), 207–238.
5. Aspland, J. R., (1992). Sulfur dyes and their application. *Textile Chemist & Colorist, 24*(3), 891–893.
6. Ball, P., & Nicholls, C. H., (1982). Azo-hydrazone tautomerism of hydroxyazo compounds – a review. *Dyes and Pigments, 3*(1), 5–26.
7. Benkhaya, S., M'rabet, S., & El Harfi, A., (2020). A review on classifications, recent synthesis and applications of textile dyes. *Inorganic Chemistry Communications, 115*, 107891.
8. Chen, X. C., Tao, T., Wang, Y. G., Peng, Y. X., Huang, W., & Qian, H. F., (2012). Azo-hydrazone tautomerism observed from UV-Vis spectra by pH control and metal-ion complexation for two heterocyclic disperse yellow dyes. *Dalton Transactions, 41*(36), 11107–11115.
9. Choi, J., Kim, S. H., Lee, W., Chang, J. B., Namgoong, J. W., Kim, Y. H., & Kim, J. P., (2014). The influence of aggregation behavior of novel quinophthalone dyes on optical and thermal properties of LCD color filters. *Dyes and Pigments, 101*, 186–195.
10. Chowdhury, M. F., Khandaker, S., Sarker, F., Islam, A., Rahman, M. T., & Awual, M. R., (2020). Current treatment technologies and mechanisms for removal of indigo carmine dyes from wastewater: A review. *Journal of Molecular Liquids, 318*, 114061.
11. Chung, K. T., & Cerniglia, C. E., (1992). Mutagenicity of azo dyes: structure-activity relationships. *Mutation Research/Reviews in Genetic Toxicology, 277*(3), 201–220.

12. Dahlen, M. A., (1939). The phthalocyanines a new class of synthetic pigments and dyes. *Industrial & Engineering Chemistry, 31*(7), 839–847.
13. Devi, L. G., & Reddy, K. M., (2010). Enhanced photocatalytic activity of silver metallized TiO_2 particles in the degradation of an azo dye methyl orange: Characterization and activity at different pH values. *Applied Surface Science, 256*(10), 3116–3121.
14. El Harfi, S., & El Harfi, A., (2017). Classifications, properties and applications of textile dyes: A review. *Applied Journal of Environmental Engineering Science, 3*(3), 01–03.
15. Fedyunyayeva, I. A., Klochko, O. P., Semenova, O. M., Khabuseva, S. U., Povrozin, Y. A., Sokolyk, O. O., & Patsenker, L. D., (2011). The synthesis, structure and spectral properties of new long-wavelength benzodipyrroleninium-based bis-styryl dyes. *Dyes and Pigments, 90*(2), 201–210.
16. Ganjoo, R., Verma, C., Kumar, A., & Quraishi, M. A., (2022). Colloidal and interface aqueous chemistry of dyes: Past, present and future scenarios in corrosion mitigation. *Advances in Colloid and Interface Science, 74*, 102832.
17. Gordon, P. F., Gregory, P., Gordon, P. F., & Gregory, P., (1987). Miscellaneous dyes. *Organic Chemistry in Color, 41*, 200–261.
18. Gray, T. H., (1985). *Picosecond Dynamics of the Solvent Dependent Ionic Photodissociation of Malachite Green Leucocyanide* (Vol. 546, pp. 3218–3220). Northwestern University.
19. Gregory, P., (1990). Classification of dyes by chemical structure. *The Chemistry and Application of Dyes, 4*(3), 17–47.
20. Hammond, P. R., Fletcher, A. N., Bliss, D. E., Henry, R. A., Atkins, R. L., & Moore, D. W., (1976). Search for efficient, near UV lasing dyes. III. Monocyclic and miscellaneous dyes. *Applied Physics, 9*, 67–70.
21. Horobin, R. W., (2020). *Nitroso and Nitro Dyes* (Vol. 1, pp. 101–105). In Conn's Biological Stains. Taylor & Francis.
22. Ince, N. H., & Tezcanli-Güyer, G., (2004). Impacts of pH and molecular structure on ultrasonic degradation of azo dyes. *Ultrasonics, 42*(1–9), 591–596.
23. Iwanaga, H., Naito, K., & Nakai, Y., (2001). The molecular structures and properties of anthraquinone-type dichroic dyes. Molecular crystals

and liquid crystals science and technology. *Section A: Molecular Crystals and Liquid Crystals, 364*(1), 211–218.

24. Jędrzejewska, B., Krawczyk, P., Pietrzak, M., Gordel, M., Matczyszyn, K., Samoć, M., & Cysewski, P., (2013). Styryl dye possessing donor-π-acceptor structure–synthesis, spectroscopic and computational studies. *Dyes and Pigments, 99*(3), 673–685.
25. Lee, H. Y., Song, X., Park, H., Baik, M. H., & Lee, D., (2010). Torsionally responsive C 3-symmetric azo dyes: Azo-hydrazone tautomerism, conformational switching, and application for chemical sensing. *Journal of the American Chemical Society, 132*(34), 12133–12144.
26. Li, H. H., Wang, Y. T., Wang, Y., Wang, H. X., Sun, K. K., & Lu, Z. M., (2019). Bacterial degradation of anthraquinone dyes. *Journal of Zhejiang University–Science B, 20*(6), 528–540.
27. Liu, J., & Sun, G., (2008). The synthesis of novel cationic anthraquinone dyes with high potent antimicrobial activity. *Dyes and Pigments, 77*(2), 380–386.
28. Mahmood, A., Tahir, M. H., Irfan, A., Al-Sehemi, A. G., & Al-Assiri, M. S., (2015). Heterocyclic azo dyes for dye sensitized solar cells: A quantum chemical study. *Computational and Theoretical Chemistry, 1066*, 94–99.
29. Mallikarjuna, N. M., & Keshavayya, J., (2020). Synthesis, spectroscopic characterization and pharmacological studies on novel sulfamethaxazole based azo dyes. *Journal of King Saud University-Science, 32*(1), 251–259.
30. Matsubara, Y., Kitano, K., Sasaki, Y., Yoshihara, M., & Maeshima, T., (1993). Crystal and molecular structure of 2-Ethylthio-3-phenyl-1, 3, 4-thiadiazolium tetrafluoroborate. *Bulletin of the Chemical Society of Japan, 66*(11), 3525–3527.
31. Mustroph, H., (2022). Bring back order in the polymethine dye medley: Classification, structure and spectra. *Dyes and Pigments, 456*(56), 110783.
32. Özen, A. S., Doruker, P., & Aviyente, V., (2007). Effect of cooperative hydrogen bonding in azo-hydrazone tautomerism of azo dyes. *The Journal of Physical Chemistry A, 111*(51), 13506–13514.
33. Patil, P. S., Gupta, P. O., Ingole, G. S., & Sekar, N., (2023). Anthrone-based carbocyclic azo dyes: Synthesis, dyeing, UV protection, anti-

microbial activity and computational study. *Fibers and Polymers, 24*(4), 1285–1296.

34. Pavlović, G., Racané, L., Čičak, H., & Tralić-Kulenović, V., (2009). The synthesis and structural study of two benzothiazolyl azo dyes: X-ray crystallographic and computational study of azo-hydrazone tautomerism. *Dyes and Pigments, 83*(3), 354–362.
35. Raval, N. P., Shah, P. U., & Shah, N. K., (2017). Malachite green "a cationic dye" and its removal from aqueous solution by adsorption. *Applied Water Science, 7*, 3407–3445.
36. Routoula, E., & Patwardhan, S. V., (2020). Degradation of anthraquinone dyes from effluents: A review focusing on enzymatic dye degradation with industrial potential. *Environmental Science & Technology, 54*(2), 647–664.
37. Shahab, S., Hajikolaee, F. H., Filippovich, L., Darroudi, M., Loiko, V. A., Kumar, R., & Borzehandani, M. Y., (2016). Molecular structure and UV-Vis spectral analysis of new synthesized azo dyes for application in polarizing films. *Dyes and Pigments, 129*, 9–17.
38. Singh, K. A. R. U. N. A., Kumar, P. A. N. K. A. J., & Srivastava, R. A. J. A. N. I., (2017). An overview of textile dyes and their removal techniques: *Indian Perspective Pollut. Res., 36*, 790–797.
39. Stasiak, N., Kukula-Koch, W., & Glowniak, K., (2014). Modern industrial and pharmacological applications of indigo dye and its derivatives – a review. *Acta Pol Pharm, 71*(71), 215–221.
40. Stoyanov, S., Antonov, L., Stoyanova, T., & Petrova, V., (1996). Ammonium-azonium tautomerism in some N,N-dialkylaminoazodyes–II. Compounds containing more than two protonation sites. *Dyes and Pigments, 32*(3), 171–185.
41. Taylor, E. C., Tseng, C. P., & Rampal, J. B., (1982). Conversion of a primary amino group into a nitroso group. Synthesis of nitroso-substituted heterocycles. *The Journal of Organic Chemistry, 47*(3), 552–555.
42. Venkatesh, G., Prabhu, A. A. M., & Rajendiran, N., (2011). Azonium-ammonium tautomerism and inclusion complexation of 1-(2, 4-diamino phenylazo) naphthalene and 4-aminoazobenzene. *Journal of Fluorescence, 21*, 1485–1497.
43. Vinodkumar, Keshavayya, J., Pushpavathi, I., Keerthikumar, C. T., Maliyappa, M. R., & Ravi, B. N., (2020). Synthesis, characterization,

computational and biological studies of nitrothiazole incorporated heterocyclic azo dyes. *Structural Chemistry, 31*, 1317–1329.

44. Waring, D. R., & Hallas, G., (2013). *The Chemistry and Application of Dyes* (Vol. 87, No. 4, pp. 218–220). Springer Science & Business Media.
45. Wöhrle, D., Schnurpfeil, G., Makarov, S. G., Kazarin, A., & Suvorova, O. N., (2012). Practical applications of phthalocyanines – from dyes and pigments to materials for optical, electronic and photo-electronic devices. *Macroheterocycles, 5*(3), 191–202.
46. Yildiz, E., & Boztepe, H., (2002). Synthesis of novel acidic mono azo dyes and an investigation of their use in the textile industry. *Turkish Journal of Chemistry, 26*(6), 897–904.
47. Zeng, W., Deng, Z., Wang, H., Zhang, H., & Zhou, Q., (2019). Benzodifuranone based color-changing epoxy-polyamine coating. *Dyes and Pigments, 164*, 198–205.
48. Zhan, H., Tian, H., Chen, K., & Zhu, W., (1999). Structural effect on photocatalytic degradation of metallized azo dyes in aqueous TiO_2 suspension. *Toxicological & Environmental Chemistry, 69*(3–4), 531–536.
49. Zhuo, J., & Sun, G., (2013). Antimicrobial functions on cellulose materials introduced by anthraquinone vat dyes. *ACS Applied Materials & Interfaces, 5*(21), 10830–10835.

CHAPTER 3

Basic Dyes

Contents

Unit Introduction 62
3.1. Properties 62
3.2. Uses of Basic Dyes 65
3.3. Application of Basic Dyes 65
3.4. Vastness Properties 68
3.5. Triphenylmethane Dyes 69
3.6. Other Basic Dyes 87
3.7. Organic Pigments from Basic Dyes 91
3.8. Manifold Record Systems 94
3.9. Summary 96
References 97

UNIT INTRODUCTION

Basic dyes are significant historically since the first synthetic dye, Mauvein (CI 50245), found in the oxidation byproducts of crude aniline by William Perkin in 1856, belongs to this class of colors. Soon after, crucially essential dyes like Malachite Green and Magenta were discovered. Almost all well-known basic dyes were found before the turn of the century, except for a few singular examples. Basic dyes are cationic dyes by definition and have substantivity for tannin-mordanted cotton. Basic dyes ionize in solution, which results in the presence of a positive-charged chromophore. Although it is likely more localized on the nitrogen atoms, the charge is generally delocalized across the chromophoric system. Methyl Green is one dye that has multivalent cationic chromophores (Dewar, 1950). The dye bases are less colorful, and color doesn't appear until after salt creation. Only a small amount of cotton is now dyed using basic colors. Most recent work in basic dyes is focused on creating new dyes for polyacrylonitrile fibers. Asfrazol Red GTL, which produces dyeings with great Vastness to light and good wet-Vastness, illustrates such a dye. The chapter on cafionic dyes for synthetic textiles covers the chemistry of these dyes. The current chapter discusses changes to traditional basic colors since 1951.

Berberine, an alkaloid, is the sole instance of a basic natural dye included in the color Index. However, its usefulness in the textile industry for coloring purposes is restricted. Basic dyes have been categorized based on their chemical composition and dyeing characteristics. Subsequent pages will review recent research on basic dyes, preceded by a concise overview of their dyeing and other pertinent characteristics (Sun & Yang, 2003).

Did You Know?

The first synthetic basic dye called mauveine or mauve was created by Sir William Henry Perkin in 1856.

3.1. PROPERTIES

The solubility of basic dyes in water is limited. It is imperative to consider the dissolution of these substances in the water to prevent the creation of an adhesive conglomerate, which, once formed, poses a challenge to dissolve. The optimal approach for dissolving a basic dye involves creating a paste by combining the dye with approximately an equivalent weight of ethanol or a diluted acetic acid solution (30%). A clear solution can be produced through the appropriate agitation of a sufficient amount of hot water. The compound Auramine, or CI 41000 or Basic Yellow 2, undergoes decomposition upon exposure to boiling water. It is recommended to dissolve it in water at temperatures that do not exceed 65°. Additional basic dyes exhibit slight decomposition when subjected to prolonged boiling with water. Basic dyes exhibit sensitivity to alkali and subsequently release dye bases with reduced color intensity upon exposure to alkali solutions. Pararosaniline (IV) can be converted into its corresponding carbinol base (V) and subsequently regenerated into the dye salt (IV) through acid treatment (Wang et al., 2008).

Subsequent acid treatment of (IV) dissolved in an alcoholic solution leads to salt formation at a different nitrogen atom, resulting in a noticeable alteration in color. The NH group in triphenylated Pararosaniline (VI) is anticipated to exhibit reduced basicity. This compound can be synthesized by heating (V) with an excess of aniline in the presence of benzoic acid at a temperature of 180°C. The carbinol bases of compounds (IV) and (VI) have been observed to produce the corresponding dye salts upon exposure to weak acids such as acetic, boric, or carbonic acids. This observation indicates that the basicities of the carbinols exhibit a comparable level, despite the relatively low basicity of the NH groups in compound (VI). Thus, it can be inferred that in acidic solutions, these dyes exist as carbonium ions, and their hue can be compared to the acid-derived salt of triphenyl carbinol (Meshko et al., 2001).

The sixth oxidation state results in the formation of a product. The pigments in printing inks are the heavy metal lakes of Ink Blue (S) or Helvetia Blue (Ciba). Many carbinol bases obtained from basic dyes exhibit limited solubility in aqueous solutions, and they undergo precipitation upon treatment of the dye solutions with alkali. Therefore, it is recommended to pre-treat water containing alkali or temporary hardness with acetic acid prior to utilizing it to dissolve basic dyes. Reducing agents facilitate the conversion of basic dyes into their respective colorless leuco bases, which can be easily reoxidized to the initial dyes. Upon reduction, Pararosaniline produces the leuco base. Similar to other aromatic compounds, the sulfonation process can be applied to basic dyes using sulfuric acid, resulting in their conversion to acid dyes. Basic dyes are characterized by their remarkable brilliance and intense color, surpassing all other synthetic dyes. These dyes are typically composed of lustrous crystalline structures that are visually appealing (Lazaridis et al., 2007).

Nevertheless, the application of basic dyes in dyeing is restricted due to their susceptibility to fading under the influence of light. Due to their cationic nature, basic dyes are susceptible to precipitation when exposed to anionic dyes, including acid and direct dyes. This characteristic has been utilized in the post-treatment process involving basic dyes and cellulosic materials that have undergone prior dyeing with direct colors. Hence, vivid hues are acquired. Azosol Fast Brilliant Red BA is a dye complex derived from Acid Yellow 99 (Wang et al., 2006) (Figure 3.1).

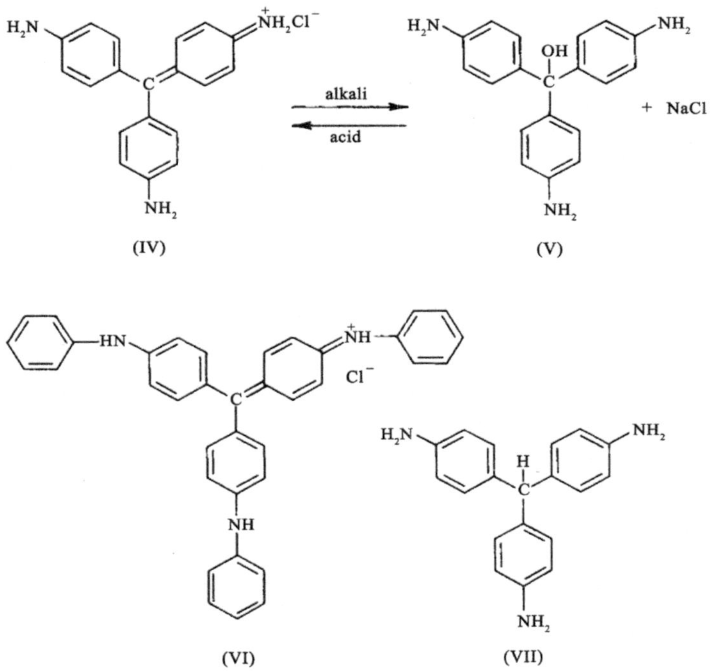

Figure 3.1. *Illustration of the structures of pararosaniline and carbinal bases.*

Source: K. Venkataraman, Creative Commons License.

The colorants employed in ink pastes for ballpoint pens are often insoluble salts derived from the combination of dye bases and acids. The red colorant for ballpoint pen inks can be derived from the insoluble salt produced through the process of heating a combination of Rhodamine 6G or 6GDN and Benzo Orange S. The resultant chromophore exhibits high resistance to photodegradation and exhibits limited solubility in ethanol or acetone. The simultaneous precipitation of cationic and anionic dyes has been harnessed in the development of novel biological stains and histochemical reagents, exemplified by Rhodanile Blue, which has been attributed with an undesirable covalently bound configuration (Asfour et al., 1985).

The pigment is synthesized by combining Rhodamine B (CI 45170, Basic Violet 10) and Nile blue. The molecular arrangements of these pigments have been debated, and it has been demonstrated that Rhodanile Blue is a mere amalgamation of Rhodamine B and Nile Blue. Basic dyes can be precipitated with tannic acid, which serves as the fundamental principle for utilizing basic dyes in the dyeing process of cellulosic fibers (Wang et al., 2004).

3.2. USES OF BASIC DYES

Basic dyes can be utilized on wool, silk, leather, and tannin-mordanted cotton. The basic dyes exhibit unparalleled tinctorial strength and color brilliance in their respective shades. Nevertheless, their fastness properties, particularly to light, are considered inadequate, and they have been almost entirely substituted by dyes belonging to other categories in terms of their application for dyeing textiles. The use of basic dyes has become widespread in dyeing textiles derived from polyacrylonitrile fiber due to the high use of these dyes for this purpose. Conventional techniques can impart vivid coloration with excellent lightfastness to synthetic fibers using basic dyes. Currently, basic dyes are predominantly utilized for non-textile applications. Basic dyes are widely employed for coloring paper, transfer sheet materials for hectographs, typewriter inks, and flexographic printing. Basic dyes have demonstrated potential interest in the field of photography. Certain fundamental dyes are also employed as biological stains and as pigments for inks (Brachet, 1953).

3.3. APPLICATION OF BASIC DYES

Under slightly acidic conditions, basic dyes can be utilized to dye protein fibers like wool, silk, and leather. The carboxylic acid groups unbound in protein fibers exhibit a proclivity to create salts with dye cations, indicating that the chemical affinity between protein fibers and cationic dyes is significant. Applying tannic acid and antimony salts, specifically tartar emetic (potassium antimony tartrate), as an after-treatment method can enhance the fastness of basic dyes on animal fibers. The strong attraction of basic dyes to protein fibers mandates the utilization of retardants during the dyeing process. A minute quantity of acetic acid fulfills the intended function (Ahmed, 2016). The presence of hydrogen ions that are unbound and available for chemical reactions can impede the absorption of dye molecules by competing for the same electronegative sites on the fiber. The uptake of Methylene Blue by wool exhibits a significant reduction below a pH of 6 and reaches a minimal level at a pH of 4.5. Crystal Violet exhibits analogous behavior. The depletion of fundamental dyes seems to be highly responsive to alterations in pH within the range of 4–6. Acrylic fibers possess acidic sulfonic acid groups, which result in a greater attraction towards basic dyes. This leads to improved wet fastness of the dyeings, albeit with slightly inferior leveling properties. In this instance, the depletion of the dyebath exhibits a lower degree of sensitivity towards alterations in pH.

Consequently, blends comprising acrylic fiber and wool can undergo dyeing with certain basic dyes at elevated temperatures, approximately at pH 4–5, while minimizing the occurrence of substantial wool staining. The exhaustion rate of basic dyes on acrylic fibers is gradual when the temperature is below 80°C. However, it experiences a significant surge beyond that temperature and peaks at 100°C. The expeditious absorption of fundamental dyes on acrylic filters has resulted in the emergence of dye retarders (Aspland, 1993).

These retarders can be classified into two categories: (i) those that interact with the dyes in the dye solution and generate a compound that disintegrates gradually at high temperatures; and (ii) those that possess cationic properties and vie with the cationic dyes for dyeing positions. The majority of quaternary ammonium salts that are soluble in water, including benzyl phenyl dimethyl ammonium chloride, exhibit a retarding effect and serve as effective leveling agents during the dyeing process of polyacrylonitrile fibers with basic dyes. The I-T Process utilizes an anionic leveling agent that creates a complex with basic dyes (El Harfi & El Harfi, 2017).

A nonionic agent is utilized to maintain them in a finely dispersed state. The dyeing process involves the gradual breakdown of the complex, resulting in uniform dye dispersion. Empirical evidence suggests that the adsorption equilibrium of a cationic dye is positively correlated with the number of sulfonic acid sites present in polyacrylonitrile filters. However, this relationship is non-linear. Polyacrylonitrile polymers are modified with acidic groups that regulate the adsorption of basic dyes by utilizing persulfates and bisulfites as polymerization catalysts. Studies have demonstrated that acrylic fibers blend potent and feeble acidic groups, with the latter exhibiting ionization constants similar to carboxylic acids. The alterations in dye absorption as pH increases can be attributed to the quantity and potency of these weakly cyclic entities (Benkhaya et al., 2020).

Basic dyes do not exhibit any affinity towards cellulosic fibers. Before dyeing with basic dye, subjecting these fibers to a mordanting process is necessary. Tannic acid is the conventional mordant employed, and it is subsequently stabilized through the administration of tartar emetic. Using synthetic mordants, namely Katanol, Katanol O, Resistone, or Tanninol BM, may sometimes serve as an alternative to circumvent the laborious process. Phenol derivatives are commonly synthesized through a process involving the application of heat to phenol in the presence of sulfur and a small amount of iron compound. The chemical composition of the substances is intricate, and their molecular structures remain unidentified. After the mordanting process, cotton fabric is subjected to dyeing with a dye solution in the presence of acetic acid. Prints produced through offset or lithographic techniques exhibit high washing and rubbing resistance (Ong et al., 2010) (Table 3.1).

Table 3.1. Fastness to Washing and Light of Some Basic Dyes on Wool and Cotton

	Fastness to			
	Washing		Light	
Dye	**Wool**	**Cotton**	**Wool**	**Cotton**
Auramine O	1–2	2	1–2	1–2
Auramine G	1–2	2	2	2
Malachite green	4–5	4–5	1	1
Brilliant green	2–3	4–5	1'	1–2'
Fuchsine (magenta)	3M	4–5	1	1
Crystal violet	2–3	3	1	1
Ethyl violet	2	2	1	1
Methyl green	1	2	1	1
Victoria pure blue BO	2	2–3	1	1
Rhodamine 6G	3–4	2–3	1	1
Rhodamine B	3	1	3	1
Meldola's blue	1–2	2	1	1
Methylene blue	1–2	3	1	3
Methylene green	4 (silk)	3'	2 (silk)	5

Source: K. Venkataraman, Creative Commons License.

The product's authenticity can be determined by applying either Continental or DEK (Deutsche Echtheitskommission) tests, while American (AATCC) tests can also be utilized for this purpose. Water-insoluble salts of basic dyes can be derived through the reaction with aliphatic carboxylic acids. These are remedied through the application of steam and soap (Gurses et al., 2016).

3.4. VASTNESS PROPERTIES

The basic dyes exhibit either subpar or moderately satisfactory washing fastness, with their primary limitation being their inadequate lightfastness. Table 3.1 presents the fastness properties of several commonly used basic dyes. Malachite Green, a diaminotriphenylmethane dye, exhibits greater wash-fastness than Crystal Violet, a triaminotriphenylmethane dye. The diethyl analog of Malachite Green, known as Brilliant Green, exhibits greater wash-fastness when compared to Ethyl Green, which is the diethylamino analog of Crystal Violet. Methyl Green, a compound possessing a dicationic nature, exhibits suboptimal wash-fastness characteristics. Rhodamine B exhibits moderate wash and lightfastness characteristics when applied to wool. Notably, Methylene Green, a nitro derivative of Methylene Blue, exhibits enhanced lightfastness compared to Methylene Blue. Nevertheless, the unappealing hue of a dull bluish green is not particularly aesthetically pleasing (Yang et al., 2008).

The photosensitivity of basic dyes is primarily attributed to their "onium" nature. The photostability of these dyes is minimally affected by modifying other aspects of the dye's chemical composition. Sulfonated basic dyes, classified as acid dyes, demonstrate enhanced lightfastness, rendering them a more desirable option for wool dyeing than basic dyes. The primary cause of the attenuation of basic dyes could be either the autoxidation of the dye or the oxidation by hydrogen peroxide, which is generated by the photochemical oxidation of minute amounts of water present in the fiber. CSD II contains a comprehensive record of the lightfastness properties of basic dyes (Freeze et al., 2019). The significance of basic dyes in the textile industry has been on the rise ever since the advent of polyacrylonitrile fibers in 1953. The fibers in question exhibit a high affinity for basic dyes, resulting in vivid coloration with excellent resistance to fading upon exposure to light. Polyacrylonitrile fibers exhibit hydrophobic characteristics, unlike natural fibers. The dye sites within the fibers are not readily accessible to moisture and oxygen, which are believed to contribute to the fading mechanisms. The observed phenomenon of basic dyes exhibiting high lightfastness on acrylic fibers could be attributed to this factor (Lacopetta et al., 2021).

In the context of dyeing basic dyes on acid-modified poly(ethylene terephthalate) (Dacron) fibers, research has indicated that certain substituted hydroxy benzophenones, including 2,2-dihydroxy-4,4'-dimethoxybenzophenone, exhibit light screening properties and enhance the lightfastness of the resulting dyeings. The application of these compounds can be carried out using conventional dyeing techniques. The fastness to light of cellulose film was enhanced by modifying it with cyanuric chloride and aminophenol treatment or by directly binding p-amino phenol, followed by dyeing with basic dyes like Rhodamine 6G or 6GDN and Crystal violet. However, the opposite outcome was observed when Rhodamine B was employed (Moosavi et al., 2020).

3.5. TRIPHENYLMETHANE DYES

The Triphenylmethane dyes constitute a significant subgroup of basic dyes. Despite the emergence of novel and superior dyes belonging to other categories, several Triphenylmethane dyes have sustained their significance (Figure 3.2).

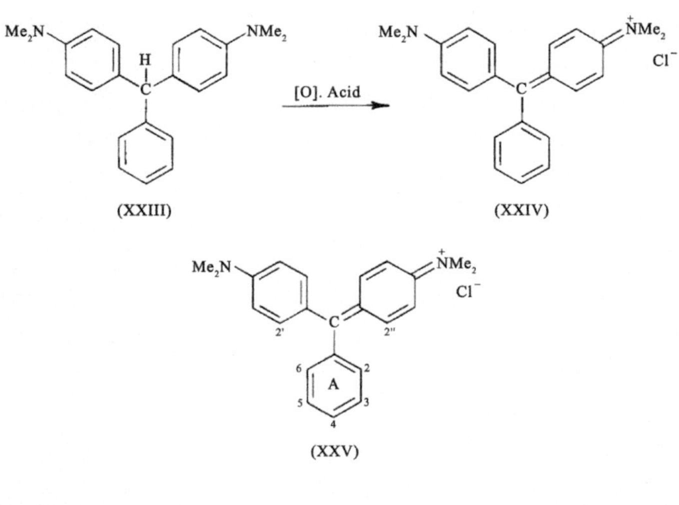

Figure 3.2. *Representation of the preparation reaction of malachite green dye.*

Source: N. R. Ayyangar, Creative Commons License.

The synthesis of Malachite Green involves the condensation reaction between two molecules of dimethylaniline and one molecule of benzaldehyde. After its production, the leuco base undergoes oxidation by utilizing oxidizing agents with high electrochemical potential, such as lead dioxide. This process results in the formation of the dye salt in an acidic environment. The study indicates that the production of Malachite Green is influenced by the speed at which the oxidizing agent is added to the 4, 4'-dimethylaminotriphenylmethane (Hameed et al., 2007).

The expeditious introduction of the oxidizing agent yielded a nearly complete conversion to compound (XXIV), whereas gradual addition led to yields of less than 35%. The reduced yields can be attributed to the creation of a hydrated structure vulnerable to oxidation and decomposition.

The preparation of derivatives of Malachite Green involves the substitution of fluorine at the 2,3-; 2,4-; 2,5-, and 2,6-positions in the phenyl ring A (refer to (XXV)). This can be achieved through condensing fluorine-substituted benzaldehydes with a molar quantity of dimethylaniline that is more than twice the required amount in the presence of zinc chloride. Subsequently, oxidation and acidification are carried out to obtain the desired product. The 2,'2"-difluoro derivative can be synthesized by reacting two fluoro dimethyl-aniline molecules with one benzaldehyde molecule in zinc chloride. Including fluorine atoms in ring A of (XXV) resulted in derivatives that exhibited blue to bluish-green dyeings with relatively favorable resistance to light. However, the derivative containing 2,'2" -difluoro demonstrated a yellowish hue and was comparatively less stable when exposed to light. The diminished lightfastness of triphenylmethane dyes can also be attributed to substituents such as chloro, nitro, and hydroxy groups in the ortho position relative to the central carbon atom (Thetford & Staff, 2000).

Water-soluble diaminotriphenylmethane derivatives, including (XXVII), can be prepared through a convenient condensation process with an aromatic amine sulfonic acid using the corresponding dimethoxy carbinol (XXVI). Applying basic dyes to polyacrylonitrile fibers has rekindled scholarly curiosity about these dyes and their preparation techniques. Methyl Violet (XXVI II) can be synthesized by oxidizing dimethylaniline using copper sulfate as a catalyst and diluents like phenol and salt. However, a more efficient method for obtaining pure Methyl Violet (XXVI II) in high yield involves conducting the oxidation reaction in the presence of an emulsifying agent. Emulsifying agents, namely sodium alkyl naphthalenesulfonate (Alkanol B), octyl alcohol, dibutyl phthalate, and dodecyl trimethylammonium chloride (Onyxsan S, Arganad 12), have the potential to be utilized for emulsification (Sani & Banerjee, 1999).

The procedural specifics are outlined as follows: A water-based solution comprising copper sulfate (160 units) and sodium chloride (53 units) was subjected to treatment with siliceous filter aid (24 units). A substance with emulsifying properties is introduced, followed by dimethylaniline. The concoction is subjected to a temperature of 60°C and agitated. A sodium chlorate solution is incrementally introduced into water and agitated vigorously until the reaction nears completion (Duxbury, 1993). Subsequently, the amalgamation is subjected to a thermal treatment at 85°C, followed by a subsequent cooling process until it reaches ambient temperature. Copper sulfide and Methyl Violet are separated from the aqueous solution through precipitation and salting. The synthesis of Crystal Violet, also known as (XXIX), has been documented through the reaction of dimethylaniline vapors with aluminum chloride in a macroscopic setting. The triphenylmethane dye, denoted as XXX, is synthesized through the condensation of Michler's hydrol and N-ethyl-N-benzyl-m-toluidine in the presence of sulfuric acid, followed by oxidation with lead dioxide. The dye exhibits a vivid blue hue after application to unbleached sulfite-treated cellulose, wood pulp, and ethyl cellulose (Cheriaa et al., 2012).

The production of dyes appropriate for dyeing and printing acrylic fibers involves the treatment of ammonia or primary aryl amines with 4,4,'4"-trihalogenotriphenyl carbinol (XXXI) (Figures 3.3 and 3.4).

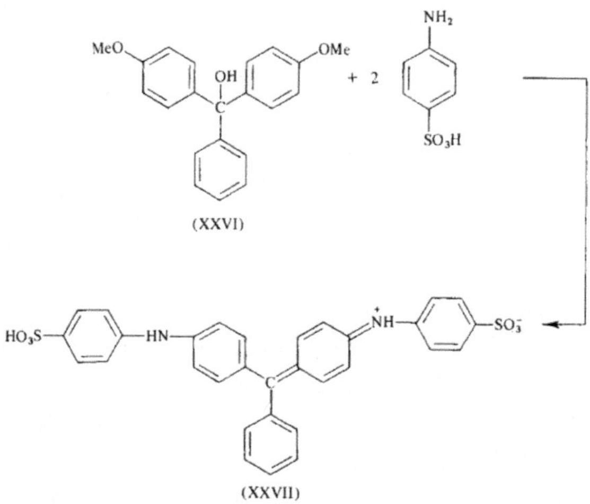

Figure 3.3. *Preparation of diaminotriphenylmethane derivative from the corresponding dimethoxy carbinol by condensation with an appropriate aromatic amine sulfonic acid.*

Source: Donald R. Baer, Creative Commons License.

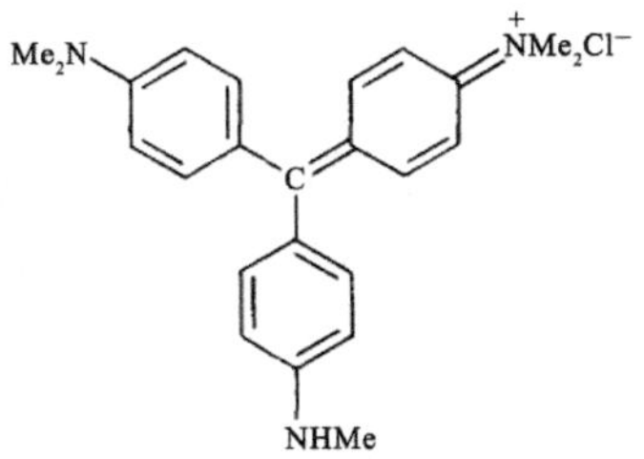

(XXVIII)

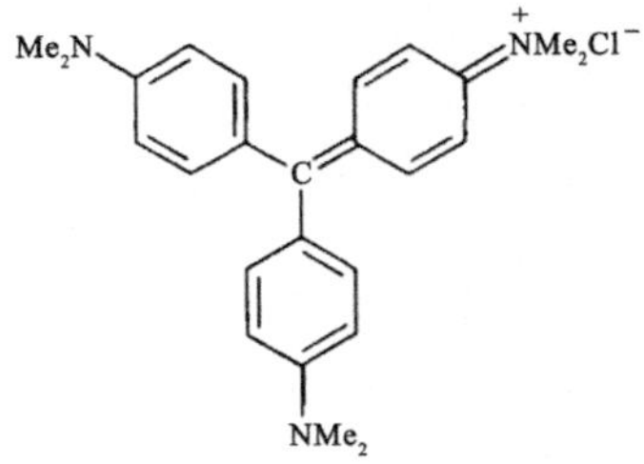

(XXIX)

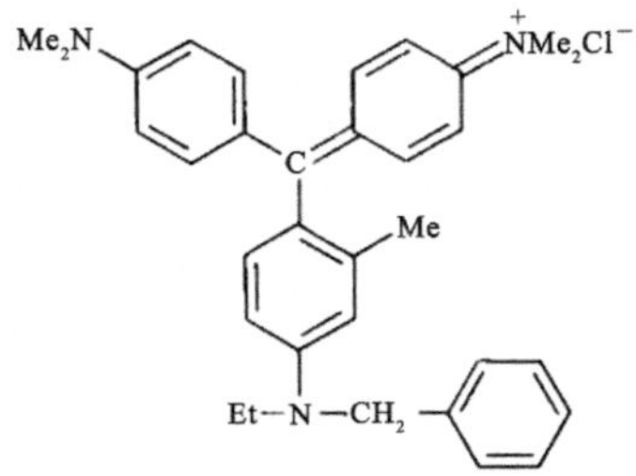

(XXX)

Figure 3.4. *Illustration of the structures of crystal violet and triphenylmethane dyes.*

Source: J. Bailey, Creative Commons License.

Carbinol (XXXI) can be synthesized by condensing p-chlorobenzotrichloride with two chlorobenzene molecules in aluminum or ferric chloride at 80° to 100°. The synthesis of diaminocarbinol (XXXII) involves subjecting a mixture of (XXXI) (182), copper powder (2), copper sulfate (10), and 25% ammonia (600 parts) to heat in an autoclave at a temperature range of 195° to 200° for 10–12 hours. The resulting product is obtained in the form of a pale colored powder. Reaction with acid produces a salt that yields a red-violet dye. When applied from a dilute acetic acid bath at high temperature, this dye imparts a Bordeaux shade on polyacrylonitrile. Triphenylrosaniline (VI) can be synthesized by reacting (XXXI) with aniline. The carbinol (XXXI) yields symmetrically substituted dyes (Przystas et al., 2012). Symmetrical and unsymmetrical halides of 4-halogeno-4,' 4-diaryl amino triphenylmethane, including compound (XXXIII), can be synthesized from the aluminum chloride complex of 4,4,'4"-trihalogenotriphenylmethyl chloride (XXXIV). In contrast to carbinol (XXXI), this complex can undergo successive single-step reactions with arylamines. The complex metal salt of 4,4'-dihalogeno-4" methyl phenylamino) triphenylmethane can be synthesized by reacting the aluminum chloride complex of (XXXIV) with one molecule of m-toluidine at a temperature range of 95–110°C in an inert organic solvent. The unsymmetrical dye (XXXII I) can be obtained by subjecting it to further reaction with an excess of a second arylamine (such as o-chloroaniline) and treating the resulting product with dilute acid (Mishra & Maiti, 2018) (Figure 3.5).

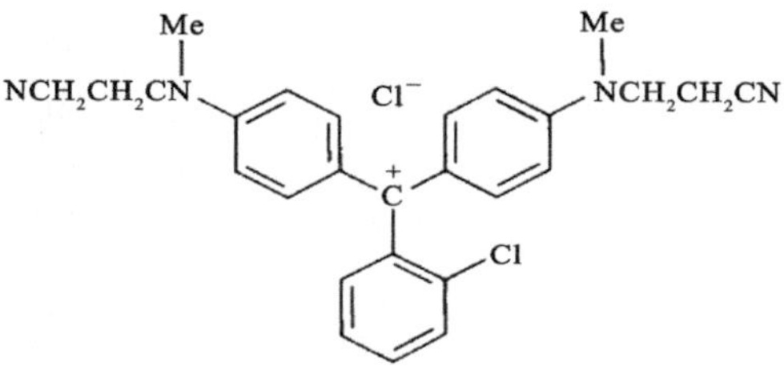

(XXXV)

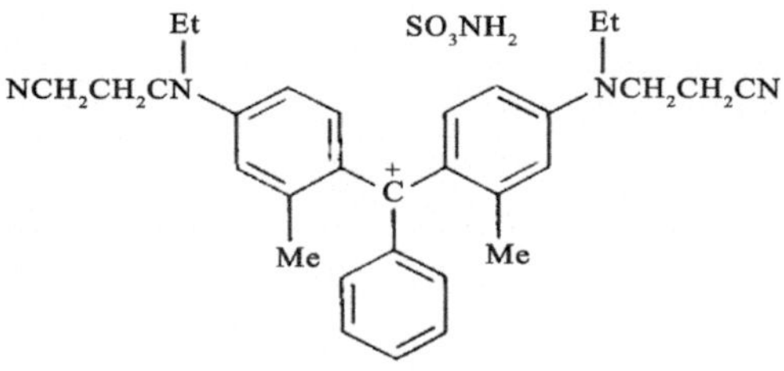

(XXXVI)

Figure 3.5. *Illustration of the structures of bright greenish blue dye for acid-modified acrylic fiber and a similar dye prepared from benzaldehyde and N-ethyl-N - (2-cyanoethy1)-m-toluidine.*

Source: G. E. Fickman, Creative Commons License.

The present study reports the synthesis of a vivid greenish blue dye (XXXV) suitable for acid-modified acrylic fiber. The dye was prepared through a reaction involving o-chlorobenzaldehyde and two molecules of A'-methyl-N-(2-cyanoethyl) aniline in isopropanol. The presence of sulfuric acid or p-toluenesulfonic acid and urea facilitated the reaction. The leuco derivative is subjected to lead dioxide oxidation, followed by acid treatment of the resulting carbinol.

The dyes used in this process exhibit pH stability. The dye denoted as XXXVI, which is synthesized through the reaction between benzaldehyde and N-ethyl-N-(2-cyanoethyl)-m-toluidine, is capable of imparting a yellowish green hue to polyacrylonitrile upon dyeing (Babendure et al., 2003) (Figure 3.6).

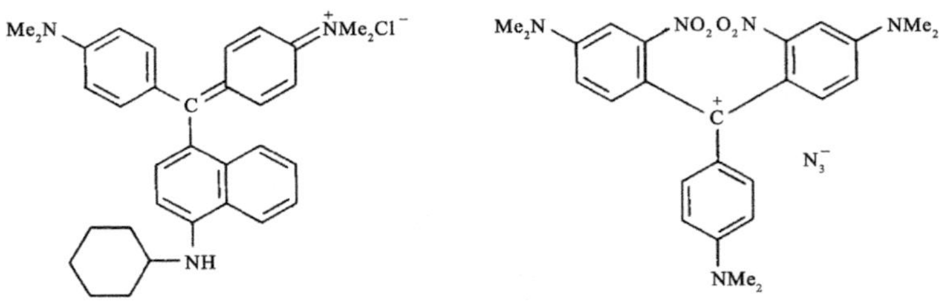

(XXXVII) (XXXVIII)

Figure 3.6. *Illustration of the structures of blue alcohol-soluble dyes and dinitro crystal violet azide.*

Source: Oskar Glenz, Creative Commons License.

Several recently developed triphenylmethane derivatives exhibit promising applications in photography, transfer sheet materials, and flexographic printing. Blue alcohol-soluble dyes, such as (XXXVII), are prevalent in producing carbon papers and offset printing inks. Additionally, these dyes are employed to tint nitrocellulose lacquers, waxes, and plastics. The dye above is synthesized through condensation between Michler's ketone and N-cyclohexyl-1-naphthylamine, facilitated by phosphorus oxychloride. Aliphatic groups can be used to substitute the cyclohexyl ring further. The reddish blue dyes exhibit superior performance compared to Victoria Blue B and demonstrate resistance to precipitation from alcoholic solutions even upon prolonged exposure. When a paper is printed with a solution of dinitro Crystal Violet azide (XXXVIII) and subsequently processed by coming into contact with either 25% water in acetone or kaolin, it produces vividly pigmented images (Sankar et al., 1999).

Tris-4 [bis (4-dimethylamino phenyl) methylamino phenylmethane (XXXIX)] solutions exhibit a colorless appearance when they are neutral or slightly alkaline. Incorporation of said solution into coatings of transfer sheet materials or typewriter ribbons, when utilized in conjunction with an acid-receptive sheet, results in the expeditious development of a dark blue hue. Under the influence of an oxidizing agent, a crimson hue is produced, which in conjunction with the blue hue, yields a violet tint. Hence, it is feasible to acquire duplicates in hues of blue and violet. Compound

(XXXIX) is synthesized through the process of condensation involving three tris (p-aminophenyl) methane (VII), which is obtained via the reduction of Pararosaniline. The photosensitive properties of cyano derivatives of dyes, such as New Magenta (XL), make them valuable for actinometry, photographic, and photoproduction applications (Goldacre & Phillips, 1949). The preparation of the XL colorless compound involves subjecting New Magenta to a temperature of 45°C in an aqueous solution (350) containing 95% sodium cyanide (5.3) in water (25). The product underwent purification through dissolution in triethyl phosphate and subsequent dilution with water at a temperature of 25°C. It is recommended that all operations be conducted in the absence of ultraviolet (UV) light. The preparation of colorless, photosensitive cyano derivatives of triphenylmethane dyes can be achieved through a heating process involving the parent dyes and an excess amount of sodium or potassium cyanide. This process is carried out in containers of a specific size and is conducted at a temperature of 100°C. Triphenylmethane derivatives, namely (XLI) and (XLII), exhibiting photosensitivity and a color range from gray to black, have practical applications in textiles and graphic arts (Combes & Haveland, 1982). These derivatives can be synthesized through the condensation reaction of piperonal with twice the molar amount of N,N-diethyl-m-toluidine in the presence of an acid catalyst. The cyano derivative designated XLII is synthesized by applying sodium cyanide to compound XLI. When a combination of (XLI), 2- (o-chlorophenyl) -4, 5-diphenyl imidazolyl dimer, methanol, and dimethylformamide is applied onto paper, dried using infrared light, and exposed to 2537–4200 A light, the resulting area is observed to have a gray to black hue. Comparable outcomes are achieved using the cyano derivative (XLII). Using heterocyclic systems in triphenylmethane derivatives, exemplified by compound (XLIII) synthesized from furfural and N,N-diethyl-m-toluidine, is advantageous in photographic applications. The preparation of the dye involves the process of oxidizing the resulting leuco derivative. The dye obtained from (XLIII) exhibits fastness properties and imparts a dark green color. The present study involved the synthesis of triphenylmethane derivatives that incorporate naphthotriazole moieties (Ashour et al., 2014). This was achieved through a series of chemical reactions, which included diazotization of a triphenylmethane derivative possessing an unbound amino group, coupling with 2-naphthylamine, oxidation with copper acetate in pyridine to obtain the triazole, and subsequent oxidation with lead dioxide to yield the carbinol. The final product was obtained by treating the carbinol with an acid to form the dye salt. The dye salt of (XLV) was provided by (XLIV).

The synthesis of a Malachite Green analog (XLVI) was achieved through the reaction of p-dimethylamino-phenyl lithium, derived from p-bromo-N,N-dimethylaniline, with methyl ferrocenoate in an ether solvent." The dye base converted to the dye salt (XLVI) through treatment with acetic acid, resulting in a melting point range of 155–156°C. The dye was found to display a rapid fading of its color, with a half-life of 10 minutes (Chen & Ting, 2015) (Figure 3.7).

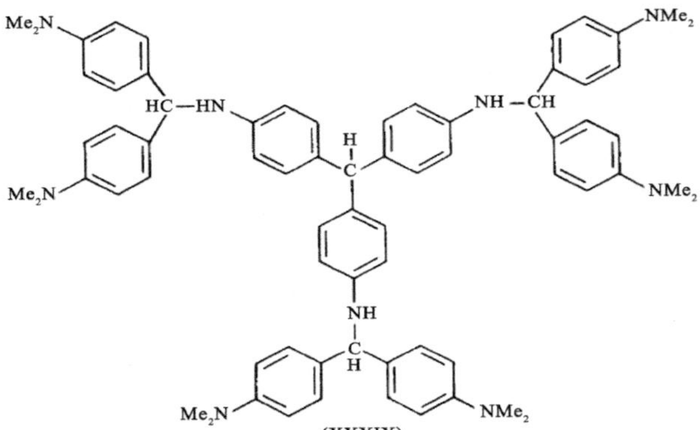

(XXXIX)

Figure 3.7. *Illustration of the structures of tris-4 [bis (4-dimethylaminophenyl) methylamino phenylmethane.*

Source: Hans Meier, Creative Commons License.

The dye stabilization was significantly enhanced by adding ethanol at 10% to the acetic acid. Substituting the phenyl group of Malachite Green (XXIV) with a ferrocenyl group induces a hypsochromic shift in the first band and a significant bathochromic shift in the second band. Incorporating a highly electron-donating methoxyl group at the para position of Malachite Green results in a comparable outcome, as demonstrated in Table 3.2. The synthesis of the thiophene derivative (XLVII) of Malachite Green involved the utilization of zinc chloride as a catalyst during the condensation reaction between 2-thenaldehyde and dimethylaniline. The leuco base that ensued was subjected to oxidation, leading to the formation of the dye (XLVII) (Ayed et al., 2010).

The triarylmethane dyes were obtained from Alkyl or halogeno-substituted 2-thenaldehydes. Substituting the phenyl group in (XXIV) with the 2-thienyl group results in a significant red shift in the dyes' absorption spectrum's second absorption band (p band). The primary absorption band of

Malachite Green, specifically the z band, is attributed to the electron cloud's oscillation across the molecule between the two auxochromes (Michaels & Lewis, 1985).

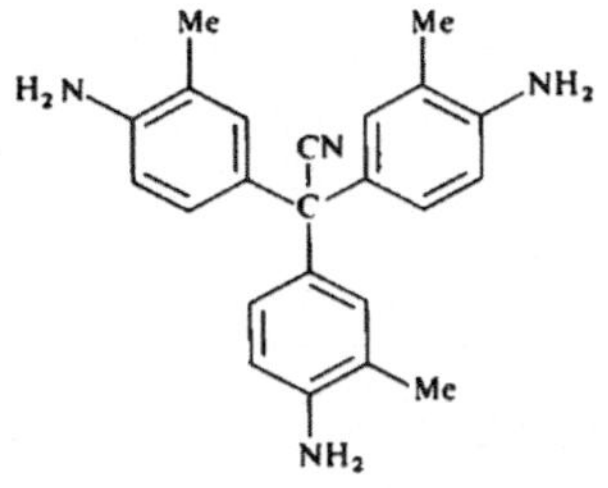

(XL)

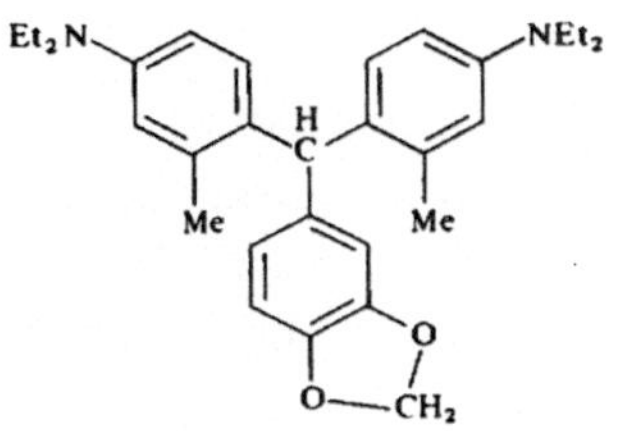

(XLI)

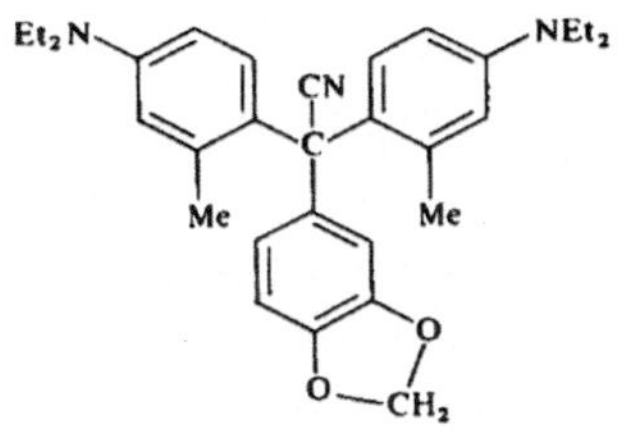

(XLII)

Table 3.2. Spectra of Malachite Green has p-Methoxy and Ferrocenyl Derivatives

Dye	)
Malachite green (XXIV)	621 (10.4); 427 (2.0)
p-OMe deriv. of (XXIV)	621 (10.4); 427 (2.0)
(XLVI)	607 (5.2); 457.5 (1.8)

Source: K. Venkataraman, Creative Commons License.

The second band, denoted as p, pertains to a vibrational mode that involves oscillation across the phenyl group. Meanwhile, the third band, I, is linked to a secondary vibrational mode that exhibits greater amplitude. The observed bathochromic effect resulting from substituting the phenyl group with the 2-thienyl group in Malachite Green is attributed to the comparatively higher electron-releasing ability of the thiophene ring compared to the benzene ring. The increased basicity of nitrogen atoms and subsequent reduction in absorption frequency can be attributed to donating electrons to the central carbon atom by the thiophene ring. The resonating structures (XLVIII, a,b,c) provide a means to visualize the electron-donating capacity of the thiophene ring. The analogs of Malachite Green, namely 3-thiophene, 2-thianaphthene, and 3-thianaphthene (designated as XLIX, L, and LI, respectively), can be synthesized by utilizing the corresponding aldehydes (Mali et al., 2000).

Numerous triarylmethane dyes have been synthesized, which feature a heterocyclic group derived from various compounds such as pyridine, quinoline, 2-methylindole, carbazole, benzothiazole, benzoselenazole, phenothiazine, phenoxazine, phenoselenazine, and sulfonyldiphenylamines, alongside two substituted phenyl groups. Many of these dyes exhibit compatibility with filter and anti-halation layers in photographic materials. The violet dyes suitable for dyeing silk and cotton were synthesized through the condensation of carbazole and N-methylcarbazole with Michler's ketone in the presence of phosphorus oxychloride. The second edition of the Color Index has included several triarylmethane dyes that incorporate heterocyclic systems. Wool Fast Blue FBL and Chromoxane Brown 5R (LIII) are two distinct chemical compounds. Wool Fast Blue FBL belongs to the class of indole derivatives. At the same time, Chromoxane Brown 5R (LIII) is a pyrasolone derivative that is synthesized through the condensation of formaldehyde with 2,3-cresotic acid, followed by co-oxidation with 3-methyl-1-phenyl-5-pyrasolone in nitrous acid. The commercial utilization of these dyes is restricted due to their high cost (Shah a et al., 2011).

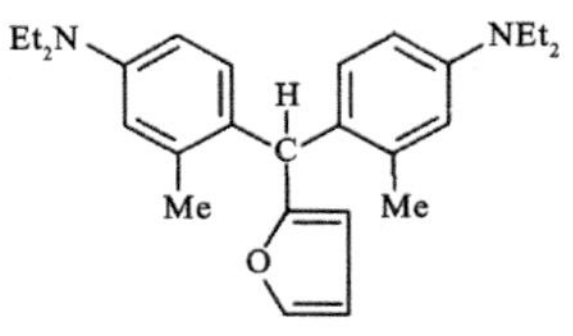

(XLIII)

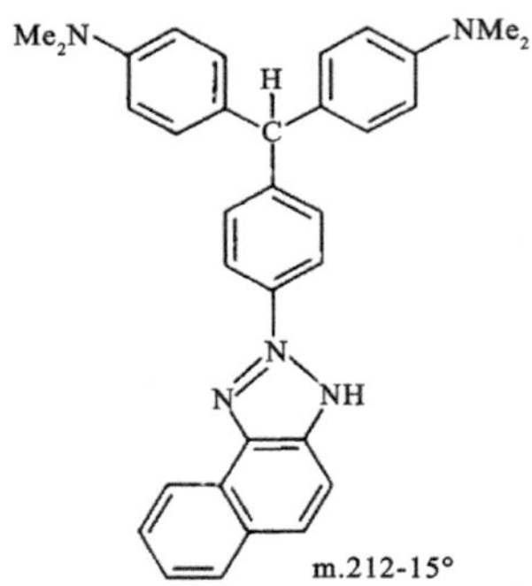

(XLV)

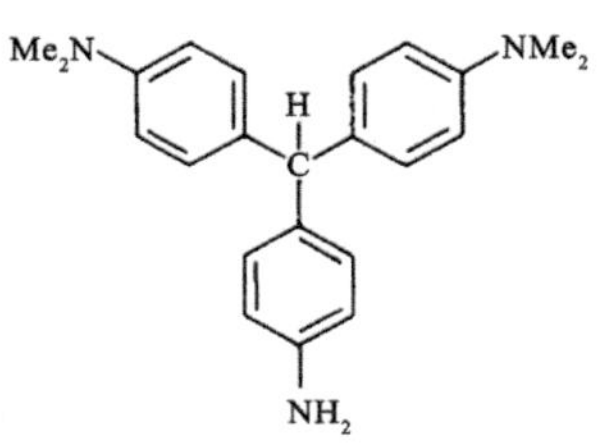

(XLIV)

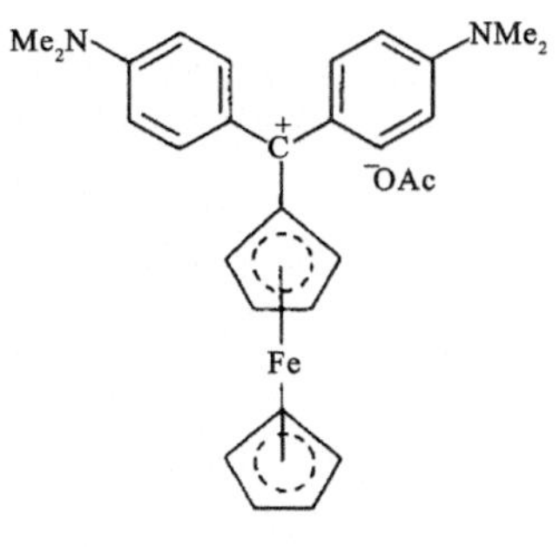

(XLVI)

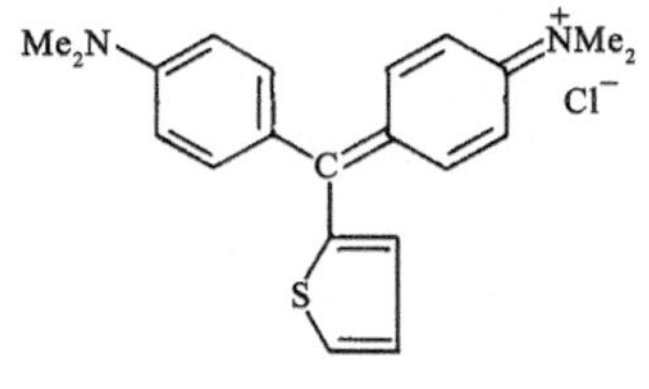

(XLVII)

Figure 3.8. *Illustration of the structures of ferrocenes analogs of malachite green.*

Source: Karl Neufang, Creative Commons License.

The silicon analog of Crystal Violet (LIV) with a melting point range of 230–232°C was synthesized through the controlled reaction of silicon chloride with p-dimethylamino phenyl lithium. The reaction between liquid sulfur dioxide and organic solvents, called LIV, resulted in chlorine displacement. However, the isolation of compounds containing the R-2Si cation was not achieved. The UV and infrared spectra analysis indicates that (LIV) exhibits significant dissimilarities compared to Crystal Violet. Additionally, the physical properties of LIV are distinct due to their non-ionic nature, which contrasts with Crystal Violet (Vogel & Retting, 1987) (Figure 3.8).

The sulfur derivative of crystal (LV) exhibited a blue hue and possessed a chemical structure resembling the triphenylmethane dye. The dye LV was synthesized by reacting sulfoxide LVI with dimethylaniline in the presence of phosphorus oxychloride. The dye produced a hue of unadulterated blue when applied to cotton and silk treated with tannin as a mordant. However, the dye's resistance to fading when applied to cotton was suboptimal, as is commonly observed. A dye base was produced upon reaction with an alkali, which could subsequently be converted back to the dye through treatment with an acid. Additional triphenyl sulfonium chlorides, such as (LVII), may be synthesized by reacting appropriately substituted phenols with thionyl chloride and anhydrous aluminum chloride. The sulfonium chlorides were observed to produce insoluble salts when combined with acid dyes and were found to be suitable for integration into bentonite (Balabanova et al., 2003). The synthesis of a Crystal Violet analog, which incorporates two methyl mercapto groups instead of dimethylamino groups, was carried out through the reaction of two thioanisylmagnesium bromide molecules with one p-dimethylamino-benzoic acid ester molecule, followed by treatment with dilute sulfuric acid. The resulting product was identified as the dye with structure (LVIII). The substance exhibited significant similarity to crystal violet (XXIX) in color and other characteristics, demonstrating notable resistance to hydrolysis. Previously, it was noted that carbinols (LIX) and (LX) exhibit the formation of pigmented salts when exposed to potent acids, and they are susceptible to hydrolysis (An et al., 2002).

The preparation of a novel triarylmethane dye (LXIII) with a melting point of 141°C has been achieved through the utilization of (LXI) and cinnamaldehyde. The resulting compound features two aryl groups separated from the central carbon atom by ethylenic linkages. The cotton, treated with tannin as a mordant, was successfully dyed to achieve a lightfast shade of brownish red. The dye (LXIV) was synthesized and applied to tannin-mordanted cotton, resulting in a blue hue. (LXV) was obtained through the condensation of benzaldehyde and (LXI), which was further condensed with dimethylaniline and phosphorus oxychloride to yield the dye (LXVI). The cotton material, treated with tannin as a mordant, was colored with a claret hue with a slight violet undertone. The stability of these dyes is observed in the presence of water and light (Vasdev et al., 1995).

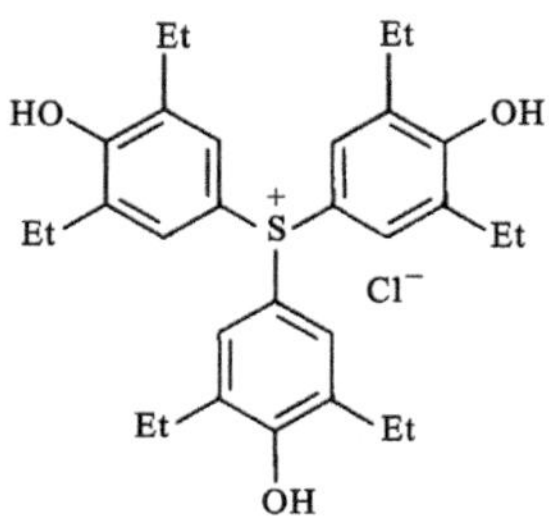

(LVII)

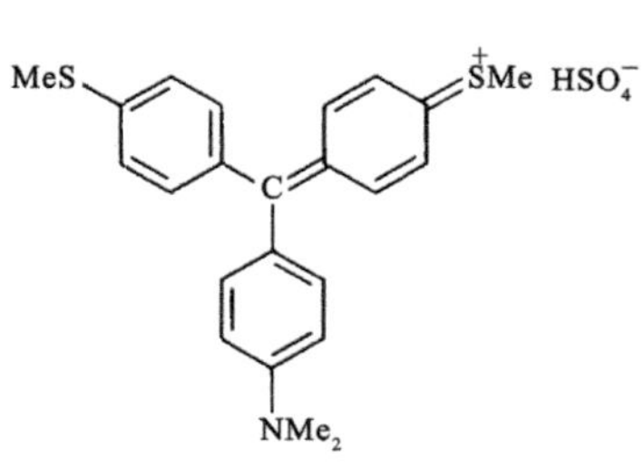

(LVIII)

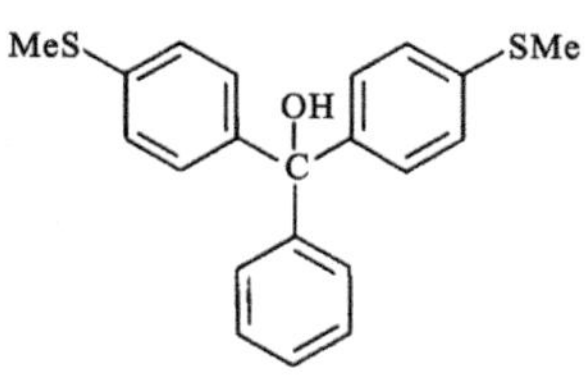

(LIX)

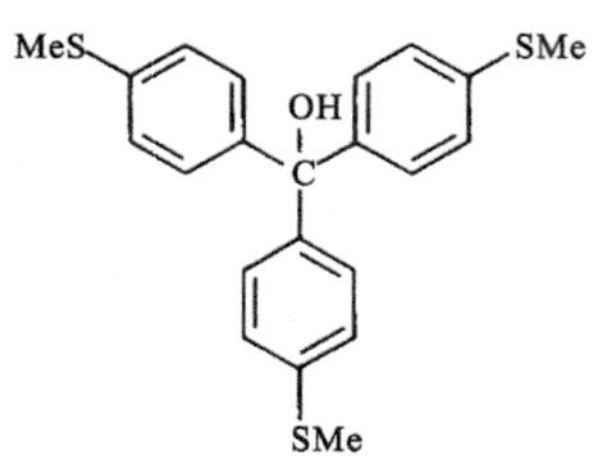

(LX)

The substitution of a methylmercapto moiety with a dimethylamino moiety in (LX) led to the stabilization of the salt due to an augmented mesomeric effect (Figure 3.9).

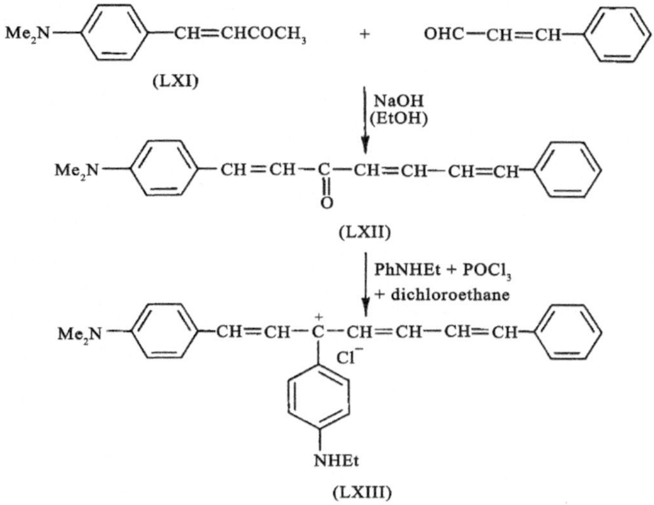

Figure 3.9. *Illustration of the preparation of triarylmethane dye, m.p. 141°.*

Source: B. D. Tilak, Creative Commons License.

It is possible to substitute the unsubstituted phenyl ring of Malachite Green with an unsaturated polymer. The Michler's ketone reaction with metalated 1, 4-cis-polybutadiene resulted in the formation of a carbinol (LXVII), which subsequently transformed into a deep red powder (LXVIII) upon treatment with iodine. The carbinol polymer's nitrogen content was 5.83%, equivalent to 25.6 units of the dye per hundred monomer units of polybutadiene (Vinod & Gowda, 2009).

The biphenyl analogs of Malachite Green and Crystal Violet, denoted as (LXIX) and (LXX), were synthesized via the reaction of (LXXI) and (LXXII) with 4'-dimethylaminobiphenyl-4-magnesium iodides, as reported in literature (Kumar et al., 2012).

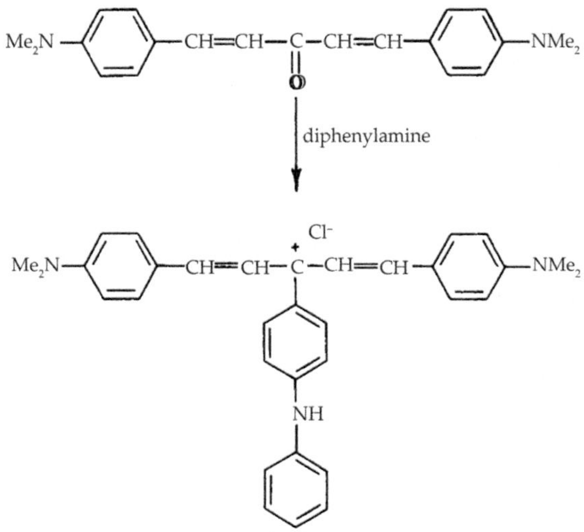

(LXIV)

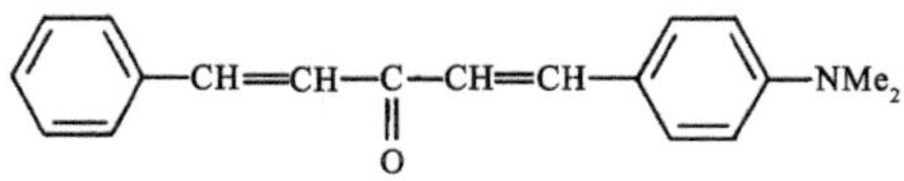

(LXV)

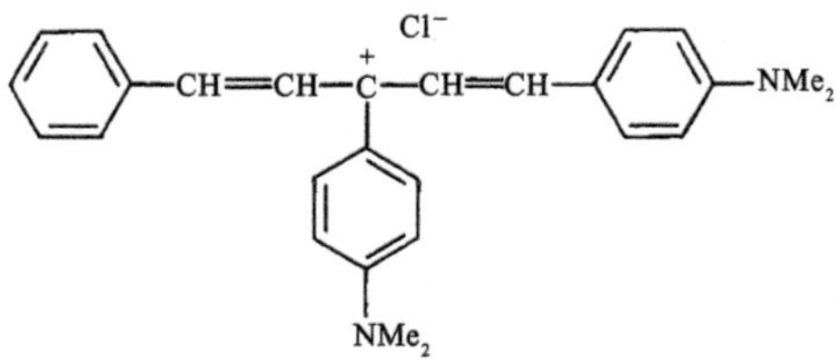

(LXVI)

The homolog LXIX of Malachite Green exhibited a green hue when subjected to hot acetic acid. In contrast, the homolog LXX of Crystal Violet displayed a deep red coloration in both sulfuric and perchloric acids. The present study involved the synthesis of Fluorene analogs (LXXIII) and (LXXIV) derived from Malachite Green and Crystal Violet bases. The synthetic procedure entailed the utilization of 3,6-bis (dimethylamino)-fluorenone (LXXV) as the starting material, which was subjected to a reaction with phenylmagnesium bromide and p- dimethylamino phenyl lithium to obtain the respective analogs (Chmelova & Ondrejovic, 2016) (Figure 3.10).

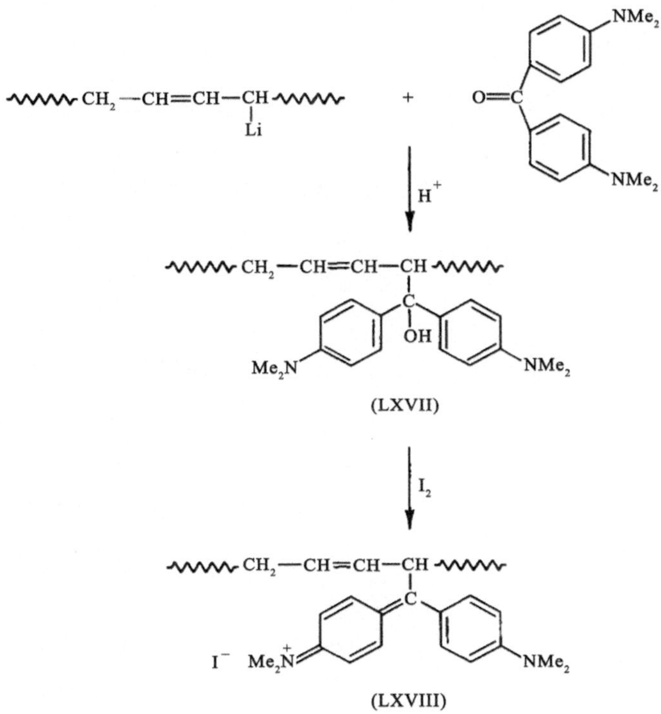

Figure 3.10. *Illustration of the reaction of Michler's ketone with metalated 1, 4-cis-polybutadiene gave a carbinol, which changed to a deep red powder with iodine.*

Source: L. A. Williams, Creative Commons License.

The Malachite Green base's unsymmetrical analog (LXXVI) was synthesized using (LXXVII). The present analogs of Malachite Green and Crystal Violet exhibit a significant bathochromic shift compared to the analogous triphenylmethane dyes. The preparation of the fluorene analog of Crystal Violet base (LXXIV) can be achieved through a convenient method involving the heating of Crystal Violet (XXIX) in 98% sulfuric acid, along with a small amount of trisodium phosphate, at a temperature range of 205–210 (Azmi et al., 1998). Subsequently, the leuco derivative acquired is subjected to oxidation utilizing lead dioxide in the presence of acetic acid. (LXXIV) dye salt is commonly used in the textile industry to produce indigo-blue shades. However, it should be noted that this dye has been reported to cause skin staining. Fluorene quinones possessing structures akin to (LXXVIII) have been synthesized. Compound (LXXVIII) was synthesized through the thermal reaction of pyrocatechol sulfonphthalein (LXXIX) and ethylamine (Wang et al., 2022) (Figure 3.11).

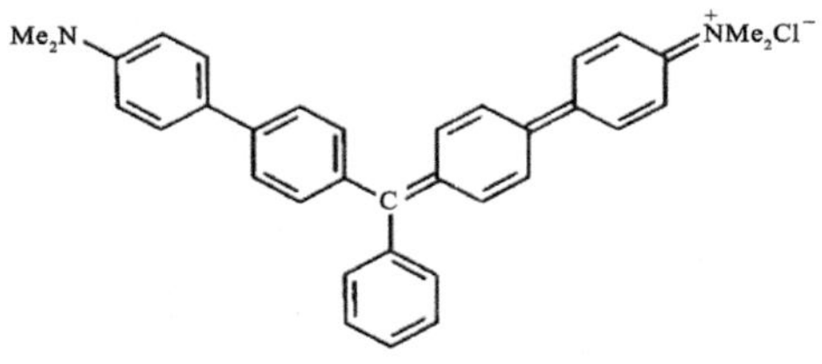

(LXIX)

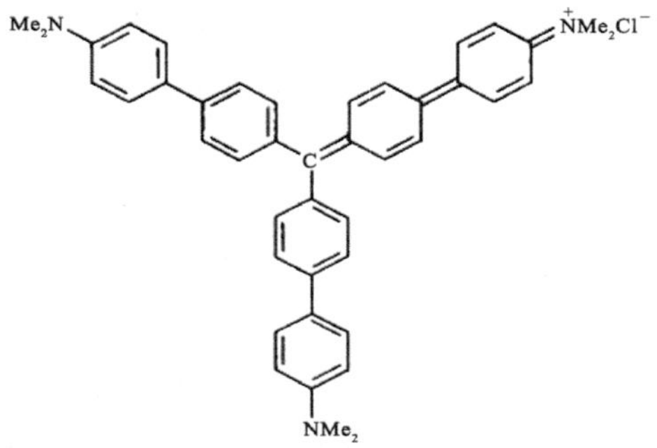

(LXX)

Figure 3.11. *Illustration of the structures of biphenyl homologs of malachite green and crystal violet dyes.*

Source: Oskar Glenz, Creative Commons License.

Sulfonphthalein (LXXIX) is synthesized via a thermal reaction between pyrocatechol and o-sulfobenzoic anhydride in equimolar proportions. The compounds in question exhibit a profound hue and demonstrate minimal chromatic variation within the pH range of 3 to 14. Triphenylmethane systems with indicator dyes with substituents such as CH_2N (CH_3COOH) located in the ortho position to the phenolic hydroxyl groups exhibit a color change upon complex formation with metal ions (Shedbalkar et al., 2008). These systems are suitable for use in both colorimetric and titrimetric determinations. The synthesis of 3,'3,"5,'5"-Tetrakis phenolphthalein (LXXX) involves the utilization of iminodiacetic acid and formaldehyde in an alkaline environment to treat a phenolphthalein solution. The solution is modified to a pH level slightly above 10 using ammonia and ammonium chloride to ascertain the presence of barium or strontium. The solution is then titrated in the presence of (LXXX) using 0.1 N disodium salt of EDTA (Cao et al., 2019).

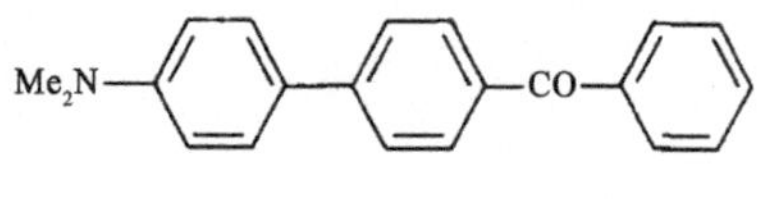

(LXXI)

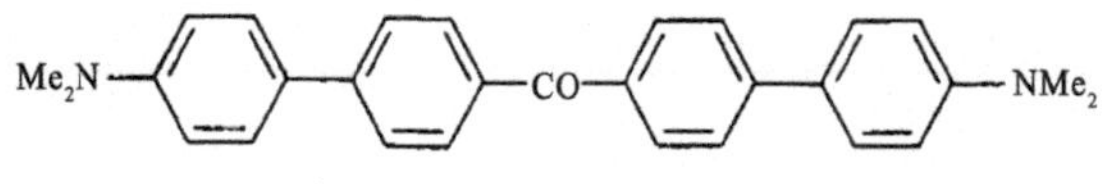

(LXXII)

The endpoint is characterized by a transition in color from violet-red to pink. Phenolphthalein, sulfonphthalein, and fluorescein are capable of producing metallochromic indicators that are highly specific.

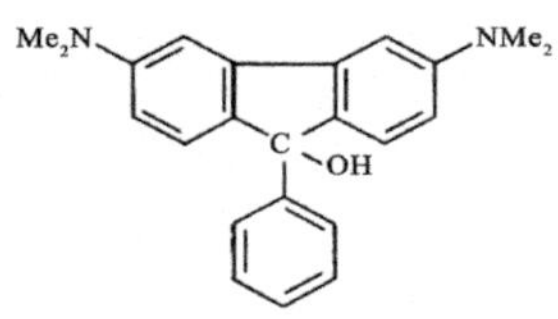

(LXXIII)

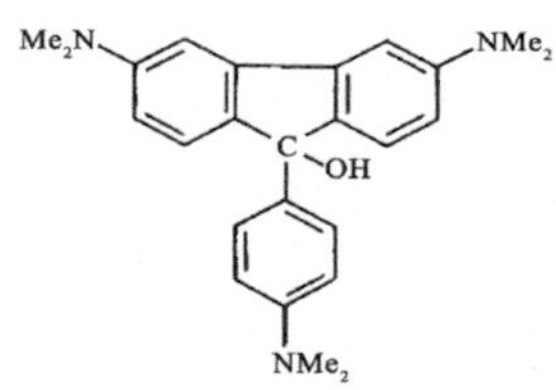

(LXXIV)

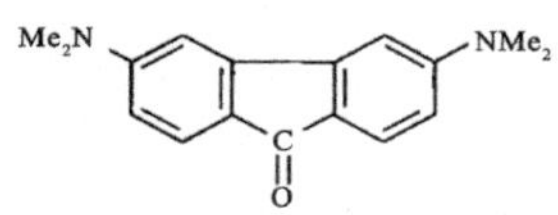

(LXXV)

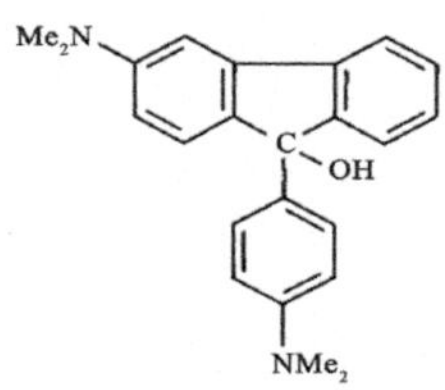

(LXXVI)

One instance is represented by the compound denoted as (LXXXI), which is synthesized through the reaction of Thymol Blue with N-methyl glycine and formaldehyde in the presence of sodium acetate and acetic acid. These substances serve as effective indicators for the presence of copper, nickel, and cadmium ions, with the ability to detect as little as mole of copper. Triphenylmethane dyes, including Malachite Green and Brilliant Green with aryl sulfonic acid as anions, exhibit enhanced solubility in alcohols, similar to Auramine dyes. These dyes are suitable for coloring inks and coating compositions with high tinctorial strength, commonly used for duplicating carbon papers. The preparation process involves the reaction of conventional basic dyes with aryl sulfonic acids, including 5-sulfo-2-hydroxybenzoic acid (Casas et al., 2009).

Did You Know?

Basic dyes have been used in the film industry for many years to dye costumes and fabrics.

The solubility of Brilliant Green nitrate in alcohols is noteworthy, as it is prepared by treating Brilliant Green crystals with aqueous sodium nitrate at a temperature of 20°C. Polyolefinic materials can be dyed using basic dyes that contain oleophilic anions, such as (LXXXII). The polyolefin can be dyed with a deep green hue by utilizing the dibutyl phosphoric acid salt of Malachite Green Base, which is produced through the aqueous solution heating of the base with dibutyl phosphoric acid (Arunprasath et al., 2019) (Figure 3.12).

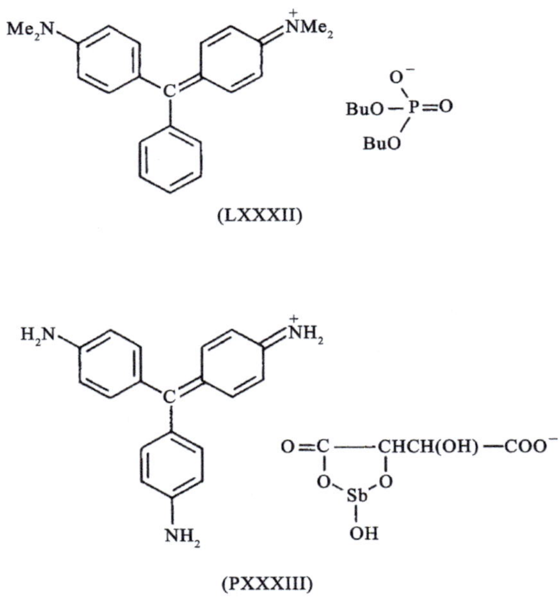

Figure 3.12. *Illustration of the structures of basic dyes containing oleophilic anions and pararosaniline antimonyl tartrate.*

Source: J. Bailey, Creative Commons License.

The Pararosaniline antimonyl tartrate (LXXXIII) has been synthesized by reacting Pararosaniline with antimony potassium tartrate in an aqueous medium. This compound has demonstrated efficacy as an anti-schistosomal agent.

The paper industry needs stable concentrated solutions of triphenylmethane dyes, including Malachite Green and Methyl Violet, due to their ease of handling and addition to the beater. The agitation of dyed-in liquids such as Q-hydroxy propionitrile can achieve these solutions (Saquib & Muneer, 2003).

3.6. OTHER BASIC DYES

In the past sixteen years, there has been a dearth of reports on novel basic dyes belonging to the xanfhene, acridine, a zine, oxazine, and thiazine categories. The chapter on cationic dyes for synthetic fibers describes several azine and oxazine dyes beneficial for polyacrylonitrile fiber.

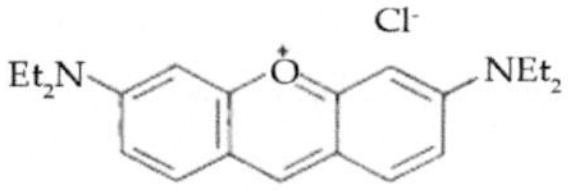

(CIII)

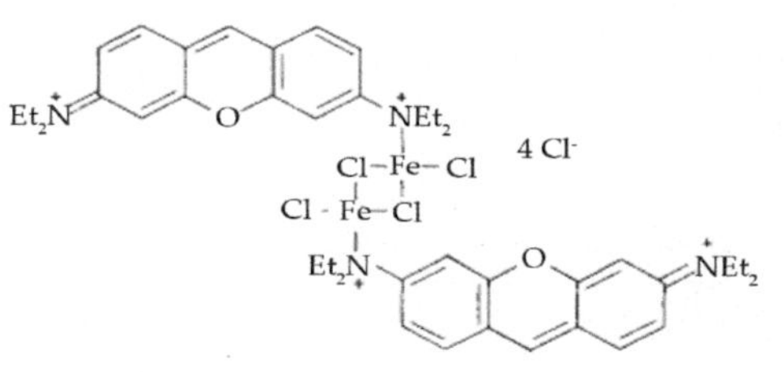

(CIV)

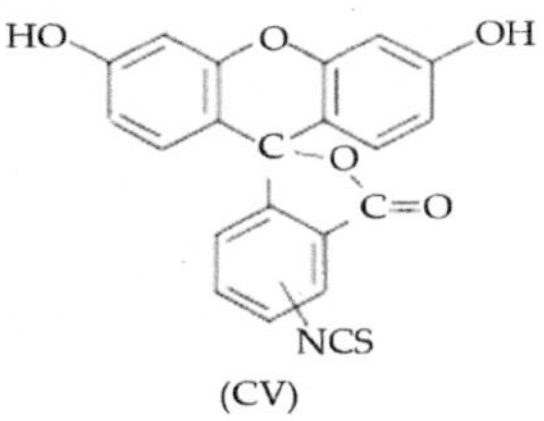

(CV)

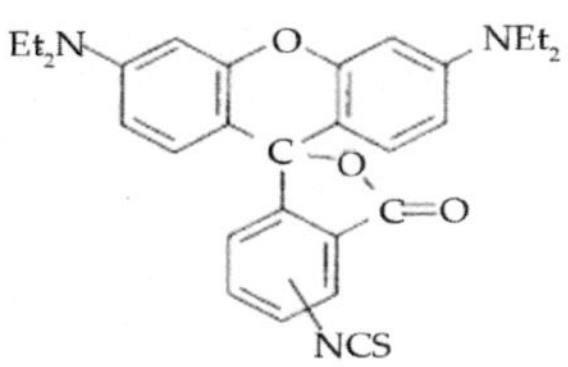

(CVI)

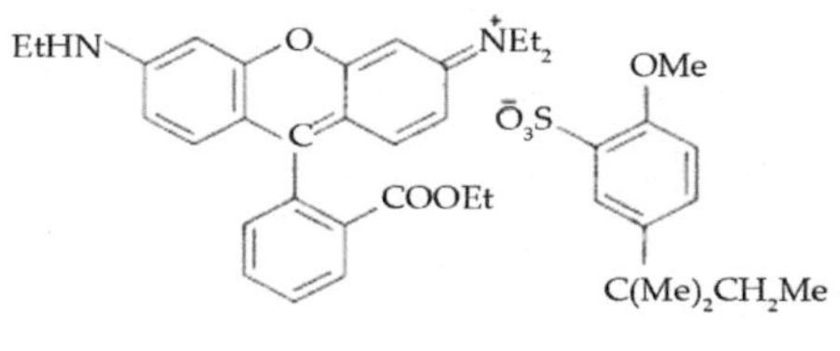

(CVII)

Pyronine B, a type of xanthene dye, is synthesized through a commercial process that involves the condensation of two molecules of in-diefhy l aminopheno l with formaldehyde, followed by oxidation with ferric chloride in hydrochloric acid. The structure (CIV) of commercial Pyronine B was determined through elemental and polarographic analyses, UV absorption spectra, and X-ray diffraction studies. The iron was likely derived from the oxidizing agent, such as ferric chloride, commonly used to produce the dye. The isothiocyanate derivatives of Fluorescein and Rhodamine B are utilized as biological staining agents to detect the existence of antigens and interacting substances (Zhou et al., 2020).

The compounds mentioned above are acquired through an aqueous alkaline carbon disulfide reaction with the corresponding amino derivatives, followed by treating the resultant dithiocarbonate with a solution containing salts of iron, mercury, copper, or lead.

The production of Rhodamine B analogs with reduced toxicity can be achieved through vinyl-substituted phthalic anhydride during the synthesis process, which is subsequently followed by polymerization or copolymerization with vinyl monomers. Rhodamine-type basic dyes that possess aryl sulfonic acid groups as anions, such as (CVII), exhibit excellent solubility in ethanol and are suitable for dyeing nitrocellulose lacquers and printing inks. Utilizing the alcohol-soluble o-nitrobenzoate derivative of Rhodamine B to create printing inks is a viable application. This derivative is synthesized by treating the basic dye with o-nitrobenzoic acid in the presence of alkali (Chu & Chen, 2002).

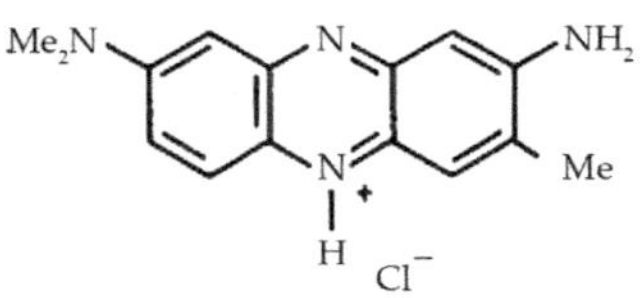

(CVIII)

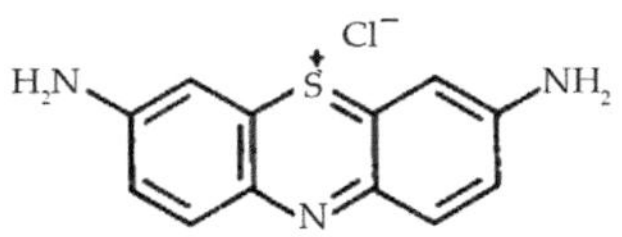

(CIX)

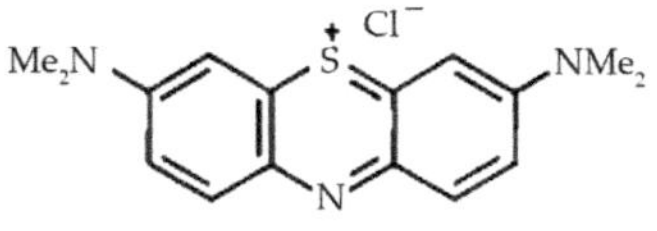

(CX)

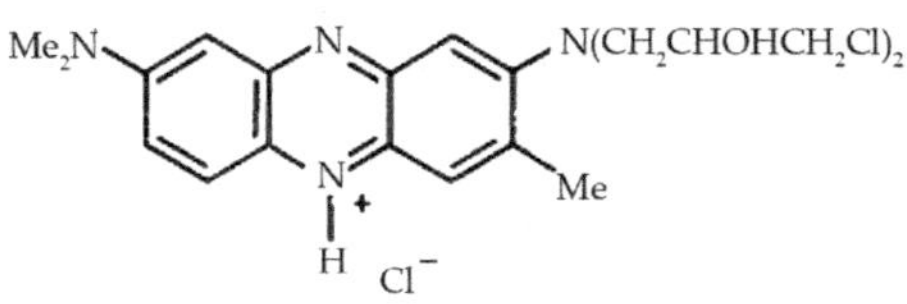

(CXI)

Loiseleur's research demonstrated that the synthesis of dyes, including Neutral Red (CVIII), Lauth's Violet (CI 52000) (CIX), and Methylene Blue (CX), can be achieved through the irradiation of suitable intermediates with UV light or X-rays. The production of (CVIII) was achieved by exposing a combination of N,N-dimethyl-p-phenylenediamine, and toluene-2,4-diamine in a solution of dilute hydrochloric acid with small amounts of copper sulfate to irradiation for thirty minutes (Singer, 1952).

The radio syntheses of Lauth's violet and Methylene Blue were accomplished through X-ray irradiation of p-phenylenediamine and N,N-dimethyl-p-phenylenediamine, respectively, in the presence of ammonium sulfide, hydrochloric acid, and copper sulfate at a pH of 7. This method was employed similarly for both compounds. The study revealed that elevating the temperature of the irradiated solution and introducing air into the reaction mixture were effective methods for enhancing the yields of the dyes. The irradiated solutions yielded pure dyes in significant quantities. Lauth's Violet and Methylene Blue were synthesized through Co y-irradiation of p-phenylenediamine and NN-bimethyl-p-phenylenediamine, respectively, using comparable procedures. It is feasible to perform radiosynthesis of various dyes, including Aniline Black, Fuchsine, Malachite Green (CI 42000; Basic Green 4), Safranine B extra (CI 50200), Acridine Yellow G (CI 46025), and Indophenol Blue (CI 49700; Vat dye) (Morthland et al., 1954).

Neutral Red (CI 50040; Basic Red 5) (CVIII) can undergo a reaction with epichlorohydrin to produce products such as (CXI). These products are capable of dyeing protein, polyamide, and cellulosic fibers by creating covalent bonds with them. Furthermore, they significantly enhance the wet-fastness of the fibers.

The electron spin resonance spectra of various azine, oxazine, and thiazine dyes, including Lauth's Violet (CIX), Methylene Blue (CX), Methylene Green (CI 52020; Basic Green 5), Gallo cyanine, and Safranine B extra, were examined after exposure to X-ray irradiation for durations ranging from 5 to 40 minutes in the presence of strong sulfuric acid, ethanol, or benzoic acid. The results indicate that the dyes were reduced to paramagnetic species identical to those obtained through chemical reduction with titanous chloride. The central nitrogen atom is attached to at least one acidic proton, and the unpaired electron is equally likely to be located on the sulfur (or oxygen) and nitrogen. The resonance structures denoting the configuration of the ions produced through the radiolytic reduction of Lauth's Violet are (CXII), (CXIII), and (CXIV) (De Souza et al., 2018).

In addition to the established absorption band located at 660μ, an additional band situated at approximately 500 μm is discernible within the spectra of solutions containing Methylene Blue in the visible range. The intensity of the latter band exhibits a significant increase in certain solid specimens, resulting in a red appearance. The red samples tend for the dimethyl-amino groups to be extruded from the molecular plane, as per the suggestion." Methylene blue (CX) dissolution in anhydrous aliphatic amines or in ethanol containing sodium ethoxide resulting in red solutions that revert to blue upon adding water, has been documented (Hao & Iqbal, 1997). The presence of the red form is contingent upon the fundamental properties of the amine. The blue dye's chromophoric system exhibits the positive charge's symmetrical distribution. The manifestation of the red hue in amine solutions is ascribed to the interplay between the dye ions and highly basic anions. The interaction mentioned above leads to a decrease in the proportion of positive charge distributed among the terminal positions, ultimately resulting in a hypsochromic spectral shift (Tsai et al., 2008) (Figure 3.13).

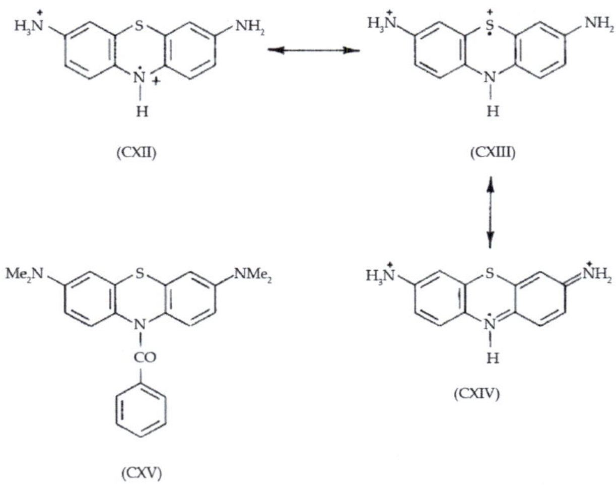

Figure 3.13. *Illustration of the structure of ions obtained by the radiolytic reduction of laugh's violet.*

Source: N. R. Ayyangar, Creative Commons License.

Phenol thiazine derivatives, specifically (CXV), have been purported to have practical applications in the photoconductive layer of electrophotographic materials and transfer sheet materials.

Tilak has proposed potential mechanisms for the formation of dyes, including Pyronine G (CI 45005) (CXVI), Methylene Blue (CX), and Meldola's Blue (CXVII), which involve hydride transfer.

Remember

Basic dyes are frequently used in tie-dyeing techniques due to their vibrant colors and affinity for fibers.

3.7. ORGANIC PIGMENTS FROM BASIC DYES

Previously, a comprehensive portrayal of the basic dye lakes was presented, which was also thoroughly examined in a commendable analysis by Williams and Conley. The utilization of intricate hetero-poly acids, namely phosphomolybdic acid (PMA), phosphotungstomolybdic acid (PTMA), and phosphotungstic acid (PTA), as precipitants for numerous basic dyes is feasible. The fundamental dyes suitable for employment in producing PMA and PTMA pigments encompass Methyl Violet, Malachite Green, Brilliant Green, Ethyl Violet, and Victoria Pure (Anliker et al., 1980).

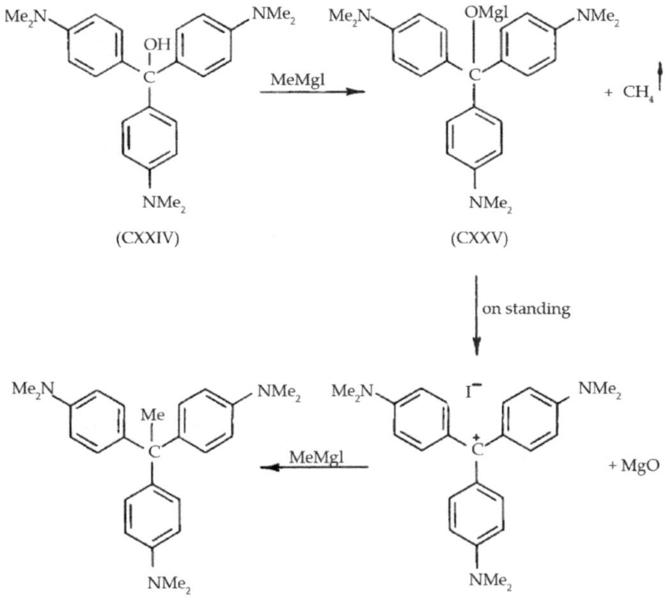

(CXXIV) (CXXV) (CXXVII) (CXXVI)

The preeminent formula for complex PMAs is the composition that comprises a ratio of 1:24 between phosphoric anhydride and molybdenum oxide, denoted $H_3P_2O_5 \cdot 24MoO_3$. According to X-ray data, the most favorable chemical formula is $H_2PO_4 \cdot 12MoO_3 \cdot XH_2O$.

The general structure of PMA pigments derived from Methyl Violet (CXXXI), Victoria Pure Blue BO (CXXXII), Brilliant Green (CXXXII I), and Rhodamine 6G (CXXXIV) can be inferred from their composition. Basic dyes are commonly found in salts, predominantly chlorides or sulfates. Producing colors through the formation of phosphomolybdate or phosphotungstate involves the substitution of highly soluble anions with significantly denser and almost insoluble anions, namely phosphomolybdate or phosphotungstate (Berradi et al., 2019).

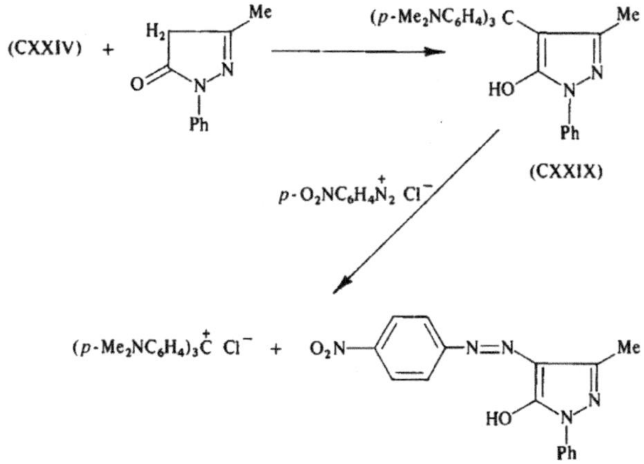

(XXIX) (CXXX)

The production process of PMA colors involves dissolving the fundamental dyestuff in water at a temperature range of 60–70°C while maintaining a dye concentration of approximately 1%. The solution's pH is modulated to approximately 3 through acetic acid. The preparation of a solution containing a complex acid (with a 1: 24 ratios of and) ratio of 1:24 involves the use of soluble phosphate and molybdate salts. The pH of the resulting solution is adjusted to fall within the range of 3 to 4 by adding mineral acid. The temperature of the intricate solution is sustained within the range of 20°C to 30°C and subsequently introduced to the heated dye solution at an appropriate pace while ensuring regulated agitation. Subsequently, the amalgamation is subjected to a heating process to stabilize the pigment (Marcelino & Muralha, 2012).

The production process of PMA colors involves the dissolution of the fundamental dyestuff in water at a temperature range of 60–70°C, with a dye concentration of approximately 1%. The solution's pH is modulated to approximately 3 through acetic acid. The preparation of the intricate acid solution involves the utilization of soluble phosphate and molybdate salts, which are subjected to adding a mineral acid to attain a pH level ranging from 3 to 4. The temperature of the intricate solution is sustained within the range of 20°C to 30°C, after which it is gradually introduced into the heated dye solution with appropriate agitation under regulated conditions. Subsequently, the concoction is subjected to heat treatment to stabilize the pigment. A small surplus of complex acid solution can be introduced to compensate for any liberated dye during the heating process. Organic pigments are recognized for their exceptional brightness, potency, and significantly enhanced resistance to high temperatures, illumination, and moisture. Regarding stability to sunlight, the PTMA colors exhibit a higher degree of superiority compared to both PMA and PTA colors. The primary application of these pigments lies within the domain of printing inks (Shindy, 2017).

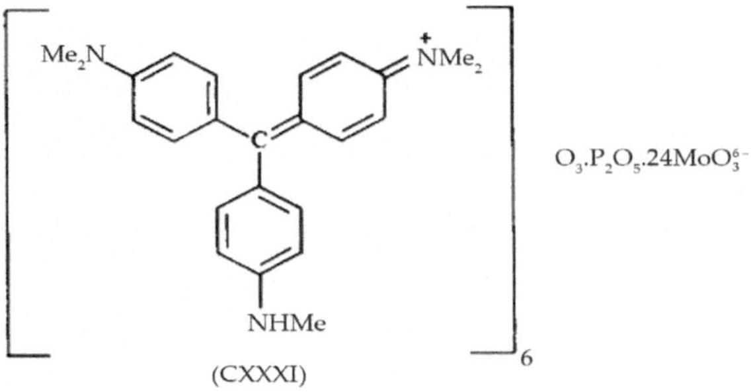

(CXXXI)

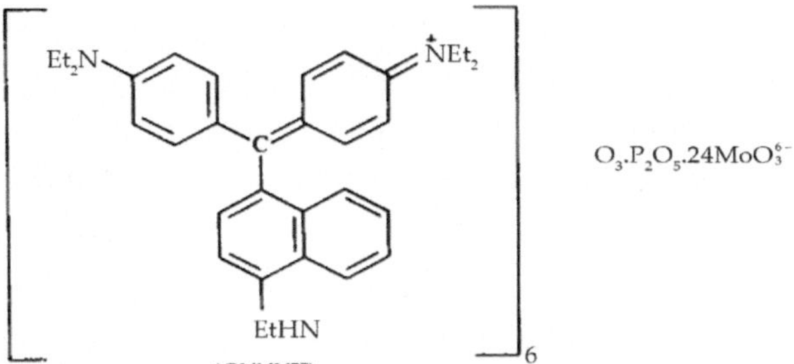

(CXXXII)

Furthermore, these substances find application in the coloration of white paper, watercolor painting, wax crayon production, colored paper manufacturing, and enamel coating for children's toys, owing to their high-water insolubility and non-toxic nature. A technique was devised to incorporate the PMA insoluble lakes into the fiber, particularly viscose rayon, resulting in favorable resistance to washing and light. The combination of a fundamental dye and an acid that forms a lake is conducted in the presence of an alkali phosphate buffer solution to ensure that the pH level remains within the range of 5 to 12.5 (Williams & Conley, 1955).

After introducing a molybdate reagent, the lake formation principle is employed in the computation of minute quantities of phosphates present in solutions. The phosphomolybdate complex can be selectively extracted from a nitric acid solution with a pH range of 0.2 to 0.5 using n-Propyl acetate. Applying acidified dye, such as Crystal Violet, Methyl Violet, Malachite Green, or Brilliant Green, to the extract results in the formation of an insoluble complex of phosphomolybdate dye. This complex can be evaluated through spectrophotometric means in a solution of tt-propyl acetate acetone (Centeno et al., 2006).

3.8. MANIFOLD RECORD SYSTEMS

"Manifold record systems" is commonly used interchangeably with "transfer sheet materials." The objective behind creating a pressure-sensitive recording mechanism is to utilize a solitary substance for both the reception and transmission of data. These materials can be utilized in a stack formation beneath a covering layer. The top sheet's inscription can be replicated onto the underlying sheets by applying force or pressure, without resorting to traditional carbon paper. The record substrate exhibits a consistent and abundant application of two distinct coatings. (Gelbwachs & Chan, 1989). One of the substances is a solid inorganic material that has been finely divided, while the other is a liquid phase that comprises a solution of an organic compound. To prevent direct contact between the two substances, the liquid phase is encapsulated in microscopic capsules made of gelatine or agar before application onto the paper. In its solid state, the inorganic phase comprises substances acting as electron acceptors or acidic agents, including attapulgite, halloysite, kaolin, magnesium trisilicate, and zeolite (Delgado et al., 2020).

The liquid organic phase comprises an oily medium solution, typically chlorinated biphenyl, containing electron donor aromatic compounds. These compounds convert into more polarized conjugated forms upon exposure to acidic substances. The exertion of pressure results in the rupture of the diminutive capsules, leading to a chemical reaction between the two phases, ultimately resulting in the manifestation of pigmented marks at the precise location of impact (Choi et al., 2022).

The organic reactants utilized for color formation must exhibit favorable solubility characteristics in nonpolar solvents. The ideal characteristic of the substance in question is its colorlessness in solution and ability to maintain stability under light exposure before its impact. The Crystal Violet lactone (CXXXV) and the Malachite Green lactone (CXXXVI) produce bluish violet and greenish markings, respectively. However, it is noteworthy that these hues tend to diminish over time. Enhanced chromatic impressions can be achieved through the utilization of leuco diphenylmethane and triphenylmethane derivatives (Redmond et al., 1997).

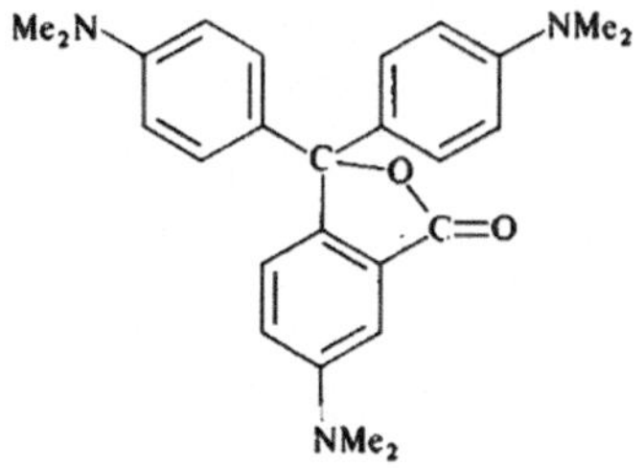

(CXXXV)

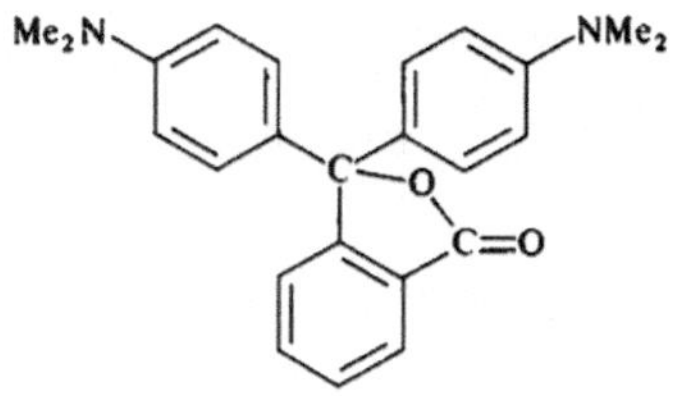

(CXXXVI)

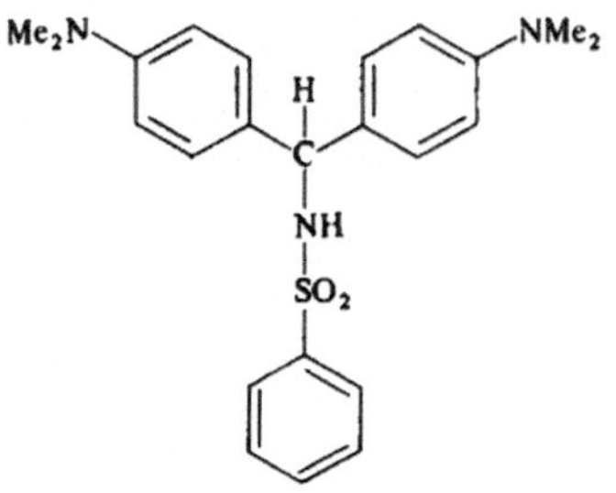

(CXXXVII)

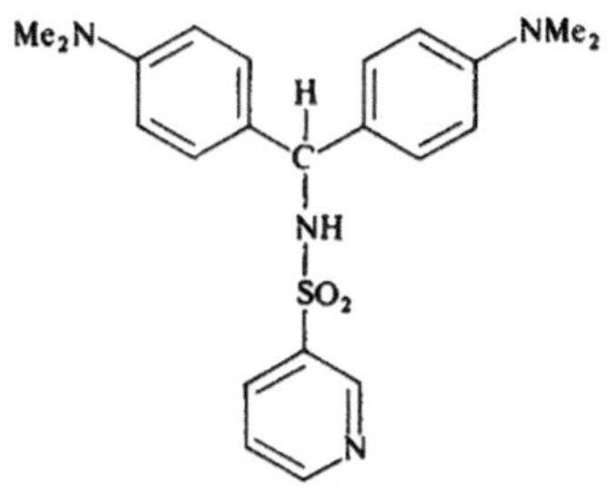

(CXXXVIII)

Several derivatives have been previously delineated in this chapter. The synthesis of sulfonamide derivatives, specifically (CXXXVII) and (CXXXVIII), involves the reaction of benzene sulfonamide and 3-pyridine sulfonamide with Michler's hydrol in the presence of a base, resulting in the formation of the desired compounds (Meng et al., 2015). The reactants are used in equimolar amounts during the condensation process. The sulfonamide derivatives that resemble (XVI) and (XVII) exhibit utility in colorless transfer sheet materials. The permanent blue markings in transfer sheet materials can be achieved by preparing the benzoyl derivative of leuco Methylene Blue (CXV). This process involves refluxing benzoyl chloride with the zinc chloride double salt of leuco Methylene Blue in pyridine (Diegel et al., 2020).

The naphthyol derivatives denoted as (CXXXIX) and (CXL) and the nitro derivatives of 10-benzoyl-3, 7-bis(dimethylamino)phenothiazine, represented by (CXLI) and (CXLII)140, exhibit superior light stability when compared to (CXV). The p-nitro derivatives (CXLII) can undergo reduction to form the corresponding p-aminobenzoic derivative. This derivative can then be condensed with Michler's hydrol to produce (CXLIII), which exhibits a colorless appearance. However, it produces lightfast blue images upon contact with acid copy sheets (Meng et al., 2017).

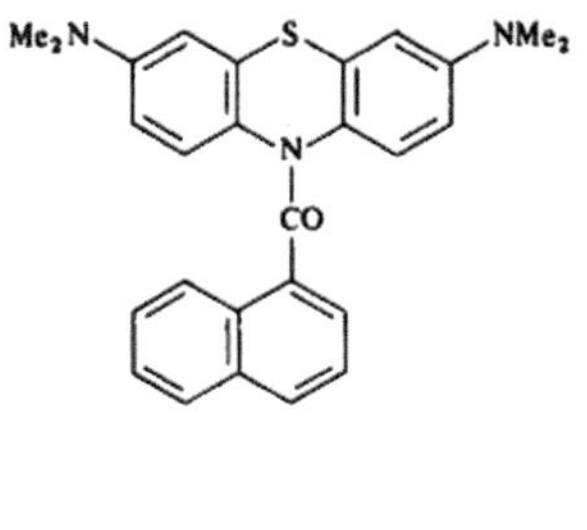

(CXXXIX)

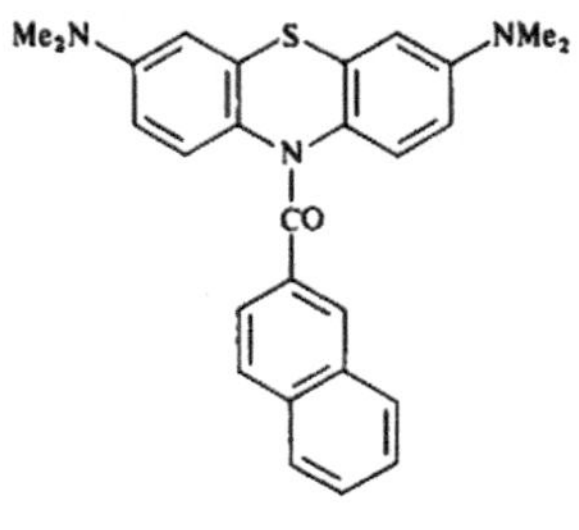

(CXL)

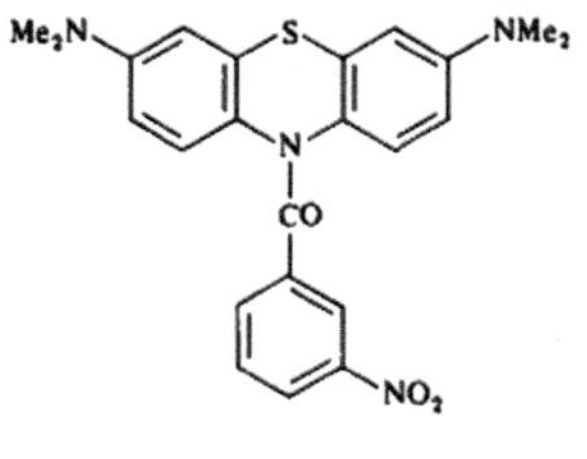

(CXLI)

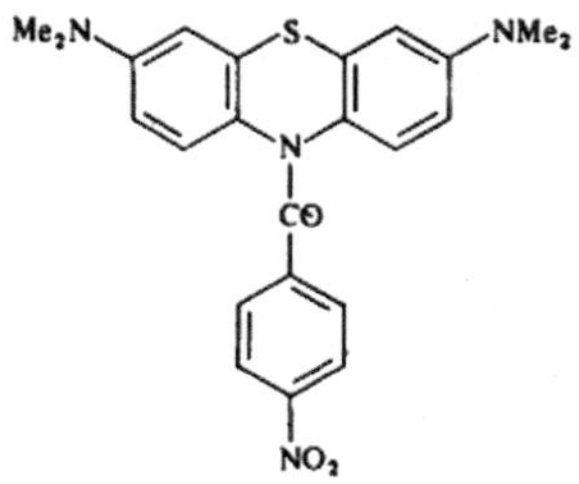

(CXLII)

3.9. SUMMARY

Basic dyes, also known as cationic dyes or basic colorants are a class of dyes that are commonly used in various applications, particularly in the dyeing of synthetic fibers, paper and leather. They are called basic dyes because their chromophore is a cationic group. Basic dyes are organic compounds with a positive charge and typically exhibit vibrant and intense colors. Basic dyes have strong affinity for negatively charged materials and are typically soluble in water. Basic dyes generally have relatively poor lightfastness and known for their low environmental compatibility. Specific properties and characteristics of basic dyes can vary depending on the particular dye.

REFERENCES

1. Ahmed, M. J., (2016). Application of agricultural based activated carbons by microwave and conventional activations for basic dye adsorption. *Journal of Environmental Chemical Engineering, 4*(1), 89–99.
2. An, S. Y., Min, S. K., Cha, I. H., Choi, Y. L., Cho, Y. S., Kim, C. H., & Lee, Y. C., (2002). Decolorization of triphenylmethane and azo dyes by *Citrobacter* sp. *Biotechnology letters, 24*, 1037–1040.
3. Anliker, R., Butler, G. C., Clarke, E. A., Förstner, U., Funke, W., Hyslop, C., & Anliker, R., (1980). Organic dyes and pigments. *Anthropogenic Compounds*, 181–215.
4. Arunprasath, T., Sudalai, S., Meenatchi, R., Jeyavishnu, K., & Arumugam, A., (2019). Biodegradation of triphenylmethane dye malachite green by a newly isolated fungus strain. *Biocatalysis and Agricultural Biotechnology, 17*, 672–679.
5. Asfour, H. M., Fadali, O. A., Nassar, M. M., & El□Geundi, M. S., (1985). Equilibrium studies on adsorption of basic dyes on hardwood. *Journal of Chemical Technology and Biotechnology. Chemical Technology, 35*(1), 21–27.
6. Ashour, A., Tony, M. A., & Purcell, P. J., (2014). Use of agriculture-based waste for basic dye sorption from aqueous solution: Kinetics and isotherm studies. *Am. J. Chem. Eng., 2*(6), 92–98.
7. Aspland, J. R., (1993). The application of basic dye cations to anionic fibers: Dyeing acrylic and other fibers with basic dyes. *Textile Chemist & Colorist, 25*(6).
8. Ayed, L., Chaieb, K., Cheref, A., & Bakhrouf, A., (2010). Biodegradation and decolorization of triphenylmethane dyes by Staphylococcus epidermidis. *Desalination, 260*(1–3), 137–146.
9. Azmi, W., Sani, R. K., & Banerjee, U. C., (1998). Biodegradation of triphenylmethane dyes. *Enzyme and Microbial Technology, 22*(3), 185–191.
10. Babendure, J. R., Adams, S. R., & Tsien, R. Y., (2003). Aptamers switch on fluorescence of triphenylmethane dyes. *Journal of the American Chemical Society, 125*(48), 14716–14717.
11. Balabanova, M., Popova, L., & Tchipeva, R., (2003). Dyes in dermatology. *Clinics in Dermatology, 21*(1), 2–6.
12. Benkhaya, S., M'rabet, S., & El Harfi, A., (2020). A review on

classifications, recent synthesis and applications of textile dyes. *Inorganic Chemistry Communications, 115*, 107891.

13. Berradi, M., Hsissou, R., Khudhair, M., Assouag, M., Cherkaoui, O., El Bachiri, A., & El Harfi, A., (2019). Textile finishing dyes and their impact on aquatic environs. *Heliyon, 5*(11), e02711.
14. Brachet, J., (1953). The use of basic dyes and ribonuclease for the cytochemical detection of ribonucleic acid. *Journal of Cell Science, 3*(25), 1–10.
15. Cao, D. J., Wang, J. J., Zhang, Q., Wen, Y. Z., Dong, B., Liu, R. J., & Geng, G., (2019). Biodegradation of triphenylmethane dye crystal violet by *Cedecea davisae. Spectrochimica Acta Part A: Molecular and Biomolecular Spectroscopy, 210*, 9–13.
16. Casas, N., Parella, T., Vicent, T., Caminal, G., & Sarrà, M., (2009). Metabolites from the biodegradation of triphenylmethane dyes by Trametes versicolor or laccase. *Chemosphere, 75*(10), 1344–1349.
17. Centeno, S. A., Buisan, V. L., & Ropret, P., (2006). Raman study of synthetic organic pigments and dyes in early lithographic inks (1890–1920). *Journal of Raman Spectroscopy: An International Journal for Original Work in all Aspects of Raman Spectroscopy, Including Higher Order Processes, and Also Brillouin and Rayleigh Scattering, 37*(10), 1111–1118.
18. Chen, S. H., & Ting, A. S. Y., (2015). Biodecolorization and biodegradation potential of recalcitrant triphenylmethane dyes by *Coriolopsis* sp. isolated from compost. *Journal of Environmental Management, 150*, 274–280.
19. Cheriaa, J., Khaireddine, M., Rouabhia, M., & Bakhrouf, A., (2012). Removal of triphenylmethane dyes by bacterial consortium. *The Scientific World Journal*, 2012.
20. Chmelová, D., & Ondrejovič, M., (2016). Purification and characterization of extracellular laccase produced by *Ceriporiopsis subvermispora* and decolorization of triphenylmethane dyes. *Journal of Basic Microbiology, 56*(11), 1173–1182.
21. Choi, G. M., Park, M., Shim, Y. H., Kim, S. Y., & Lee, H. S., (2022). Mass production of 2D manifolds of graphene oxide by shear flow. *Advanced Functional Materials, 32*(1), 2107694.
22. Chu, H. C., & Chen, K. M., (2002). Reuse of activated sludge biomass: I. Removal of basic dyes from wastewater by biomass. *Process*

Biochemistry, 37(6), 595–600.

23. Combes, R. D., & Haveland-Smith, R. B., (1982). A review of the genotoxicity of food, drug and cosmetic colors and other azo, triphenylmethane and xanthene dyes. *Mutation Research/Reviews in Genetic Toxicology, 98*(2), 101–243.
24. de Souza, T. N. V., de Carvalho, S. M. L., Vieira, M. G. A., da Silva, M. G. C., & Brasil, D. D. S. B., (2018). Adsorption of basic dyes onto activated carbon: Experimental and theoretical investigation of chemical reactivity of basic dyes using DFT-based descriptors. *Applied Surface Science, 448*, 662–670.
25. Delgado-Blanca, I., Llorent-Martínez, E. J., Ruiz-Medina, A., & Pilar, O. B., (2020). Automated on-line liquid-liquid extraction in a multisyringe flow injection analysis manifold for migration studies in food-contact materials: Analysis of 4,4′-dihydroxybiphenyl. *Food Additives & Contaminants: Part A, 37*(1), 174–182.
26. Dewar, M. J. S., (1950). Color and constitution. Part I. Basic dyes. *Journal of the Chemical Society (Resumed)*, 2329–2334.
27. Diegel, O., Schutte, J., Ferreira, A., & Chan, Y. L., (2020). Design for additive manufacturing process for a lightweight hydraulic manifold. *Additive Manufacturing, 36*, 101446.
28. Duxbury, D. F., (1993). The photochemistry and photophysics of triphenylmethane dyes in solid and liquid media. *Chemical Reviews, 93*(1), 381–433.
29. El Harfi, S., & El Harfi, A., (2017). Classifications, properties and applications of textile dyes: A review. *Applied Journal of Environmental Engineering Science, 3*(3), 1–3.
30. Freeze, J. G., Kelly, H. R., & Batista, V. S., (2019). Search for catalysts by inverse design: Artificial intelligence, mountain climbers, and alchemists. *Chemical Reviews, 119*(11), 6595–6612.
31. Gelbwachs, J. A., & Chan, Y. C., (1989). CW laser spectroscopy of the triplet manifold of magnesium. In: *Laser Spectroscopy* (pp. 113–116). Academic Press.
32. Goldacre, R. J., & Phillips, J. N., (1949). The ionization of basic triphenylmethane dyes. *Journal of the Chemical Society (Resumed)*, 1724–1732.
33. Gürses, A., Açıkyıldız, M., Güneş, K., Gürses, M. S., Gürses, A., Açıkyıldız, M., & Gürses, M. S., (2016). Classification of dye and

pigments. *Dyes and Pigments*, 31–45.

34. Hameed, B. H., Ahmad, A. L., & Latiff, K. N. A., (2007). Adsorption of basic dye (methylene blue) onto activated carbon prepared from rattan sawdust. *Dyes and Pigments, 75*(1), 143–149.
35. Hao, Z., & Iqbal, A., (1997). Some aspects of organic pigments. *Chemical Society Reviews, 26*(3), 203–213.
36. Iacopetta, D., Ceramella, J., Catalano, A., Saturnino, C., Bonomo, M. G., Franchini, C., & Sinicropi, M. S., (2021). Schiff bases: Interesting scaffolds with promising antitumoral properties. *Applied Sciences, 11*(4), 1877.
37. Kumar, V. V., Kumar, M. P., Thiruvenkadaravi, K. V., Baskaralingam, P., Kumar, P. S., & Sivanesan, S., (2012). Preparation and characterization of porous cross linked laccase aggregates for the decolorization of triphenyl methane and reactive dyes. *Bioresource Technology, 119*, 28–34.
38. Lazaridis, N. K., Kyzas, G. Z., Vassiliou, A. A., & Bikiaris, D. N., (2007). Chitosan derivatives as biosorbents for basic dyes. *Langmuir, 23*(14), 7634–7643.
39. Liu, K., Fan, Y., Zhang, K., Yang, J., Yao, Y., & Liu, Y. (2021). Generative manifold learning thermography for non-destructive evaluation of defects in composite materials. In: *2021 IEEE 10th Data Driven Control and Learning Systems Conference (DDCLS)* (pp. 432–437). IEEE.
40. Mali, P. L., Mahajan, M. M., Patil, D. P., & Kulkarni, M. V., (2000). *Biodecolorisation of Members of Triphenylmethane and Azo Group of Dyes 3*(1), 7–14.
41. Marcelino, M. D. R., & Muralha, V. S., (2012). Synthetic organic pigments in contemporary Balinese painting: A Raman microscopy study. *Journal of Raman Spectroscopy, 43*(9), 1281–1292.
42. Meng, L., Breitkopf, P., & Le Quilliec, G., (2017). An insight into the identifiability of material properties by instrumented indentation test using manifold approach based on Ph curve and imprint shape. *International Journal of Solids and Structures, 106*, 13–26.
43. Meng, L., Breitkopf, P., Raghavan, B., Mauvoisin, G., Bartier, O., & Hernot, X., (2015). Identification of material properties using indentation test and shape manifold learning approach. *Computer Methods in Applied Mechanics and Engineering, 297*, 239–257.

44. Meshko, V., Markovska, L., Mincheva, M., & Rodrigues, A. E., (2001). Adsorption of basic dyes on granular activated carbon and natural zeolite. *Water Research, 35*(14), 3357–3366.
45. Michaels, G. B., & Lewis, D. L., (1985). Sorption and toxicity of azo and triphenylmethane dyes to aquatic microbial populations. *Environmental Toxicology and Chemistry: An International Journal, 4*(1), 45–50.
46. Mishra, S., & Maiti, A., (2018). The efficacy of bacterial species to decolorize reactive azo, anthroquinone and triphenylmethane dyes from wastewater: A review. *Environmental Science and Pollution Research, 25*, 8286–8314.
47. Moosavi, S., Lai, C. W., Gan, S., Zamiri, G., Akbarzadeh Pivehzhani, O., & Johan, M. R., (2020). Application of efficient magnetic particles and activated carbon for dye removal from wastewater. *ACS Omega, 5*(33), 20684–20697.
48. Morthland, F. W., De Bruyn, P. P. H., & Smith, N. H., (1954). Spectrophotometric studies on the interaction of nucleic acids with aminoacridines and other basic dyes. *Experimental Cell Research, 7*(1), 201–214.
49. Ong, S. T., Keng, P. S., Lee, S. L., Leong, M. H., & Hung, Y. T., (2010). Equilibrium studies for the removal of basic dye by sunflower seed husk (*Helianthus annuus*). *International Journal of the Physical Sciences, 5*(8), 1270.
50. Przystaś, W., Zabłocka-Godlewska, E., & Grabińska-Sota, E., (2012). Biological removal of azo and triphenylmethane dyes and toxicity of process by-products. *Water, Air, & Soil Pollution, 223*, 1581–1592.
51. Redmond, R. W., Kochevar, I. E., Krieg, M., Smith, G., & McGimpsey, W. G., (1997). Excited state relaxation in cyanine dyes: A remarkably efficient reverse intersystem crossing from upper triplet levels. *The Journal of Physical Chemistry A, 101*(15), 2773–2777.
52. Sani, R. K., & Banerjee, U. C., (1999). Decolorization of triphenylmethane dyes and textile and dye-stuff effluent by *Kurthia* sp. *Enzyme and Microbial Technology, 24*(7), 433–437.
53. Sankar, M., Sekaran, G., Sadulla, S., & Ramasami, T., (1999). Removal of diazo and triphenylmethane dyes from aqueous solutions through an adsorption process. *Journal of Chemical Technology & Biotechnology: International Research in Process, Environmental & Clean Technology,*

74(4), 337–344.

54. Saquib, M., & Muneer, M., (2003). TiO_2-mediated photocatalytic degradation of a triphenylmethane dye (gentian violet), in aqueous suspensions. *Dyes and Pigments, 56*(1), 37–49.
55. Shah, B. A., Patel, H. D., & Shah, A. V., (2011). Equilibrium and kinetic studies of the adsorption of basic dye from aqueous solutions by zeolite synthesized from bagasse fly ash. *Environmental Progress & Sustainable Energy, 30*(4), 549–557.
56. Shedbalkar, U., Dhanve, R., & Jadhav, J., (2008). Biodegradation of triphenylmethane dye cotton blue by *Penicillium ochrochloron* MTCC 517. *Journal of Hazardous Materials, 157*(2–3), 472–479.
57. Shen, Y., Béreux, N., Frangi, A., & Touzé, C., (2021). Reduced order models for geometrically nonlinear structures: Assessment of implicit condensation in comparison with invariant manifold approach. *European Journal of Mechanics-A/Solids, 86*, 104165.
58. Shindy, A., (2017). Problems and solutions in colors, dyes and pigments chemistry: A review. *Chem. Int., 3*(2), 97–105.
59. Singer, M., (1952). Factors which control the staining of tissue sections with acid and basic dyes. In: *International Review of Cytology* (Vol. 1, pp. 211–255). Academic Press.
60. Sun, Q., & Yang, L., (2003). The adsorption of basic dyes from aqueous solution on modified peat–resin particle. *Water Research, 37*(7), 1535–1544.
61. Thetford, D., & Updated by Staff., (2000). Triphenylmethane and related dyes. *Kirk Othmer Encyclopedia of Chemical Technology*, 1–19.
62. Tsai, W. T., Hsu, H. C., Su, T. Y., Lin, K. Y., & Lin, C. M., (2008). Removal of basic dye (methylene blue) from wastewaters utilizing beer brewery waste. *Journal of Hazardous Materials, 154*(1–3), 73–78.
63. Vasdev, K., Kuhad, R. C., & Saxena, R. K., (1995). Decolorization of triphenylmethane dyes by the bird's nest fungus *Cyathus bulleri. Current Microbiology, 30*, 269–272.
64. Vinod, K. N., & Gowda, K. N., (2009). Oxidative decolorization of triphenylmethane dyes by chloramine-T in alkaline medium catalyzed by Pd (II): A comparative spectrophotomeric kinetic and mechanistic approach. *Journal of Molecular Catalysis A: Chemical, 298*(1–2), 60–68.

65. Vogel, M., & Rettig, W., (1987). Excited state dynamics of triphenylmethane-dyes used for investigation of microviscosity effects. *Berichte der Bunsengesellschaft für physikalische Chemie, 91*(11), 1241–1247.
66. Wang, C. C., Juang, L. C., Hsu, T. C., Lee, C. K., Lee, J. F., & Huang, F. C., (2004). Adsorption of basic dyes onto montmorillonite. *Journal of Colloid and Interface Science, 273*(1), 80–86.
67. Wang, H., Le, Y., & Sun, J., (2022). Light-driven bio-decolorization of triphenylmethane dyes by a *Clostridium thermocellum*-CdS biohybrid. *Journal of Hazardous Materials, 431*, 128596.
68. Wang, S., Li, H., & Xu, L., (2006). Application of zeolite MCM-22 for basic dye removal from wastewater. *Journal of Colloid and Interface Science, 295*(1), 71–78.
69. Wang, X. S., Zhou, Y., Jiang, Y., & Sun, C., (2008). The removal of basic dyes from aqueous solutions using agricultural by-products. *Journal of Hazardous Materials, 157*(2–3), 374–385.
70. Williams, W. W., & Conley, J. W., (1955). Organic pigments. *Industrial & Engineering Chemistry, 47*(8), 1507–1510.
71. Yang, N., Zhu, S., Zhang, D., & Xu, S., (2008). Synthesis and properties of magnetic Fe_3O_4-activated carbon nanocomposite particles for dye removal. *Materials Letters, 62*(4–5), 645–647.
72. Zhou, G. L., Xu, T., Zhu, W. C., Konietzky, H., Heng, Z., & Yu, X., (2020). A damage-based numerical manifold approach to crack propagation in rocks. *Engineering Analysis with Boundary Elements, 117*, 76–88.

CHAPTER 4

Cationic Dyes for Synthetic Fibers

Contents

Unit Introduction 106
4.1. Classification of Cationic Dyes for Synthetic Fibers 107
4.2. Pendant Cationic Dyes 108
4.3. Delocalized Cationic Dyes 119
4.4. Amine Salts 134
4.5. Trends in Cationic Dyes for Synthetic Fibers 135
4.6. Summary 136
References 137

UNIT INTRODUCTION

The commercial dye industry experienced a significant shift in 1953 with the du Pont Company's introduction of Orlon Type 42 polyacrylonitrile fiber. The fiber in question possesses acidic dye sites. It exhibits a high affinity towards classical basic dyes, resulting in remarkable light and wash-fastness properties surpassing similar dyes on natural fibers (Parvinzadeh, 2012). While the lightfastness of basic dyes on acrylics has been enhanced compared to their performance on silk and acetate, further enhancements are still required due to the more rigorous fastness requirements imposed by the long-term reliability of acrylic fibers.

The primary commercial dyes utilized for acrylic fibers comprised carefully chosen and altered polymethinez, oxazine, and triarylmethane varieties. In a brief period, alterations were made to the dyes mentioned above to enhance their lightfastness to a greater extent. Acrylics have been widely utilized in the sweater industry and conjunction with wool blends, primarily due to the fiber's high bulk and thermal properties (Hou et al., 2020). These circumstances resulted in an increased need for improved dye light fastness. Dye manufacturers directed their research towards innovative dye varieties to enhance this characteristic. The phenomenon mentioned above has resulted in a novel area of study known as "pendant" cationic dyes. This involves the introduction of a cationic charge onto a molecule at a location that is effectively isolated from the chromophore. Thus, it is now feasible to modify a majority of lightfast chromophores for application on acrylic fibers. The size of certain molecules, such as copper phthalocyanine and benzanthrone derivatives, poses a limitation as they are unable to permeate the fiber due to their large size (Raval et al., 2017). The industrialization and expansion of a wide-ranging field of diazahemicyanine dyes have emerged as the second significant development, providing an unprecedented combination of brightness and fastness.

The progress made in the field of dye development has resulted in recognition of dyed acrylic fibers as possessing the most optimal blend of lightfastness, brightness, and wash-fastness among all textile fibers (Kim & Sun, 2002). As a result, they have established a benchmark for the industry to follow.

The market potential of the acrylic fiber industry has been significantly impacted by advancements in dyeing technology, making it a unique case in this regard. Polyacrylonitrile fibers gained significant market traction only after 1954, following the introduction of cationic-dyeable fibers. The worldwide manufacturing of acrylic and modacrylic fibers has experienced a significant increase since 1954, with an anticipated production capacity of 2.8 billion pounds by the 1970s (Bouatay et al., 2016). Additionally, the saturation level of each fiber for basic dyes is provided, indicating the number of acidic dye sites present. Most acrylic fibers exhibit dyeability with cationic dyes, except for a select few products of relatively small volume, such as Orlon Type 44, which can be dyed using acid dyes.

Remember

Due to their strong affinity for synthetic fibers, cationic dyes can create interesting color effects such as gradient or variegated patterns of fabrics.

The triumph of acrylic fibers capable of accommodating cationic dye has prompted the adaptation of alternative synthetic fibers to enable the acceptance of cationic dyes to enhance their luminosity. In general, acrylic fibers are unparalleled in providing a unique blend of brightness and fastness, which is not found in any other acid-modified fibers (Aminoddin, 2012). The mere incorporation of acidic sites within a polymer does not ensure the manifestation of any of the abovementioned characteristics. Generally, the lightfastness of cationic dyes exhibits inferior performance on substrates other than their own. The fluorescence emitted by dyes tends to exhibit a lower level of brightness. The degree of hydrophilicity of a fiber is inversely proportional to its wash-fastness.

4.1. CLASSIFICATION OF CATIONIC DYES FOR SYNTHETIC FIBERS

Cationic dyes are characterized by the presence of a positive charge attributed to the compound's chromophoric moiety. Typically, a concomitant anion exerts minimal influence on the dyeing characteristics. Notwithstanding, it might significantly impact the solubility properties of a dye and the feasibility of obtaining a pure dye. The existing cationic dyes can be categorized into three discrete classes.

4.1.1. Pendant Cation (Non-Resonating Charge)

The dyes in question exhibit structural similarity to disperse dyes, with the notable distinction of possessing an added cationic charge isolated from the chromophore. This modification facilitates the absorption of the dye by acid-modified fibers (El Harfi & El Harfi, 2017). These products have been specifically engineered to provide superior resistance to fading caused by exposure to light. These dyes' chromatic, mechanical, and colorfast characteristics are akin to those of the dispersed dyes from which they are extracted. As a result, using these dyes with high lightfastness properties does not provide the same level of vividness in color and potent coloring ability as the conventional triaryl methane and associated dyes.

4.1.2. Delocalized Positive Charge (Resonating Charge)

Dyes that possess a delocalized positive charge, known as cationic dyes, are typically the most vibrant and potent in terms of their coloring ability (Gürses et al., 2016). Various dyes have been observed, such as modified oxazine, triarylmethane, aza cyanine, polymethine, and polymethine.

4.1.3. Amine Salts

Acrylic fibers are commonly dyed under acidic conditions (Affat, 2021). However, basic dyes that lack a formal positive charge, except in the presence of an acid, can also be utilized for dyeing acrylic fibers. Under either neutral or alkaline conditions, these particular products will exhibit characteristics that are more akin to those of dispersed dyes.

4.1.4. Structures of Commercial Cationic Dyes

Due to the recent and ongoing development of cationic dyes for synthetic fibers, there is significant competition among dye manufacturers (Benkhaya et al., 2020). Consequently, there is a shortage of published technical information about these dyes' precise structures. Most recently developed cationic dyes are catalogued in the color Index using a non-specific nomenclature. The ensuing discourse on dye structures delineates customary commercial dye structures and technology. However, particular commercial dye designations shall be referenced solely if documented in published literature.

4.2. PENDANT CATIONIC DYES

Pendant cationic dyes are categorized as dyes with a localized positive charge not combined with the chromophore system. Typically, the cationic functionality comprises a substituted quaternary ammonium salt, although there exist instances of sulfonium, hydrazinium, and thiouronium salts. The quaternary ammonium salt is favored due to its ease of preparation and stability across a wide pH spectrum (She et al., 2016).

Before the emergence of acrylic fibers, a restricted range of pendant cationic dyes had been synthesized. The Janus color range included derivatives of (in-aminophenyl) tri-methylammonium chloride, including (I) (Mahdavinia et al., 2014). These derivatives exhibited restricted applicability on rayon, silk wool unions, and cotton that had undergone treatment with fixing agents.

Isothiouronium salts can be synthesized by heating a chloromethylated dye with tetramethyl thiourea. Subsequently, the salts mentioned above are administered onto fibers composed of cellulose and rendered insoluble through heat and an acid acceptor, ultimately creating dyeings that exhibit resistance to washing. The chemistry mentioned above is the fundamental component for the Alcian X (II) dye series developed by ICI (Mahdavinia et al., 2012). Recently, ICI has obtained patents for using monoisothiouronium salts of specific azo compounds as dyes for polyacrylonitrile.

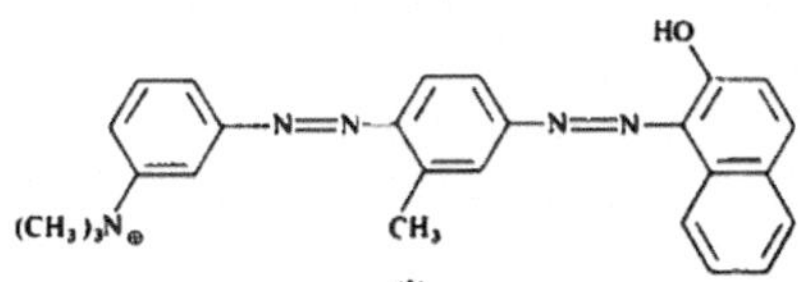

(I)

Janus Red B, Basic Red (CI 26115)

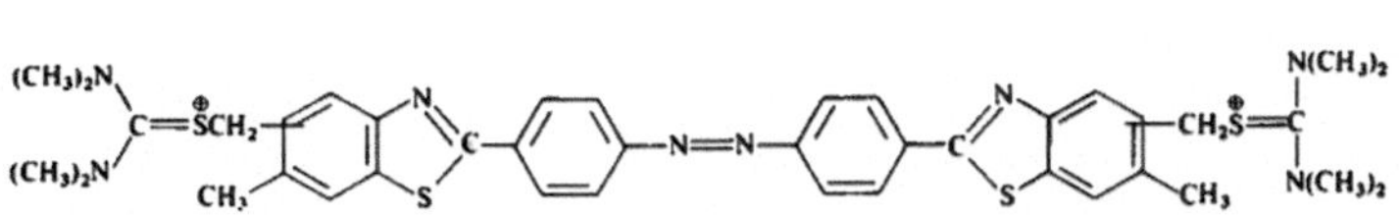

(II)

Alcian Yellow GXS

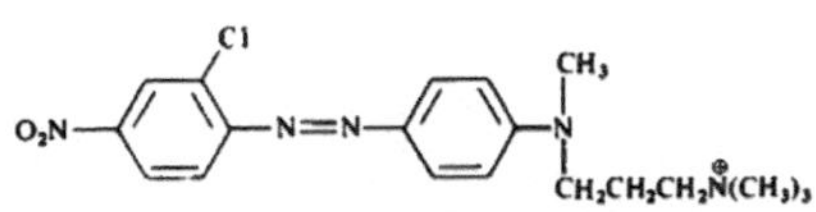

(III)

Basic Red 18 (CI 11085)

Figure 4.1. *Illustration of the structures of Janus red B, Alcian yellow GXS, and basic red 18 dyes.*

Source: Donal R. Baer, Creative Commons License.

The category of pendant ammonium dyes has been recognized as a viable option for coloring leather. It is worth noting that a particular red pendant cationic azo dye (III) of significance today was initially introduced in the patent literature as a leather dye (Mahdavinia & Asgari, 2013) (Figure 4.1).

4.2.1. Azo Dyes

The cationic group in a pendant azo dye is most conveniently prepared by performing it into the radical containing the diazonium group or into the coupler with subsequent dye formation (Yun et al., 2020). Generally speaking, there are more pendant cationic dyes based on a cationic coupling component than on the diazo compound due to the ease of synthesis.

Early evaluation of the Janus-type dyes containing a trimethylammonium group directly on the diazo component showed them deficient in lightfastness on acrylic fibers. Boyd developed a new diazo component (IV) and a practical synthesis (Chart 1), leading to lightfast pendant cationic azo dyes in the yellow to red range (Ali et al., 2017) (Figure 4.2).

CHART 1

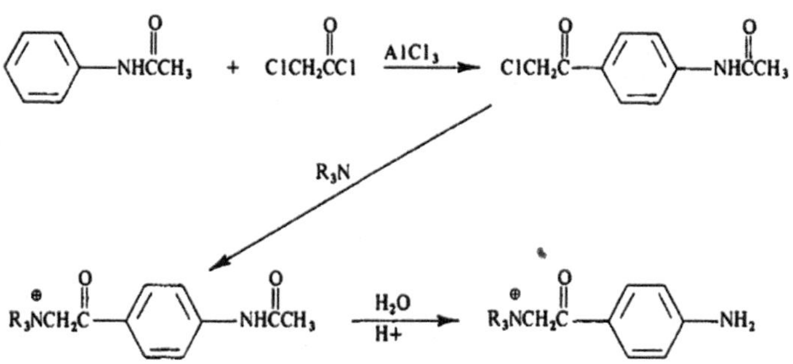

(IV)

Figure 4.2. *Representation of synthesis reaction leading to lightfast pendant cationic azo dyes.*

Source: J. Bailey, Creative Commons License.

Diazotization and subsequent coupling of compound (IV) with aromatic primary, hindered secondary, or tertiary aromatic amines produces dyes that exhibit high resistance to fading upon exposure to light. The shades acquired exemplify the following: The chemical compounds m-toluidine, o-chloro-N-(2 cyanoethyl) aniline, and N,N-diethyl aniline exhibit orange, yellow, and red colors, respectively (Zhou et al., 2018). The impact of R group variations on the shade and fastness properties is negligible; however, it can affect additional physical attributes such as dyeing rate and solubility.

The coupling reaction with pyrazolones results in the formation of green-yellow hues (V).

The disazo dyes synthesis involves diazotizing the m-toluidine above derivative and coupling with a diverse range of couplers. The desulfonation of the coupler occurs during the dye-forming process upon coupling of (IV) to Tobias acid, producing the 2-naphthyl amine derivative (VI) (Pakdel et al., 2022). The scarlet dye exhibits remarkable pH stability and demonstrates unexpected lightfastness. The compound bearing the 6-(methylsulfonyl) moiety in its structure exhibits an orange hue.

Given that most acrylics undergo dyeing within the pH range of 3 to 6, the dyes must exhibit shade stability within this range. Azo dyes that possess highly basic amine groups, particularly the p-dimethylamino group, are typically characterized by significant prototropic behavior, manifesting as a variation in hue in response to changes in acidity (Hosseinzadeh & Khoshnood, 2016).

As a result, most dyes that fall under this category are rendered impractical due to the presence of basic amines such as (cyanoethyl) or (hydroxyethyl) anilines or their coupling with active methylene compounds. Regrettably, it can be observed that as the amine becomes less basic, the resulting shade of the dye shifts towards the hypsochromic end of the spectrum.

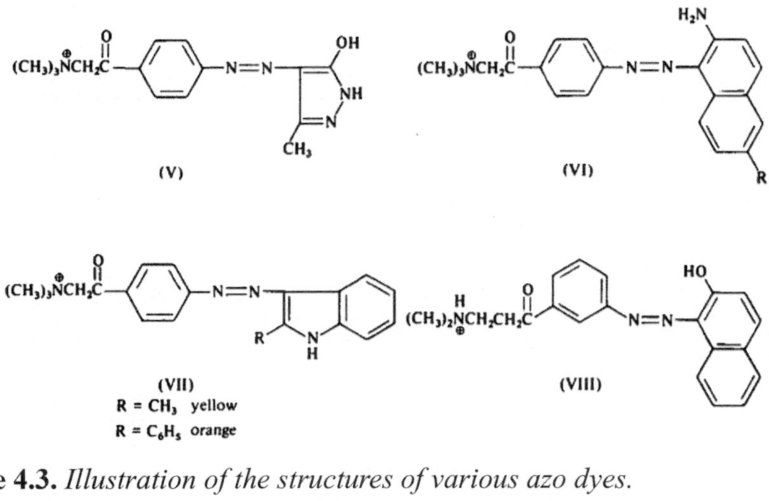

Figure 4.3. *Illustration of the structures of various azo dyes.*

Source: N. R. Ayyangar, Creative Commons License.

Consequently, the applicable range of hues for dyes synthesized from (IV) is limited to the green-yellow to the scarlet range. The most environmentally friendly yellow varieties are obtained through linkages to 1,3-indandione or 2,4-quinoline diol (Lewis & Ho, 1996). The introduction of couplings to heterocyclic rings, such as 2-methylindole or 2-phenylindole, produces yellow and orange dyes (VII) that exhibit superior lightfastness and pH stability. An example of such a dye is Deorlene Yellow 4RL (Figure 4.3).

A viable pathway towards cationic diazo components involves the utilization of in-amino-acetophenone. A modified Mannich base featuring a diazotizable amino group can be obtained from the CCO through treatment with formaldehyde and a secondary amine. The orange dye commonly used is identified as (VIII). The technical significance of this particular dye type has not been attained to the same extent as its predecessors due to the reversible nature of the Mannich reaction, which poses challenges regarding dye stability (Tsai et al., 2016).

The generation of the cationic moiety in the coupling agent can occur through various mechanisms, as illustrated in Chart 2 (Shukla et al., 2012). Similar to the preceding dyes, the R group exhibits minimal impact on hue and durability but directly impacts dyeing characteristics. The brownish-red dye of significant importance is obtained through the diazotization of 2-chloro-4-nitroaniline followed by its coupling to a molecule resembling (IX) (Figure 4.4).

CHART 2

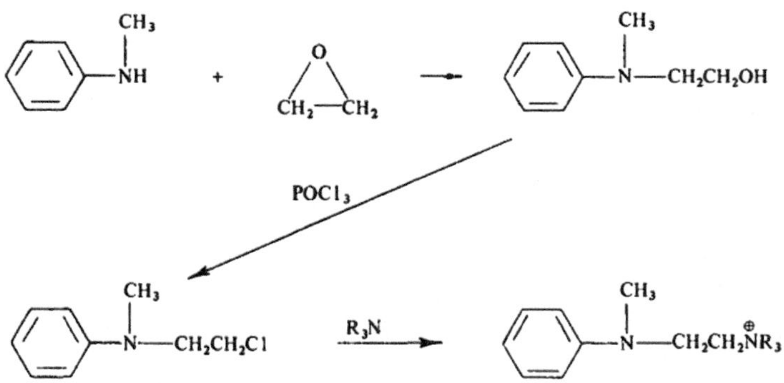

(IX)

Figure 4.4. *Formation of the cationic group.*

Source: Hans Meier, Creative Commons License.

As mentioned earlier, several patents address alternative cationic groups in the type of couplers. Several instances include the period from the 10th century to the 19th century. All of these couplers generate comparable hues under identical conditions. The primary distinction lies in the cationic charge's stability under the specific temperature and pH conditions necessary for the dyeing process (Hosseinzadeh et al., 2015).

The majority of azo dyes that are synthesized through coupling with a standard "quaternised dialkyl aniline (IX)–(XIX)" exhibit hues ranging from dull yellow-orange to red-brown. The coupling reaction of aniline typically yields p-nitro- or p-cyano- anilines in yellow color, 2,6-dihalo-4-nitroanilines in orange to red hues, and brown shades (Lei et al., 2010). Disazo dyes have been synthesized; however, their utility appears to be limited compared to monoazo dyes. Disazo dyes exhibit reduced dyeing rates and have the propensity to cause significant discoloration on alternative fibers.

The utilization of heterocyclic diazo constituents has been employed to enhance the luminosity of cationic dyes comprising quaternised dialkyl aniline couplers. Bayer and du Pont have obtained patents for producing red and violet dyes derived from thiadiazole and 2-amino-5-nitrothiasole (Shirai et al., 1978). Additionally, du Pont has patented cationic dyes derived from 2-amino-5-(alkylsulfonyl) thiazoles. Efforts to generate vivid red hues using naphthol have created couplers belonging to the naphthalene category, namely XXIa, and XXIb. The naphthol hue can typically be synthesized with favorable resistance to fading and heightened luminosity using cyanine-class chromophores. The chromophores mentioned above possess minimal significance in the realm of commerce (Figure 4.5).

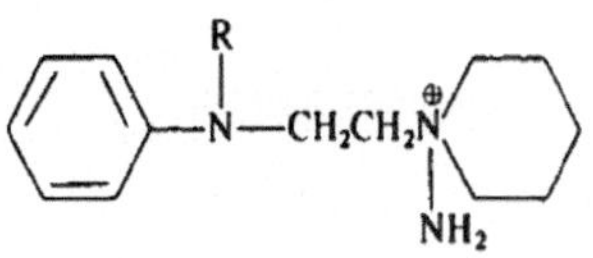

(X)[24]

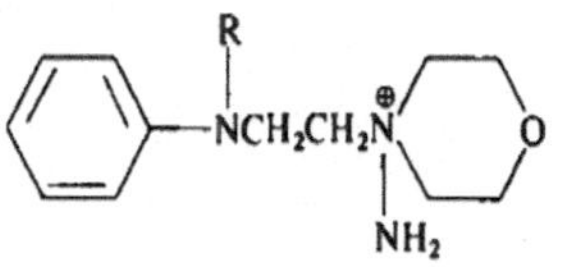

(XI)[24]

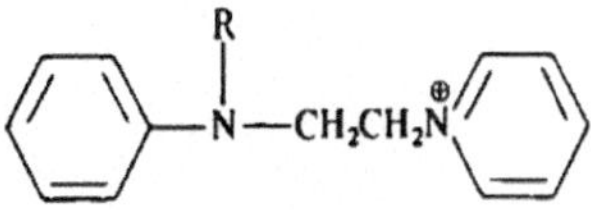

(XII)[25]

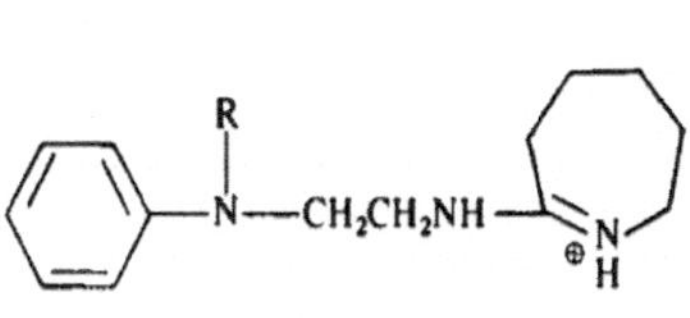

(XIII)[26]

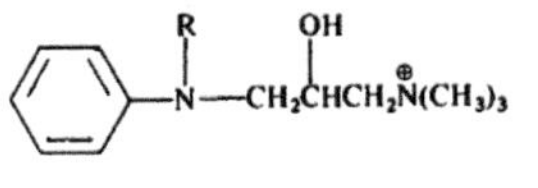

(XIV)[27]

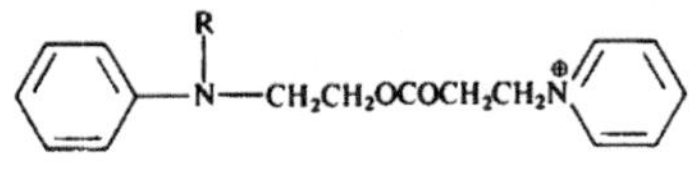

(XV)[28]

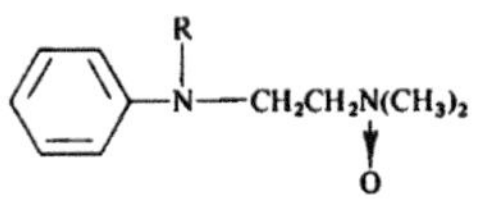

(XVI)[29]

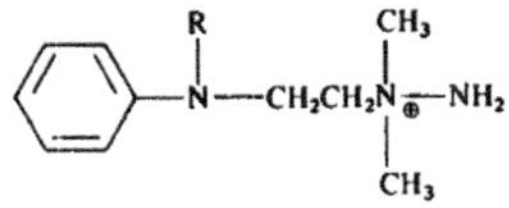

(XVII)[30]

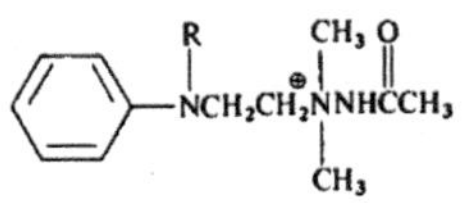

(XVIII)[31]

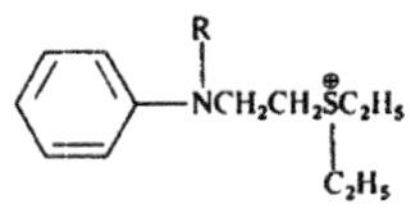

(XIX)[32]

Figure 4.5. *Illustration of examples of alternate cationic groups.*

Source: B. D. Tilak, Creative Commons License.

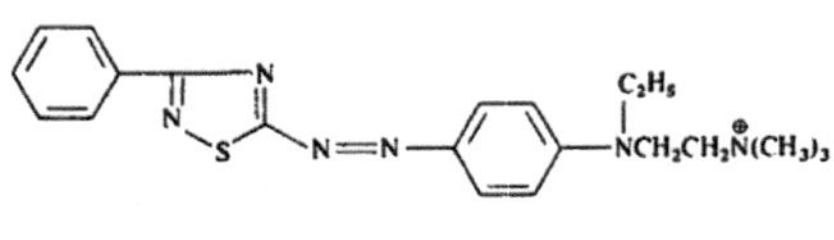

(XX)

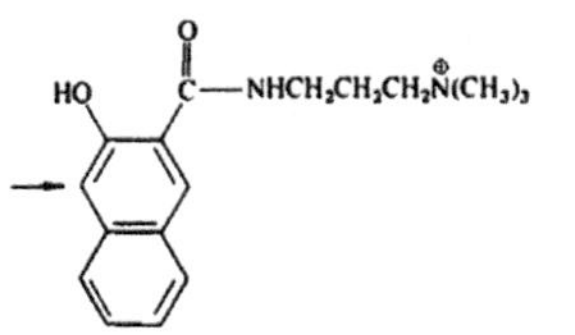

(XXIa)

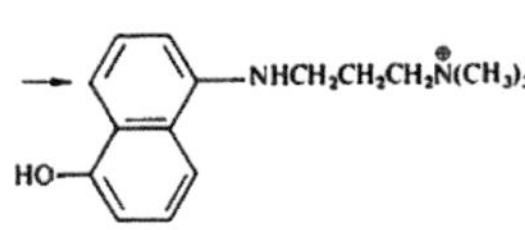

(XXIb)

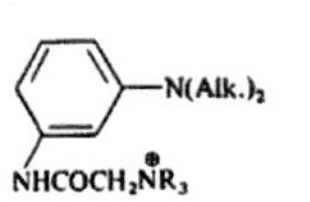

(XXII)

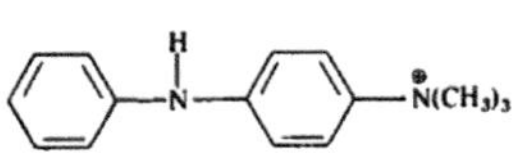

(XXIII)

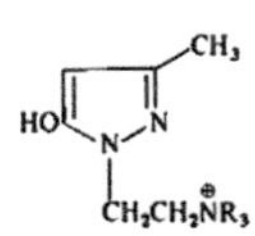

(XXIV)

Figure 4.6. *Illustration of the sampling tools.*

Source: Meyers, Creative Commons License.

Additional pathways for the production of cationic couplers in azo dyes involve the utilization of intermediates such as (XXII), (XXIII), and (IY). Pre-metalized mono-cationic azo dyes can be prepared by utilizing naphthol couplers (XXIa) and (XXIb). Metals with a valence of two, such as copper or nickel, are known to produce hues ranging from subdued red to deep violet (Dhodapkar et al., 2007). The pyrazolones present in these dyes exhibit satisfactory yellow lightfastness as anticipated; however, they tend to result in undue discoloration of adjacent fibers, such as wool, when employed in composite textiles (Figure 4.6).

4.2.2. Anthraquinone Dyes

During the initial stages of cationic dye development for acrylic fibers, it was acknowledged that the conventional blue basic dyes could provide vividness but lacked the essential level of lightfastness that was imperative for blends, particularly those involving wool. Novel dyes needed to be formulated. The research findings indicate that applying blue anthraquinone disperses dyes on Orlon 42 fabric yields pastel hues that exhibit commendable resistance to fading upon exposure to light (Vaid & Jindal, 2023). Introducing a cationic pendant charge to these molecules produced dyes with equivalent fastness properties but significantly improved buildup. To achieve satisfactory lightfastness for acetate and polyesters using anthraquinone dyes, it is necessary to establish hydrogen bonding between the carbonyl group and amino or hydroxy auxochromes in the n-positions. Therefore, it is imperative that the positioning of the charged group in lightfast pendant cationic anthraquinones does not disrupt the essential n-hydrogen bonds of the amino or hydroxyl groups within the molecule. The rapidity of fuming exhibited by anthraquinone dyes on acrylics and modacrylics does not pose a significant issue due to the fiber's limited permeability to nitrogen oxides (Li et al., 2022). Conversely, the impact of oxides of nitrogen or ozone on the identical dyes utilized in acid-modified polyamides can be significant.

The anthraquinone dyes exhibit a molar extinction coefficient that is approximately lower than that of the azo dyes and approximately lower than that of the delocalized cationic dyes. Hence, most pendant cationic anthraquinones utilized in commercial applications are derived from established anthraquinone intermediates, as the utilization of specialized intermediates leads to the production of dyes that possess limited commercial value in use.

The production of lightfast blues for acrylics in the early stages involved the reaction of a 4-halo-1-aminoanthraquinone with a dialkylaminoalkylamine, which was subsequently followed by quaternization, resulting in the formation of types (XXV)–(XXVIII) (Kyzas et al., 2015).

The acquisition of reddish violet dyes (XXIX) is possible when the quaternary group is integrated into a 1,2,4-triazole side chain. Du Pont has developed green-blue dyes (XXX) that exhibit high lightfastness, utilizing 1,4-diamino-2,3-anthraquinone dicarboximides as the basis for their development.

The introduction of cationic groups into the anthraquinone nucleus at the 2-position through an oxygen or sulfur bridge has been observed to maintain favorable fastness properties. The application of dialkylamino alcohols in the presence of a base to 1-amino-2-bromo-4-hydroxyanthraquinone produces pink hues (XXXI) resistant to fading. The production of pink shades with a slightly increased blue hue is achieved through the process of chloromethylation of 1-amino-2-phenoxy-4-hydroxyanthraquinones, which is subsequently followed by quaternization with either an amine or hydrazine (XXXII) (Goel & Jacob, 2020) (Figure 4.7).

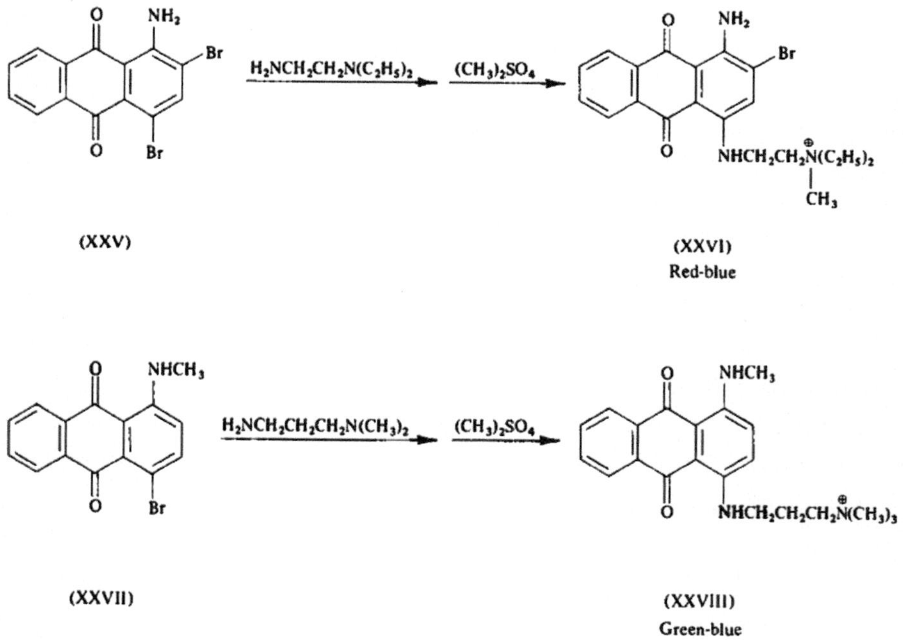

Figure 4.7. *Representation of the formation of red-blue and green-blue dyes.*

Source: Oskar Glenz, Creative Commons License.

The preparation of blue dyes can be achieved through analogous techniques, wherein an amino or substituted amino group is present in the 4-position of the anthraquinone. Commercial blue dyes (XXXIII) and (XXXIV) have been established.

Dyes incorporating a sulfur bridge are synthesized through the reaction between dialkylaminoethanethiols and 1-amino-2-anthraquinonesulfonic acids, which is followed by quaternization (Bayramoglu et al., 2009). The structure (XXXV) of Basic Blue 49 has been disclosed.

Various techniques have been revealed for producing cationic anthraquinone dyes that hang down, including the process of treating a sulfonyl halide with a dialkylaminoalkylamine and then quaternizing it, the reaction of primary aminoanthraquinones with paraformaldehyde and N,N-dimethylaniline followed by quaternization, and the creation of carboxamides on either the anthraquinone nucleus or a side chain by transforming a carboxylic acid (El Kaoutit et al., 2013) (Figure 4.8).

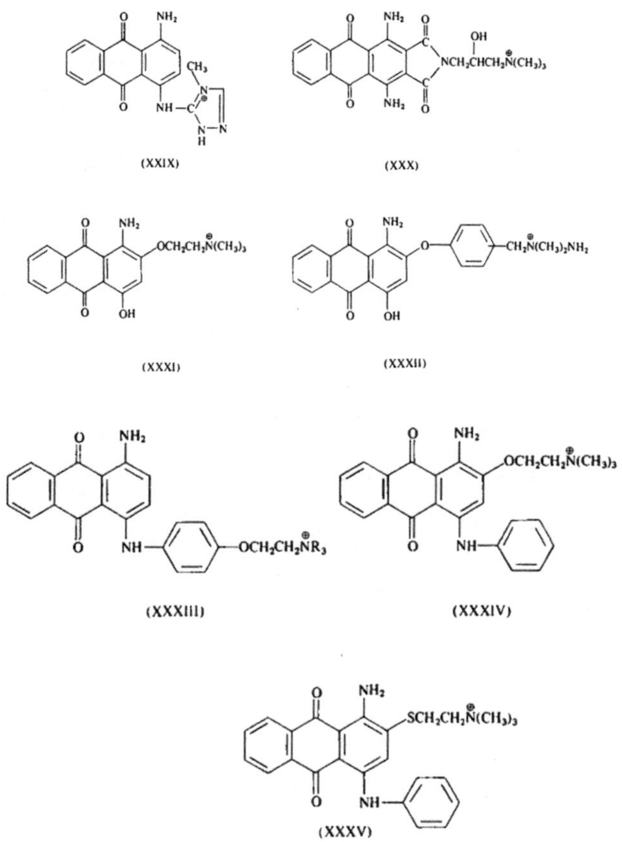

Figure 4.8. *Illustration of the structures of commercial blue and basic blue 49 dyes.*

Source: L. A. Williams, Creative Commons License.

Numerous variations of the methods mentioned above have been documented in the patent literature. The utilization of alicyclic amines as charge carriers and the possibility of employing various linkages to the chromophore are feasible options (Singh et al., 2021). This particular technology has the potential to be applied to a wide range of stable chromophores as long as the resulting dye molecule is of a size that is capable of penetrating the fiber.

4.2.3. Miscellaneous Pendant Dyes

Bright yellow, fluorescent methine dyes (XXXVI) were prepared by condensing a dialkylaminobensaldehyde containing a cationic group with an active methylene compound (Ying et al., 2022).

Lightfast quinophthalone yellows (XXXVII) have been prepared by condensation of trimellitic anhydride with a hydroxyquinaldine followed by the formation of the cationic ester or amide derivatives. Yellow dyes having excellent lightfastness on acid-modified polyesters have been obtained from nitrodiphenylamine dyes (XXXVIII) (Lan et al., 2017) (Figures 4.9–4.11).

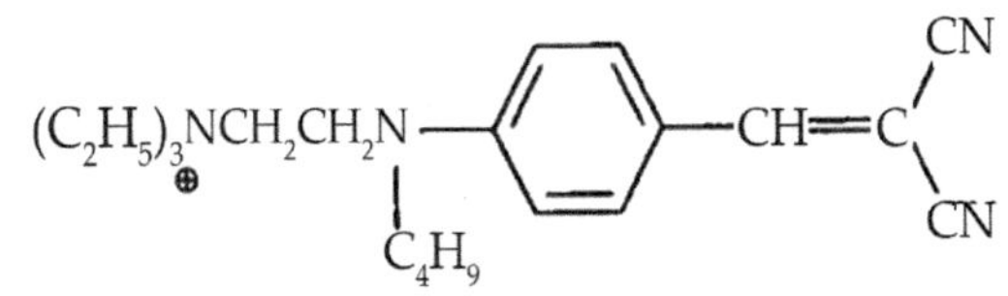

(XXXVI)

Figure 4.9. *Structure of the fluorescent methane dye.*

Source: Donald R. Baer, Creative Commons License.

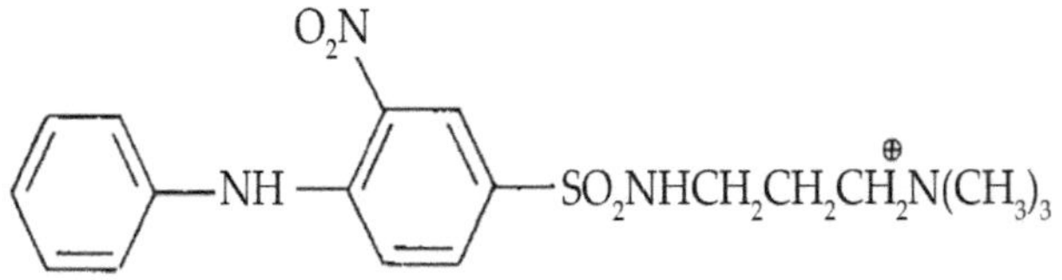

(XXXVIII)

Figure 4.10. *Structure of the lightfast quinophthalone yellows dye.*

Source: G. E. Fickman, Creative Commons License.

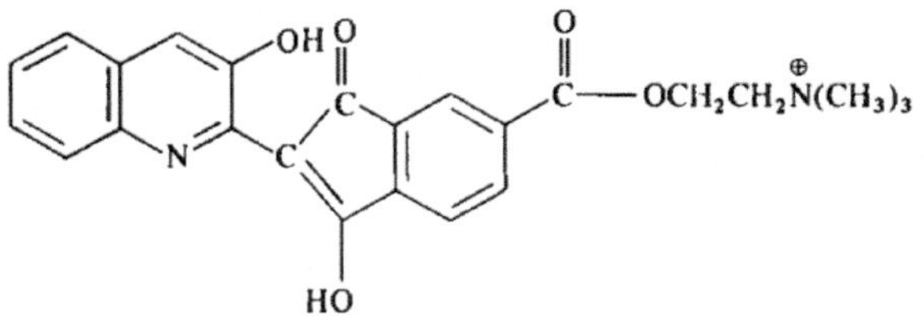

(XXXVII)

Figure 4.11. *Structure of nitrodiphenylamine dyes.*

Source: L. A. Williams, Creative Commons License.

4.3. DELOCALIZED CATIONIC DYES

Delocalized or distributed dyes are those in which the chromophoric system involves the participation of the positive charge in the molecule. Numerous patents and publications have been dedicated to studying diverse delocalized cationic dyes for synthetic fibers. However, triarylmethane, oxazine, cyanine, and azacyanine groups are the most significant commercially available categories. Most other cationic dyes with delocalized properties exhibit inadequate hydrolytic stability for acrylic fibers or are excessively susceptible to light fading.

Classical basic dyes are highly significant in producing deep shades, particularly blacks, due to their exceptional tinctorial value. Nevertheless, most of these necessitate alteration to enhance their lightfastness, which frequently decreases their tinctorial strength. Nevertheless, a significant anomaly exists in which conventional triarylmethane dyes are employed in the commercial sector. Malachite Green (XXXIX) and Fuchsin (XL) exhibit a lightfastness of fewer than 5 hours when used as self-shades on Orlon (Ehrenberg et al., 1988). The combination of these dyes produces a black hue that exhibits a lightfastness of 20 hours, owing to the compensatory fading properties of the dyes. The amalgamation in question holds paramount significance for acid-modified fibers owing to its exceptional tinctorial potency and value-in-use. Enhanced fastness properties for black coloration in acrylic fibers can be achieved by using dyes specifically formulated for this purpose. However, this approach incurs a cost premium of 3 to 6 times that of the conventional dye mixture. Except for a limited number of cyanines and a black blend of (XXXIX) and (XL), most fundamental dyes that existed before 1954 have not gained significant recognition on synthetic fibers without undergoing chemical alterations (Figure 4.12).

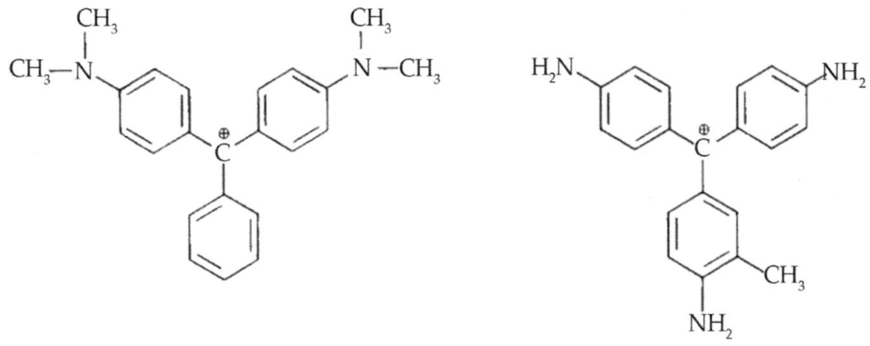

(XXXIX) Malachite Green

(XL) Fuchsin

Figure 4.12. *Structure of malachite green and fuchsin dyes.*

Source: Oskar Glenz, Creative Commons License.

4.3.1. DI and Triphenylmethanes

Auramine is the primary commercial representative of the diplienylmethane class. Auvamine is a yellow dye with high tinctorial strength (Shakeela et al., 2020). However, its utility on acrylic fibers is limited due to its imine nature, which makes it susceptible to hydrolysis in a refluxing dye bath, leading to the formation of Michler's ketone and ammonia. As of now, there is no viable method for stabilizing this particular dye variant, hence its exclusion from usage on synthetic materials.

Triarylmethanes remain a significant category of pigments, especially in the spectral range from red to green, where the highest chromaticity is sought. Nevertheless, these have various drawbacks, and recent research endeavors have focused on mitigating these limitations. The issues encompass inadequate lightfastness, substantial wool staining in blends, and the retrogression of specific dyes, such as Malachite Green, to a colorless leuco or carbinol form upon exposure of fibers to the elevated temperatures necessary for permanent creases or wrinkles.

The advancement of enhanced triarylmethane dyes for acrylics has encountered constraints due to contradictory factors. Dyes that exhibit the highest degree of symmetry and possess more fundamental amino groups are deemed the most potent (Kim & Choi, 2011). Nevertheless, these two factors diminish the likelihood of a feasible resolution. The diminution of basicity results in the imposition of a penalty on the preeminent characteristic of triaryl methanes, namely their elevated tinctorial potency. The enhancement of fastness through substitution often results in the deterioration of symmetry, diminishing the tinctorial value.

Wegmann analyzed the composition and properties of basic dyed polyacrylonitrile fibers in the context of these aspects. In general, it can be observed that the lightfastness of a dye molecule tends to decline as its basicity increases. In contrast, the affinity of the dye for wool fibers tends to increase.

Despite the favorable lightfastness properties exhibited by (XLI) (soluble sky-blue base), which sets it apart from other triaryl methanes, it is not a suitable dye for acrylic fibers due to its insolubility in water and its inability to produce uniform and deep penetrating dyeing effects (Bui et al., 2017).

Enhancement of dyeing properties can be achieved by introducing asymmetry in the aforementioned dyed material. The application of a Lewis acid, such as aluminum chloride, in the presence of an inert solvent to treat 4,4,'4"-trichlorotriphenylmethane with 4 moles of a substituted aniline at 100°C results in the production of dyes that exhibit lightfastness on acrylic fibers (Figure 4.13).

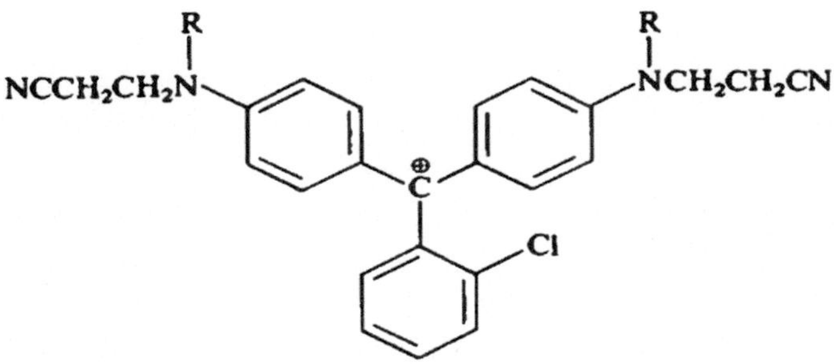

(XLV)

Figure 4.13. *Illustration of bright greenish blue dyes prepared from o-chloro-bensaldehyde.*

Source: Donald R. Bear, Creative Commons License.

The synthesis of stable, bright greenish blue dyes (XLV) with favorable pH stability involves the utilization of o-chlorobensaldehyde and two moles of N-alkyl-N- (2-cyanoethyl) -anilines (Dallel et al., 2018).

Indole derivatives of diphenylmethanes have been found to generate dyeings on polyacrylonitrile that exhibit high resistance to fading under light exposure. Blue-green-hued dyes can be synthesized by the reaction of 4-chlorobenzophenone with 1-methyl-2-phenylindole and p-anisidine. A range of hues from green to violet can be achieved by altering the amine and benzophenone derivative (XLVI) (Baer, 1971) (Figure 4.14).

CHART 4

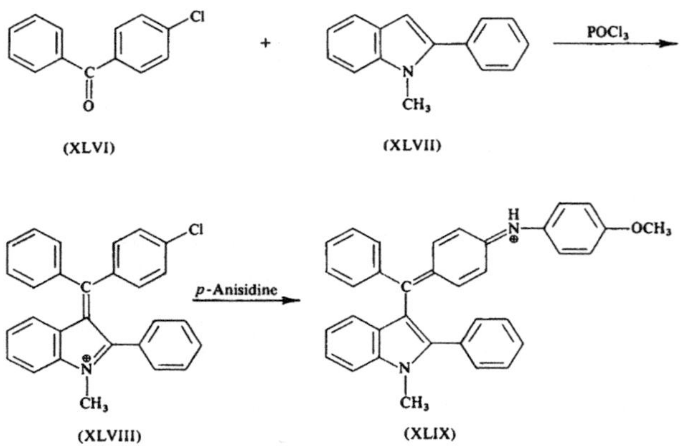

Figure 4.14. *Representation of reaction between 4-chlorobenzophenone with 1-methyl-2-phenylindole and p-anisidine.*

Source: Karl Neufang, Creative Commons License.

4.3.2. Azines

The indamine types and their cyclic derivatives, including acridines, pyrans, azines, oxazines, and thiazines, are closely associated with diphenylmethane dyes. These derivatives are widely recognized in the field. Wegmann has demonstrated the correlation between shade and lightfastness in this series. This is presented in a tabular format in Table 4.1 (Roulia & Vassiliadis, 2008).

Table 4.1. Shade and Fastness of Azines and Related Structures

X	Y	Dye Type	Shade	Lightfastness
–CH=	–H, –H	Diphenylmethane	Blue	Good
–CH=	–N–	Acridine	Orange	Poor
–CH=	–O–	Pyran	Red	Fair

–CH=	–S–	Thiopyran	Voilet	Fair
–N=	–H, –H	Indaniline	Green	(Unstable)
–N=	–N–	Azine	Violet	Poor
–N=	–O–	Oxazine	Blue	Good
–N=	–S–	Thiazine	Blue-green	Fair

Source: Karl Neufang, Creative Commons License.

The oxazines are the sole significant commercially viable dyes within this series, as determined by their shade and lightfastness properties. Ciba has characterized Basic Blue 4, CI 51004, as a valuable bright blue hue for application on polyacrylonitrile fibers (El-Shishtawy & Melegy, 2001). Using the 2-cyanoalkyl group in the oxazine has enhanced lightfastness, similar to other basic dyes.

Basic Blue 3, CI 51005, shares similarities with the abovementioned compound. However, it is distinguished by the presence of an ethoxyl group in the α position peri to the oxazine nitrogen. The abovementioned procedure is founded upon the synthetic methodology that entails the nitrosation of N,N-diethyl-in-anisidine, followed by its condensation with m-diethylamino-phenol in an acidic environment. Recent findings from Germany and the United States have revealed that ethanol is eliminated during the condensation process and that the structure of Basic Blue 4 is indistinguishable from that of Basic Blue 3 (Polit et al., 2012).

Most classical basic azine dyes exhibit inadequate light stability on acrylic fibers, rendering them unremarkable. The process of diazotization of Safranine T, subsequent coupling to N,N-dialkylanilines, produces deep blues (LII) that exhibit superior lightfastness on acrylic and polyester fibers (Zucca et al., 2016). Phenols or cresols can be used for couplings to obtain violet shades with equivalent fastness. The diazotization of diethylsafranine and subsequent coupling to phenol results in the formation of a dye that exhibits excellent lightfastness and imparts shades of charcoal and black on acrylic and polyester fibers. The lightfastness of azo safranines is commendable; however, their applicability is restricted to 100% artificial fibers and their amalgamations. This is attributed to the fact that the dye undergoes decomposition to the red shade of safranine in the presence of wool or potent acids. The absence of stability is a common characteristic observed in cationic azo dyes that contain an azo group as a constituent of the resonance conjugation system, as evidenced by the diazahem cyanines.

The expeditious blue dyeings can be generated by the quaternary salts of 5,12-diphenylfluorindine upon being subjected to dimethyl sulfate (Berneth, 2000).

Rhodamine 3B and its derivatives exhibit remarkable characteristics, yielding vivid and intense fluorescent pink and red shades. The lightfastness of acrylics is deemed inadequate, with a duration of only 5 hours. However, using acid-modified polyesters is possible to achieve light pink hues with a lightfastness of 20 hours. In contrast to the conventional principles of lightfastness, it has been observed that augmenting the dye concentration on the fiber decreases lightfastness in this particular instance.

4.3.3. Cyanine Dyes and Their Derivatives

Cyanine dyes have been historically recognized and utilized within the photographic sector as photosensitizers (El Amri et al., 2022). The prior discourse encompassed the chemistry and nomenclature of said dyes. The discovery that certain cyanine dyes exhibit exceptional colorfastness on polyacrylonitrile fibers while simultaneously imparting vivid and intense hues was an unexpected finding.

Several cyanine classes exhibit insufficient photostability or hydrolytic stability for synthetic fibers at or near boiling temperatures.

Extensive research has been conducted in the past decade and a half to explore the diverse properties of cyanine dyes. This has resulted in significant advancements in the development of highly luminous and durable dyes and in the synthesis technology of these dyes. The ensuing discourse will primarily focus on the alterations in the chemistry of cyanine dyes that have resulted in the development of novel commercial dyes.

4.3.3.1. Hemicyanines (Styryl Dyes)

Astrazon Red FG was one of the initial products utilized to produce acrylic fibers. The chemical compounds denoted as CI Basic Red 13 and CI Basic Violet 16, also known as Genacryl Red 4B and "Sevron" Brilliant Red 2B, respectively, exhibit a characteristic property of remaining in a state of stasis when subjected to a holding condition, wherein the values of R and R' are held constant (Joshi et al., 2022). The color spectrum of cationic dyes for synthetic materials can be altered from scarlet to blue-red by substituting R groups on the aniline ring. This modification is commonly observed in numerous product lines. Typically, amines with lower levels of complexity exhibit a greater tendency towards blue hues and lower resistance to fading when exposed to light. When both R' and R" are equal, acrylic fibers exhibit brilliant reddish orange shades, while polyester fibers display scarlet shades with good lightfastness. Typically, dyes of this nature elicit greater bathochromic hues on polyester fibers than acrylics.

The synthesis of bright orange dyes involves the condensation of Fischer's aldehyde (l,3,3-trimethyl-Δ2, α-indolineacetaldehyde) with substituted indoles (Dapson, 2005).

Two techniques are employed in the preparation of these dyes. The choice of methodology is contingent upon the simplicity of obtaining the uncontaminated dyes and the accessibility of the initial substances.

4.3.3.2. Carbocyanines

The technical efficacy of these products, characterized by the presence of a trimethine bridge connecting the ring systems, has been limited on acrylic fibers owing to inadequate lightfastness. One prominent illustration is Basic Red 12, synthesized through the reaction between Fischer's aldehyde and Fischer's base in acetic acid (Xiao et al., 2016).

4.3.3.3. Azacarbocyanines

Substituting a carbon atom with nitrogen in a carbocyanine molecule reduces the length of the conjugated system. The majority of the dyes within this particular series exhibit a yellow hue. The initial synthetic dyes in this sequence were produced through the condensation of Fischer's base and primary aromatic amines. The abovementioned products yield vivid shades ranging from bright green-yellow to yellow, exhibiting exceptional resistance to fading under light exposure, with a lightfastness of up to 80 hours on acrylic fibers.

Basic Yellow 11 (CI 48055) and 12 (CI 48065) represent commercial dyes. When an aromatic amine undergoes N-alkylation within a ring system, the resulting coloration is minimally altered, and the condensation process becomes more challenging.

The dyeing properties of aza cyanines have been evaluated and compared with those of the parent cyanine in the symmetrical benzothiazole series by Volts (Gregory, 1990). The study involved a thorough analysis of shade and fastness, focusing on acrylic fibers. The results indicate that the triazatrimethine cyanines and the unsymmetrical diazatrimethine cyanines exhibit superior lightfastness and dyebath stability compared to other azacyanine dyes.

The production of stable triazatrimethines involves the alkylation of linear triazene groups, which are diazoamino compounds that link an aromatic ring with a heterocyclic radical comprising nitrogen. The production of a pure yellow shade with excellent lightfastness can be achieved through the diazotization of a p-substituted aniline coupling to 2-aminobenzothiazole in glacial acetic acid, followed by the dimethylation of the isolated triazene (Asai et al., 2016). The configuration of the yellow dye likely comprises a blend of isomers (LXVI) and (LXVII), with their respective proportions contingent upon the electronic properties of R and the degree of steric hindrance.

The synthesis of unsymmetrical diazatrimethine cyanines involves the application of oxidative coupling methods to couple heterocyclic hydrazine with an active methylene heterocycle (Wainwright, 2010). Thus, the yellow chromophores (LXVIII) and (LXIX) are acquired (Figure 4.15).

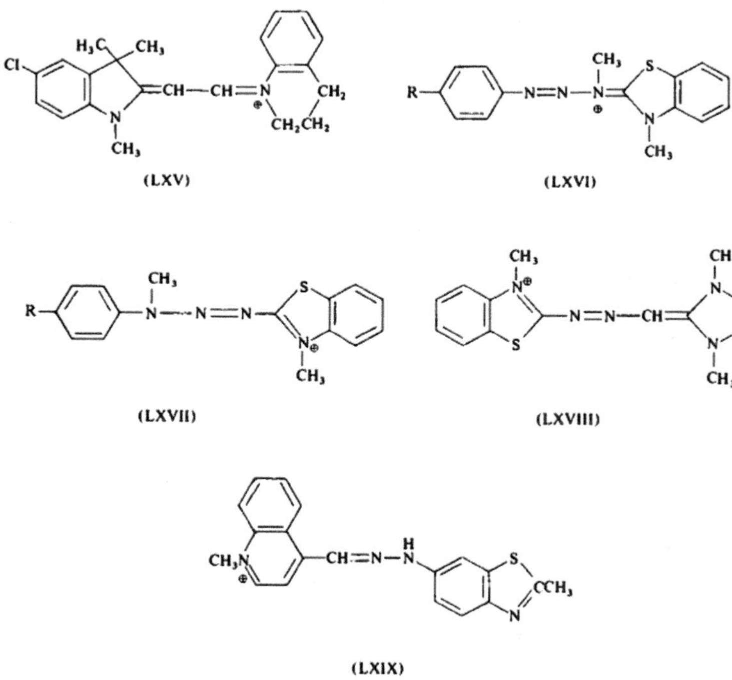

Figure 4.15. *Representation of structure of yellow chromophores.*

Source: J. Bailey, Creative Commons License.

It is possible to perform additional alkylation on the nitrogen of the chain in Diazatrimethine cyanines. In comparison to unsubstituted dyes, it can be observed that they exhibit lower stability under conditions of exposure to light and alkaline pH. The production of cationic dyes in yellow and orange hues, specifically the alkylated diazatrimethine variants, can be achieved through a coupling reaction between a diazonium salt and an active methylene located on a heterocyclic ring, which is subsequently followed by quaternization (Ramesha et al., 2011).

4.3.3.4. Diazahemicyanines (Diazastyryl Dyes)

The discovery of the efficacy of diazahemicyanines on acrylic fibers for generating deep shades with superior lightfastness and brightness, made almost simultaneously in Germany, Switzerland, and the United States, stands out as a significant breakthrough in the realm of cationic dyes for synthetic fibers.

Prior knowledge of Diazahemicyanines existed; however, the synthetic procedures were arduous, and the resulting products were acquired in suboptimal yields. The application of normal diazotization to heterocyclic amines is limited to specific cases. Amino compounds with heterocyclic structures, such as 2-aminopyridine, pose challenges in the diazotization process due to their resistance to the reaction (Kawakami et al., 1997). This resistance is attributed to the requirement of strong acidic conditions that lead to the protonation of the hetero atom, thereby hindering the diazotization process or causing instability in the diazonium salt. Consequently, this results in a low yield.

The process of oxidative coupling – Hunig's (1957) findings revealed that nitrogen heterocyclic hydrazones could be effectively linked with an aromatic coupling component by utilizing a mild oxidizing agent, including sodium or potassium ferricyanide, cupric salts, mercuric salts, ferric chloride, lead dioxide, hydrogen peroxide, ferrous salts, lead tetraacetate, or other comparable compounds. Typically, the reaction medium comprises water and an organic solvent, such as dimethylformamide, methanol, or glycol ethers.

The intricate details of this chemistry have been expounded upon in a comprehensive collection of 29 scholarly articles that were published during the period spanning from 1957 to 1968. A comprehensive analysis of the chemical process of oxidative coupling is required. When subject to a conventional azo coupling reaction, aromatic amines, phenols, or active methylene compounds tend to generate azo dyes with hydrazones in the presence of oxidizing conditions. The facile oxidation of specific couplers in the presence of an oxidizing agent impedes their coupling ability, as their susceptibility to oxidation surpasses that of the hydrazone. The coupling of pH conditions exhibits similarities to conventional principles. Amines necessitate a milieu of weakly acidic nature, with a pH range of 2 to 5, whereas phenols and enols mandate a milieu of mildly alkaline nature, with a pH of 9 (Yousif et al., 2009). It is advisable to refrain from utilizing strongly basic media due to the propensity of hydrazones to undergo decomposition under high pH conditions. N-sulfonylhydrazones can be utilized when highly alkaline conditions are necessary.

Zollinger compared oxidative coupling and classical electrophilic substitution in the context of the diazo coupling reaction (Andreiuk et al., 2021). The key distinction between the two lies in how the electrophilic component is formed. The reaction between phenylenediamines, aromatic amines, and phenols results in indamine and indophenol dyes, which exhibit a parallel formation process.

The following equations demonstrate that the primary chemical process implicated in oxidative coupling entails four-electron oxidation.

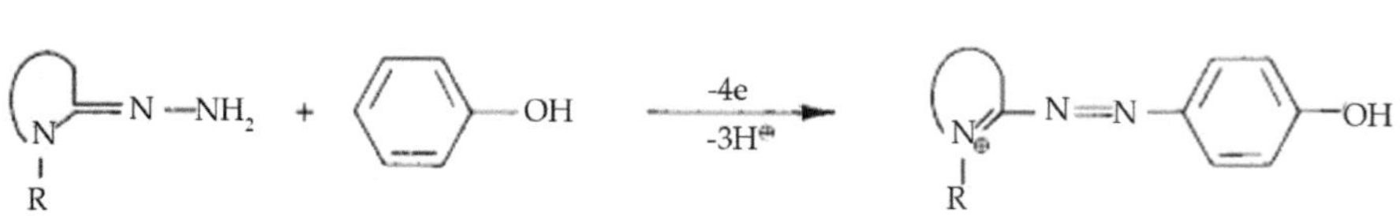

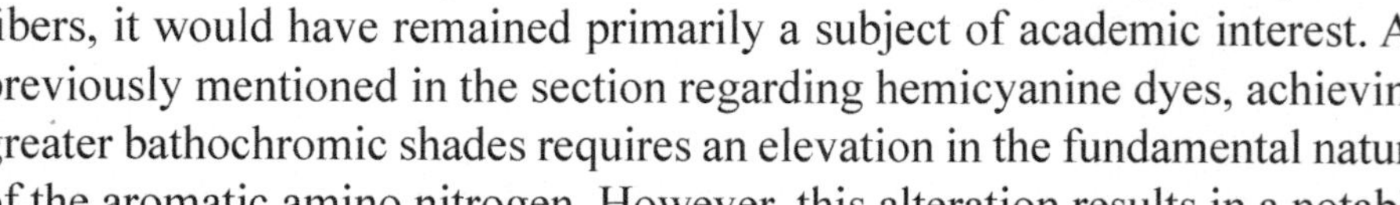

Characteristics of diazahemicganine dyes – If Hunig's technology had not demonstrated exceptional color shades and durability on synthetic fibers, it would have remained primarily a subject of academic interest. As previously mentioned in the section regarding hemicyanine dyes, achieving greater bathochromic shades requires an elevation in the fundamental nature of the aromatic amino nitrogen. However, this alteration results in a notable reduction in lightfastness. The substitution of the -CH-CH- moiety in hemicyanine dyes with -N-N- results in a redshift of approximately 100 mp and enhanced photostability. The pyridine series exhibits a range of hues, transitioning from yellow-orange to blue-red.

The benzothiazole series exhibits a range of hues, transitioning from red hemicyanine dyes to a vivid and durable blue-violet diazahemicyanine (Kiernan, 2001).

Introducing an arylene moiety between the heterocyclic ring and the azo group leads to a considerable hypsochromic shift in the absorption spectrum, causing a transition in color from blue-violet to yellow (Toumi et al., 2018).

A comparatively lesser degree of chromatic alteration is observed upon the dispersion of conjugation across a quinoline moiety as opposed to a pyridine moiety. Several commercial dyes have been synthesized using this chemical process.

Two examples of 6-methoxy-2-aminobenzothiazole derivatives are Basacryl Blue and Maxilon Blue (CI basic blue 41). Maxilon Red BL (CI basic red 22) belongs to the class of 1,2,4-triazolium dyes, while Calcozine Acrylic Violet 3R (CI basic violet 18) and Calcozine Acrylic Red 3G (CI basic red 30) are categorized as pyridinium dyes.

The shade of the hemicyanine class is deepened by increasing the basicity of the coupling component. Baumann & Dehnert (1961) demonstrated this phenomenon in the imidazole series.

The bensothiasole series exhibits a similar pattern of deepening shade as the basicity of the coupling component increases. Specifically, a shift in shade from red-violet at 563 mμ is observed when R = R' – H to blue at 616 mμ at R ==, R' = (You et al., 2009).

The diasahemicyanine dyes exhibit a wide range of hues that effectively span the entire visible spectrum, owing to their ability to be synthesized from a diverse array of heterocyclic compounds.

Alternative pathways leading to Diazahemicyanine D have been identified. Apart from the significant breakthrough achieved by Hunig et al. regarding the pragmatic synthesis of these chromophores, several other methods have been devised during subsequent explorations of this chromophoric system.

The process of diazotization and coupling is subsequently followed by quaternization. This approach can be effectively employed provided that the amine intended for diazotization is situated at a distance of a minimum of one vinyl group from the hetero nitrogen, as observed in specific heterocyclic compounds or in cases where the hetero atom does not exhibit significant basicity. Direct diazotization of 4-aminopyridines or quinolines, 2-aminothiazo1es, or benzothiazoles is feasible (Jiang et al., 2020). It is imperative to ensure meticulous pH, time, and temperature regulation. The condensation of amine with aromatic nitroso compounds. A diverse range of heterocyclic dyes were synthesized by the chemists at Geigy (1962) using this approach. The sodio salt is obtained by treating a toluene solution 2-aminopyridine with sodium amide. The reaction mixture was subjected to reflux for three hours after adding freshly prepared p-nitroso-N,N-dimethylaniline. The resultant mixture was then subjected to evaporation until complete dryness was achieved. The dye extraction is done by utilizing hot acetic acid and crystallization (Muthyala et al., 2001). The process of producing cationic dye involves the alkylation of dry dye. The process of condensing halogen compounds with hydrazines followed by subsequent oxidation. The compound 2-B aminopyridine has the potential to undergo condensation with p-nitro phenylhydrazine. The desired dye can be obtained upon oxidation of the resulting hydrazone, followed by reduction of the nitro group and alkylation. The practical utility of this approach is limited by the restricted accessibility of the initial substances and the substantial number of reaction stages involved.

The process of quaternization involves acrylamides. Most synthetic procedures necessitate the quaternization of the hetero nitrogen as a crucial step during the synthesis. Commonly employed alkylating agents include dialkyl sulfates, alkyl halides, and alkyl toluenesulfonates. The researchers Suzuki et al. have discovered that the alkylation process can be facilitated by heating azo dyes generated through the coupling of diazotized 2-aminothiazoles or 2-aminobenzothiazoles with different N,N-dialkylanilines or 2-phenylindoles, along with acrylamide in glacial acetic acid, at a temperature of 95°C in the presence of a mineral acid. The cationic dye that was obtained exhibits a side chain of 2-carbamoylethyl (Studenyak et al., 2012).

The present study concerns the generation of diazahemicyanines through the utilization of N-o: side dyes. In contrast to 2-aminopyridine, which exhibits resistance to diazotization, the diazotization process can be easily carried out on 2-aminopyridine-1-oxide using conventional methods. After Y-oxide formation, methylation may be carried out to yield Y-alkoxy quaternary dye, exhibiting analogous characteristics to those of the Y-alkyl derivative (Powers, 1988). Brody has demonstrated that the Y-alkoxy compound can undergo demethoxylation through refluxing in alcoholic ammonium hydroxide. After its formation, the resulting dye base may be subjected to re-alkylation procedures utilizing standard conditions, yielding the diasahemicyanine dye. Liss (1967) has demonstrated that the diasahemicyanine dye can be synthesized directly from the Y-oxide, subject to the quaternization conditions.

Remember

Some cationic dyes can be used in combination with fabric finishes such as water repellents or flame retardants, without significantly affecting their performance.

The anticipated Y-methoxy derivative was obtained through alkylation with dimethyl sulfate in toluene. The Y-methyl derivative was obtained unexpectedly when excessive methyl iodide was utilized in refluxing chloroform. The reagent yielded the o-methyl product at standard temperature and pressure. The reversibility of alkylation through methyl iodide was demonstrated, and it was observed that demethylation occurs as a result of the oxidation of iodide by Y-oxide. This oxidation leads to the formation of a base that can be easily alkylated with excess methyl iodide to produce the triiodide salt (Modica-Napolitano et al., 2003).

Table I presents the physical properties of the Y-oxide derivatives. The present study investigates the displacement of active halogen or alkosyl on preformed diazahemi-cyanine dyes. Amines that are weakly basic or contain oxidizable groups exhibit poor coupling performance when subjected to the oxidative conditions of the Hünig reaction (Cristian et al., 2004). Baumann & Dehnert (1961) discovered an alternative method for producing this dye category, which entails displacing an activated halogen atom or alkoxy group on the pre-existing diasahemicyanine. The reddish yellow dye undergoes a heating process with 4-amino acetanilide and sodium acetate at a temperature of 50°C until the yellow hue dissipates, and a claret dye emerges. This approach exhibits a high degree of versatility in generating diverse derivatives.

4.3.3.5. Dyeing Properties of the Diazahemicyanine Dyes

Diazahemicyanine dyes can be structurally varied to provide a wide spectrum of fast-fasting colors. Due to their tiny size and great color intensity, they dye at a lower temperature and transfer more easily than triarylmethane or cationic pendant dyes. They are crucial for modacrylic fibers dyed at 80–90°.

Wool or strong acids destroy diazahemicyanine dyes like azosafranines. Diaza-hemicyanine dyes can use this apparent drawback to color wool mixes. The initial dyeing procedure for major wool blends with acrylic or acid-modified polyesters absorbs the most cationic dye (Puchtler et al., 1985). The dye slowly travels from wool to synthetic fibers during dyeing. Due to minimal transfer and residual dye affinity for wool, triarylmethane, oxazine, and hemicyanine varieties generally retain most of the dye. The blend's brightness and wash fastness suffer. In wool, extended diazahemicyanine coloring decomposes to colorless compounds. The combination has optimal contrast and fastness after dyeing the wool with acid or pre-metalized colors.

4.3.3.6. Naphthostyril Dyes

Bayer's naphthostyril (per (-naphthasolone or 8-amino-1-naphfhoic lactam) cationic dyes provide an alternate way to achieve lightfast and deep-shade hues on synthetic fabrics. The study created brilliant acrylic fiber colors from deep red to green-blue that are light-resistant. The dyes also stain wool minimally. As said, these colors may be prepared using hemicyanine dye techniques.

The equation shows that heating N-ethylnaphthostyril (CXVII I) with 1-methylindole (CXIX) at 120°C in POC produces the dye (CXX). This creates red hues (Salvador et al., 2008) (Figure 4.16).

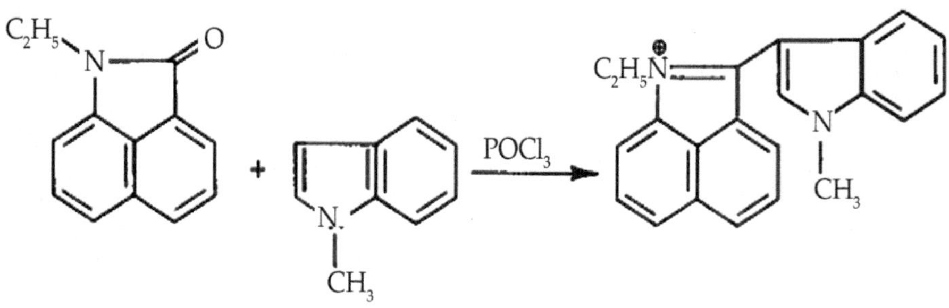

(CXVIII) (CXIX) (CXX)

Figure 4.16. *Representation of reaction between N-ethylnaphthostyril with 1-methylindole in the presence of $POCl_3$ to produce red dyes.*

Source: Karl Neufang, Creative Commons License.

With N-methyl-N-phenyl-*p*-phenetidine, a red violet is obtained (Basic Violet 20). Blue dyes are obtained by condensing the quaternities of naphtholactam with dialkylaminobensaldehydes in the presence of acetic acid and acetic anhydride at 100° (Weissig & Torchilin, 2001). Compound, the 1, 2-dimethylbens [c,d] indolium cation, is prepared from

N-methylnaphthostyril via a Grignard reaction, followed by elimination of water on hydrolysis.

4.3.3.7. Miscellaneous Delocalized Cationic Dyes

Several positively charged dyes have been examined for their applicability to artificial fibers. Only a limited number of options exhibit the requisite blend of rapidity, cost-effectiveness, and chromaticity to rival the aforementioned varieties in the marketplace effectively. In numerous instances, the alkylation of heterocyclic nitrogen atoms on a chromophore can create a dye that exhibits a favorable affinity towards acid-modified fibers. This section delves into the discussion of certain types outlined in the literature. Hiiusermann & Voltsl (1962) have discovered a cluster of highly intelligent fluorescent fiavine shade dyes within the bensimidasolium derivatives of dialkylaminocoumarins (Salvador et al., 2009). The dyes presented below can produce highly fluorescent shades in the green-yellow spectrum, which can be utilized independently or in combination with other dyes. The lightfastness property exhibits only a moderate level of performance; however, it is deemed sufficient to create fashion-oriented hues.

The production of vivid green-yellow compounds through the quaternization of a ring-closed naphthoylbensimidasole [11-methoxy-7H-benzo [de] [1,2] benzimidazo [2,1] o-benzimidazole] has been documented (Modica-Napolitano & Aprille, 2001).

Polycyclic derivatives have been employed as colorants for polyacrylonitrile. Quaternary coeramidonine is a type of self-shade polycyclic green that comprises a blend of isomers and bis-cationic products. Hoechst (1963) has obtained patents for the vivid hues of the novel polycyclic series. Variations of the hetero atom from oxygen to sulfur to nitrogen have produced shades ranging from yellow to blue that exhibit favorable fastness properties.

The infrequent utilization of hetero atoms apart from nitrogen can be attributed to either the challenges encountered during synthesis or the vulnerability of the resultant dye to hydrolysis in the boiling dye bath, which is necessary for achieving intense hues on synthetic fibers.

Sulfonium dyes exhibiting shades ranging from deep blue to green and exhibiting favorable fastness properties within the thioflavonium category have been characterized by BASF. American Cyanamid has obtained a patent for the 1,2-Benzodithiolium dyes. Bright blue dyes are synthesized through the condensation of a tertiary aniline with a 1,2-benzodithiole-3-thione in the presence of thiophosgene (Ngen et al., 2009).

The thiolium group has been granted a patent by Ciba (Chemical Industries Basel - Switzerland). The compounds mentioned above are synthesized via condensation, whereby 1,2-dithiols are reacted with substituted aromatic carboxylic acids in phosphorus oxychloride. Numerous structural variations are feasible.

Literature is scarce regarding phosphonium dyes. The derivatives of diazotized (m-aminophenyl) trimethylphosphonium chloride have been patented by ICI. The orange dye obtained from coupling N,N-diethylaniline has been reported to exhibit favorable light and wash fastness properties when applied to polyacrylonitrile.

4.4. AMINE SALTS

The cationic dye in this particular category exhibits no discernible formal charge when subjected to neutral or alkaline conditions (Kreuz et al., 1966). Under acidic dyeing conditions necessary for synthetic fibers, the molecule undergoes a positive charge and exhibits the characteristics of a cationic dye. Chrysoidine and Acridine Orange are considered classical illustrations of amine dyes.

Amine dyes exhibit distinct dyeing characteristics compared to conventional cationic dyes, owing to their tendency to assume a more dispersed dye character. This can be attributed to the fact that an equilibrium reaction takes place under normal dyebath conditions. This leads to a significant level of consistency in the dyeing process. Conversely, the dispersed dye's properties allow it to have an attraction to various fibers, leading to significant discoloration of wool, nylon, polyesters, and other materials when utilized alongside acid-modified fibers.

A diverse range of amine-based dyes have been synthesized and developed for application on synthetic materials. The pendant aminoalkyl disperse dyes hold significant technical significance, imparting superior lightfastness and multifiber affinity (Akkus et al., 2019).

The 7V-oxides represent a distinctive category of amine dyes. Under acidic dyebath conditions, the N-oxide exhibits properties similar to those of an amine dye, despite lacking a formal cationic charge. The process of synthesizing amine oxide dyes involves the utilization of oxidizing agents, including but not limited to hydrogen peroxide, peracids, and per-salts (such as persulfate), in a solvent medium at temperatures that are considered moderate. The compound denoted as CXXXVII elicits a red chromophore on polyacrylonitrile when acetic acid is present (Smith, 2019). The experimental results demonstrate that the amine oxide functions as a cationic dye. The saturation of dyesites within the fiber is achieved through treatment with an excessive cationic agent that lacks color. The obstructed fiber can undergo dyeing with dispersed dyes; however, it lacks the potential to be dyed with cationic dyes. The dyes containing amine oxide exhibit similar behavior to the conventional cationic or amine dyes as they do not possess the ability to dye the blocked fiber.

4.5. TRENDS IN CATIONIC DYES FOR SYNTHETIC FIBERS

Since 1953, textile fibers have used cationic dyes. Optimism prevails. Cationic dyes are the most powerful textile colors due to their intensity and brightness. Acrylic fibers that resist light-induced fading dispute that such colors are photodegradable. The synthetic organic chemist has shown for 15 years that chemically modifying dye molecules may improve brightness and lightfastness (El-Ola, 2008). Novel reactions and dye classes achieved this. In 15 years, this tendency will continue. Cationic dyes are predicted to grow as customers embrace color and become less concerned about its poor fastness. Cationic dyes will benefit from the textile and carpet industries' trend toward fashion-forward designs and multi-color dyeing. Coloration has become more important in textile production as customer color preferences change. Piece dyeing should replace yarn dyeing. Synthetic fibers receptive to various dyes, such as acid-modified fibers interlaced with basic-modified fibers, allow for multicolored dye baths using basic and acid dyes (Vinu et al., 2010). Due to their great brightness, basic dyes are used extensively to provide excellent contrast. In modern markets, weavers blend textile fibers to combine their best qualities to create clothing with desirable aesthetic and physical traits. High-production runs require pieces and continuous dying. Luminosity and little discoloration of the blended fibers are important. Cationic dyes can color many materials that accept dye. Staining reduces contrast and compromises blends' light and wet fastness (Yunusa et al., 2021). Organic and textile chemists must create new dyes and dyeing methods to quickly color fiber blends without affecting their characteristics. Over the past 15 years, cationic dyes have advanced, but there is always space for improvement.

4.6. SUMMARY

Cationic dyes are a type of dye that is designed to bond specifically with synthetic fibers that carry a positive charge, such as nylon, acrylic and polyester. These dyes contain positively charged molecules, known as cations, which are attracted to the negatively charged sites on the surface of synthetic fibers. These dyes require a different dyeing process compared to traditional dyeing methods. These dyes are not compatible with natural fibers such as cotton, silk, or wool, as these fibers do not carry a significant positive charge. Cationic dyes find applications in various industries including textile, automotive and carpet manufacturing.

REFERENCES

1. Affat, S. S., (2021). Classifications, advantages, disadvantages, toxicity effects of natural and synthetic dyes: a review. *University of Thi-Qar Journal of Science, 8*(1), 130–135.
2. Akkus, B., Kiskan, B., & Yagci, Y., (2019). Counterion effect of amine salts on ring-opening polymerization of 1,3-benzoxazines. *Macromolecular Chemistry and Physics, 220*(1), 1800268.
3. Ali, S. A., Rachman, I. B., & Saleh, T. A., (2017). Simultaneous trapping of Cr (III) and organic dyes by a pH-responsive resin containing zwitterionic aminomethylphosphonate ligands and hydrophobic pendants. *Chemical Engineering Journal, 330*(1), 663–674.
4. Aminoddin, H. A. J. I., (2012). Antibacterial dyeing of wool with natural cationic dye using metal mordants. *Materials Science, 18*(3), 267–270.
5. Andreiuk, B., Aparin, I. O., Reisch, A., & Klymchenko, A. S., (2021). Bulky barbiturates as non-toxic ionic dye insulators for enhanced emission in polymeric nanoparticles. *Chemistry–A European Journal, 27*(50), 12877–12883.
6. Asai, K., Fukazawa, A., & Yamaguchi, S., (2016). Stable red-emissive cationic dithienotropylium dyes. *Chemistry–A European Journal, 22*(49), 17571–17575.
7. Baer, D. R., (1971). Cationic dyes for synthetic fibers. In: *The Chemistry of Synthetic Dyes* (Vol. 1, pp. 161–210). Academic Press.
8. Baumann, H., & Dehnert, J., (1961). New synthesis of diazastyryl dyes. *Chimia, 15*, 163–168.
9. Bayramoglu, G., Altintas, B., & Arica, M. Y., (2009). Adsorption kinetics and thermodynamic parameters of cationic dyes from aqueous solutions by using a new strong cation-exchange resin. *Chemical Engineering Journal, 152*(2–3), 339–346.
10. Benhalima, T., & Ferfera-Harrar, H., (2019). Eco-friendly porous carboxymethyl cellulose/dextran sulfate composite beads as reusable and efficient adsorbents of cationic dye methylene blue. *International Journal of Biological Macromolecules, 132*(1), 126–141.
11. Benkhaya, S., M'rabet, S., & El Harfi, A., (2020). A review on classifications, recent synthesis and applications of textile dyes. *Inorganic Chemistry Communications, 115*(1), 107891.
12. Berneth, H., (2000). Cationic dyes. *Ullmann's Encyclopedia of*

Industrial Chemistry, 1, pp. 4–7.

13. Bouatay, F., Meksi, N., Adeel, S., Salah, F., & Mhenni, F., (2016). Dyeing behavior of the cellulosic and jute fibers with cationic dyes: Process development and optimization using statistical analysis. *Journal of Natural Fibers, 13*(4), 423–436.
14. Braun, G., Lengyel, M., Enyedi, P., & Czirják, G., (2015). Differential sensitivity of TREK-1, TREK-2 and TRAAK background potassium channels to the polycationic dye ruthenium red. *British Journal of Pharmacology, 172*(7), 1728–1738.
15. Bui, T. P. Q., Kim, S., Doan, T. Q. M., Nguyen, T. M. D., Pham, V. T., & Nguyen, T. T., (2017). Magnetic $NiFe_2O_4$ decorated-exfoliated graphite for adsorptive removal of anionic dyes and cationic dyes from aqueous solution. *Desalination and Water Treatment, 82*(1), 101–113.
16. Candido, I. C. M., Pires, I. C. B., & de Oliveira, H. P., (2021). Natural and synthetic fiber-based adsorbents for water remediation. *CLEAN–Soil, Air, Water, 49*(6), 2000189.
17. Cristian, L., Sasaki, I., Lacroix, P. G., Donnadieu, B., Asselberghs, I., Clays, K., & Razus, A. C., (2004). Donating strength of azulene in various azulen-1-yl-substituted cationic dyes: Application in nonlinear optics. *Chemistry of Materials, 16*(18), 3543–3551.
18. Dallel, R., Kesraoui, A., & Seffen, M., (2018). Biosorption of cationic dye onto" Phragmites australis" fibers: Characterization and mechanism. *Journal of Environmental Chemical Engineering, 6*(6), 7247–7256.
19. Dapson, R. W., (2005). Dye–tissue interactions: mechanisms, quantification and bonding parameters for dyes used in biological staining. *Biotechnic & Histochemistry, 80*(2), 49–72.
20. Dhodapkar, R., Rao, N. N., Pande, S. P., Nandy, T., & Devotta, S., (2007). Adsorption of cationic dyes on Jalshakti®, super absorbent polymer and photocatalytic regeneration of the adsorbent. *Reactive and Functional Polymers, 67*(6), 540–548.
21. Ehrenberg, B., Montana, V. A. L. E. R. I. E., Wei, M. D., Wuskell, J. P., & Loew, L. M., (1988). Membrane potential can be determined in individual cells from the Nernstian distribution of cationic dyes. *Biophysical Journal, 53*(5), 785–794.
22. El Amri, A., Bensalah, J., Idrissi, A., Lamya, K., Ouass, A., Bouzakraoui, S., & Lebkiri, A., (2022). Adsorption of a cationic dye (Methylene

bleu) by *Typha latifolia*: Equilibrium, kinetic, thermodynamic and DFT calculations. *Chemical Data Collections, 38*(1), 100–834.

23. El Harfi, S., & El Harfi, A., (2017). Classifications, properties and applications of textile dyes: A review. *Applied Journal of Environmental Engineering Science, 3*(3), 00000–3.
24. El Kaoutit, H., Estévez, P., Ibeas, S., García, F. C., Serna, F., Benabdelouahab, F. B., & García, J. M., (2013). Chromogenic and fluorogenic detection of cations in aqueous media by means of an acrylic polymer chemosensor with pendant Rhodamine-based dyes. *Dyes and Pigments, 96*(2), 414–423.
25. El-Ola, S. A., (2008). Recent developments in finishing of synthetic fibers for medical applications. *Designed Monomers and Polymers, 11*(6), 483–533.
26. El-Shishtawy, R. M., & Melegy, A. A., (2001). Geochemistry and utilization of montmorillonitic soil for cationic dye removal. *Adsorption Science & Technology, 19*(8), 609–620.
27. Goel, S., & Jacob, J., (2020). D-galactose-based organogelator for phase-selective solvent removal and sequestration of cationic dyes. *Reactive and Functional Polymers, 157*(1), 104766.
28. Gregory, P., (1990). Dyes for polyacrylonitrile. *The Chemistry and Application of Dyes, 1*, 165–201.
29. Gürses, A., Açıkyıldız, M., Güneş, K., Gürses, M. S., Gürses, A., Açıkyıldız, M., & Gürses, M. S., (2016). Classification of dye and pigments. *Dyes and Pigments, 1*, 31–45.
30. Hosseinzadeh, H., & Khoshnood, N., (2016). Removal of cationic dyes by poly(AA-co-AMPS)/montmorillonite nanocomposite hydrogel. *Desalination and Water Treatment, 57*(14), 6372–6383.
31. Hosseinzadeh, H., Zoroufi, S., & Mahdavinia, G. R., (2015). Study on adsorption of cationic dye on novel kappa-carrageenan/poly(vinyl alcohol)/montmorillonite nanocomposite hydrogels. *Polymer Bulletin, 72*(1), 1339–1363.
32. Hou, A., Chen, H., Xu, Y., Yang, X., Zhang, Y., Xie, K., & Gao, A., (2020). Rapid and environmental-friendly continuous gel-dyeing of polyacrylonitrile fiber with cationic dyes. *Journal of Cleaner Production, 274*(1), 122935.
33. Hünig, S., & Fritsch, K. H., (1957). Azofarbstoffe Durch Oxydative Kupplung I. Farbstoffe aus N-Methyl-benzthiazolon-hydrazon. *Justus*

Liebigs Annalen der Chemie, 609(1), 143–160.

34. Jia, Q., Song, C., Li, H., Huang, Y., Liu, L., & Yu, Y., (2017). Construction of polycationic film coated cotton and new inductive effect to remove water-soluble dyes in water. *Materials & Design, 124*(1), 1–15.
35. Jiang, Z., Liu, H., He, H., Yadava, N., Chambers, J. J., & Thayumanavan, S., (2020). Anionic polymers promote mitochondrial targeting of delocalized lipophilic cations. *Bioconjugate Chemistry, 31*(5), 1344–1353.
36. Joshi, P., Sharma, O. P., Srivastava, M., & Khatri, O. P., (2022). Fibrous aerogel-templated nanostructured boron carbon oxynitride networks for preferential adsorptive separation of cationic dyes. *ACS Applied Nano Materials, 5*(12), 18093–18105.
37. Kawakami, M., Koya, K., Ukai, T., Tatsuta, N., Ikegawa, A., Ogawa, K., & Chen, L. B., (1997). Synthesis and evaluation of novel rhodacyanine dyes that exhibit antitumor activity. *Journal of Medicinal Chemistry, 40*(20), 3151–3160.
38. Kiernan, J. A., (2001). Classification and naming of dyes, stains and fluorochromes. *Biotechnic & Histochemistry, 76*(5–6), 261–278.
39. Kim, E. M., & Choi, J. H., (2011). Dyeing properties and color fastness of 100% meta-aramid fiber. *Fibers and Polymers, 12*(1), 484–490.
40. Kim, Y. H., & Sun, G., (2002). Functional finishing of acrylic and cationic dyeable fabrics: Intermolecular interactions. *Textile Research Journal, 72*(12), 1052–1056.
41. Kreuz, J. A., Endrey, A. L., Gay, F. P., & Sroog, C. E., (1966). Studies of thermal cyclizations of polyamic acids and tertiary amine salts. *Journal of Polymer Science Part A-1: Polymer Chemistry, 4*(10), 2607–2616.
42. Kyzas, G. Z., Siafaka, P. I., Pavlidou, E. G., Chrissafis, K. J., & Bikiaris, D. N., (2015). Synthesis and adsorption application of succinyl-grafted chitosan for the simultaneous removal of zinc and cationic dye from binary hazardous mixtures. *Chemical Engineering Journal, 259*(1), 438–448.
43. Lan, Q., Zhang, Z. M., Li, Y. G., Lu, Y., & Wang, E. B., (2014). Synthesis of a poly-pendant 1-D chain based on 'trans-vanadium'bicapped, Keggin-type vanadtungstate and its photocatalytic properties. *Dalton Transactions, 43*(43), 16265–16269.
44. Lei, W., Xie, J., Hou, Y., Jiang, G., Zhang, H., Wang, P., & Zhang, B., (2010). Mitochondria-targeting properties and photodynamic

activities of porphyrin derivatives bearing cationic pendant. *Journal of Photochemistry and Photobiology B: Biology, 98*(2), 167–171.

45. Lewis, D. M., & Ho, Y. C., (1996). Improved fixation of dyes on polyamide fibers—IV. The use of a nucleophilic aminoethylsulphonyl cationic dye for dyeing nylon followed by a fixation aftertreatment with 2-chloro-4, 6-di (aminobenzene-4′-sulphatoethylsulphone)-s-triazine [XLC]. *Dyes and Pigments, 31*(2), 111–129.
46. Li, J., Yang, Q., Chen, S., McPhedran, K., Gu, Y., Huang, R., & Yan, B., (2022). Sustainably closed loop recycling of hierarchically porous polymer microbeads for efficient removal of cationic dyes. Environmental Science: *Water Research & Technology, 8*(3), 575–585.
47. Mahdavinia, G. R., & Asgari, A., (2013). Synthesis of kappa-carrageenan-g-poly (acrylamide)/sepiolite nanocomposite hydrogels and adsorption of cationic dye. *Polymer Bulletin, 70*(1), 2451–2470.
48. Mahdavinia, G. R., Massoudi, A., Baghban, A., & Shokri, E., (2014). Study of adsorption of cationic dye on magnetic kappa-carrageenan/PVA nanocomposite hydrogels. *Journal of Environmental Chemical Engineering, 2*(3), 1578–1587.
49. Mahdavinia, G. R., Massoumi, B., Jalili, K., & Kiani, G., (2012). Effect of sodium montmorillonite nanoclay on the water absorbency and cationic dye removal of carrageenan-based nanocomposite superabsorbents. *Journal of Polymer Research, 19*(1), 1–13.
50. Modica-Napolitano, J. S., & Aprille, J. R., (2001). Delocalized lipophilic cations selectively target the mitochondria of carcinoma cells. *Advanced Drug Delivery Reviews, 49*(1–2), 63–70.
51. Modica-Napolitano, J. S., Nalbandian, R., Kidd, M. E., Nalbandian, A., & Nguyen, C. C., (2003). The selective in vitro cytotoxicity of carcinoma cells by AZT is enhanced by concurrent treatment with delocalized lipophilic cations. *Cancer Letters, 198*(1), 59–68.
52. Muthyala, R., Watanabe, D., Asato, A. E., & Liu, R. S., (2001). The nature of the delocalized cations in azulenic bacteriorhodopsin analogs. *Photochemistry and photobiology, 74*(6), 837–845.
53. Ngen, E. J., Rajaputra, P., & You, Y., (2009). Evaluation of delocalized lipophilic cationic dyes as delivery vehicles for photosensitizers to mitochondria. *Bioorganic & Medicinal Chemistry, 17*(18), 6631–6640.
54. Pakdel, P. M., Peighambardoust, S. J., Arsalani, N., & Aghdasinia, H., (2022). Safranin-O cationic dye removal from wastewater using

carboxymethyl cellulose-grafted-poly (acrylic acid-co-itaconic acid) nanocomposite hydrogel. *Environmental Research, 212*(1), 113201.

55. Parvinzadeh, M., (2012). Surface modification of synthetic fibers to improve performance: Recent approaches. *Global J. Phys. Chem., 3*(2), 3–8.
56. Polit, W., Exner, T., Wuttke, E., & Winter, R. F., (2012). Vinylruthenium-triarylamine conjugates as electroswitchable polyelectrochromic NIR dyes. *BioInorganic Reaction Mechanisms, 8*(3–4), 85–105.
57. Powers, S. K., (1988). Cationic dyes with mitochondrial specificity for phototherapy of malignant tumors. In: *New Directions in Photodynamic Therapy* (Vol. 847, pp. 74–89). SPIE.
58. Puchtler, H., Meloan, S. N., & Spencer, M., (1985). Current chemical concepts of acids and bases and their application to anionic ("acid") and cationic ("basic") dyes. *Histochemistry, 82*(1), 301–306.
59. Ramesha, G. K., Kumara, A. V., Muralidhara, H. B., & Sampath, S., (2011). Graphene and graphene oxide as effective adsorbents toward anionic and cationic dyes. *Journal of Colloid and Interface Science, 361*(1), 270–277.
60. Raval, N. P., Shah, P. U., & Shah, N. K., (2017). Malachite green "a cationic dye" and its removal from aqueous solution by adsorption. *Applied Water Science, 7*(1), 3407–3445.
61. Roulia, M., & Vassiliadis, A. A., (2008). Sorption characterization of a cationic dye retained by clays and perlite. *Microporous and Mesoporous Materials, 116*(1–3), 732–740.
62. Salvador, M. A., Almeida, P., Reis, L. V., & Santos, P. F., (2009). Near-infrared absorbing delocalized cationic azo dyes. *Dyes and Pigments, 82*(2), 118–123.
63. Salvador, M. A., Reis, L. V., Almeida, P., & Santos, P. F., (2008). Delocalized cationic azo dyes containing a thiazole moiety. *Tetrahedron, 64*(2), 299–303.
64. Shakeela, K., Guru, S., & Rao, G. R., (2020). Cationic dye adsorption by phosphomolybdate nanoclusters immobilised on polyelectrolyte matrix. *Journal of Chemical Sciences, 132*(1), 1–10.
65. She, N., Moncelet, D., Gilberg, L., Lu, X., Sindelar, V., Briken, V., & Isaacs, L., (2016). Glycoluril-derived molecular clips are potent and selective receptors for cationic dyes in water. *Chemistry–A European Journal, 22*(43), 15270–15279.

66. Shirai, M., Nagatsuka, T., & Tanaka, M., (1978). Interaction between dyes and polyelectrolytes, 8. Effect of polyanions on the mixed dimer formation of cationic dyes. *Die Makromolekulare Chemie: Macromolecular Chemistry and Physics, 179*(1), 173–179.

67. Shukla, N. B., Rattan, S., & Madras, G., (2012). Swelling and dye-adsorption characteristics of an amphoteric superabsorbent polymer. *Industrial & Engineering Chemistry Research, 51*(46), 14941–14948.

68. Singh, R., Munya, V., Are, V. N., Nayak, D., & Chattopadhyay, S., (2021). A biocompatible, pH-sensitive, and magnetically separable superparamagnetic hydrogel nanocomposite as an efficient platform for the removal of cationic dyes in wastewater treatment. *ACS Omega, 6*(36), 23139–23154.

69. Smith, B., (2019). Organic nitrogen compounds V: Amine salts. *Spectroscopy, 34*(9), 30–37.

70. Studenyak, Y., Fershal, M., Kushnir, L., & Gomonnai, A. V., (2012). Tetrafluoroborate selective electrodes on the basis of cations with delocalized charge. *Electroanalysis, 24*(7), 1621–1629.

71. Toumi, K. H., Benguerba, Y., Erto, A., Dotto, G. L., Khalfaoui, M., Tiar, C., & Amrane, A., (2018). Molecular modeling of cationic dyes adsorption on agricultural Algerian olive cake waste. *Journal of Molecular Liquids, 264*(1), 127–133.

72. Tsai, C. H., Chang, W. C., Saikia, D., Wu, C. E., & Kao, H. M., (2016). Functionalization of cubic mesoporous silica SBA-16 with carboxylic acid via one-pot synthesis route for effective removal of cationic dyes. *Journal of Hazardous Materials, 309*(1), 236–248.

73. Vaid, V., & Jindal, R., (2023). Biodegradable tamarind kernel powder/kappa-carrageenan hydrogel for efficient removal of cationic dyes from effluents. *Water Environment Research, 2*(1), 10–97.

74. Vinu, R., Akki, S. U., & Madras, G., (2010). Investigation of dye functional group on the photocatalytic degradation of dyes by nano-TiO_2. *Journal of Hazardous Materials, 176*(1–3), 765–773.

75. Wainwright, M., (2010). Dyes, trypanosomiasis and DNA: A historical and critical review. *Biotechnic & Histochemistry, 85*(6), 341–354.

76. Weissig, V., & Torchilin, V. P., (2001). Cationic bolasomes with delocalized charge centers as mitochondria-specific DNA delivery systems. *Advanced Drug Delivery Reviews, 49*(1–2), 127–149.

77. Xiao, J., Lv, W., Xie, Z., Tan, Y., Song, Y., & Zheng, Q., (2016).

Environmentally friendly reduced graphene oxide as a broad-spectrum adsorbent for anionic and cationic dyes via π–π interactions. *Journal of Materials Chemistry A, 4*(31), 12126–12135.

78. Ying, T., Su, J., Jiang, Y., Ni, L., Jiang, X., Ke, Q., & Xu, H., (2022). Superhydrophobic MOFs decorated on hierarchically micro/nanofibrous membranes for high-performance emulsified oily wastewater separation and cationic dyes adsorption. *Journal of Materials Chemistry A, 10*(2), 829–845.
79. You, Y., Ngen, E. J., & Rajaputra, P., (2009). Conjugate systems using delocalized cationic dyes as a carrier of photosensitizers to mitochondria. In: *Photodynamic Therapy: Back to the Future* (Vol. 7380, pp. 930–935). SPIE.
80. Yousif, L. F., Stewart, K. M., & Kelley, S. O., (2009). Targeting mitochondria with organelle-specific compounds: strategies and applications. *ChemBioChem, 10*(12), 1939–1950.
81. Yue, X., Huang, J., Jiang, F., Lin, H., & Chen, Y., (2019). Synthesis and characterization of cellulose-based adsorbent for removal of anionic and cationic dyes. *Journal of Engineered Fibers and Fabrics, 14, 1558925019828194, 2*(1), 6–10.
82. Yun, J., Wang, Y., Liu, Z., Li, Y., Yang, H., & Xu, Z. L., (2020). High efficient dye removal with hydrolyzed ethanolamine-polyacrylonitrile UF membrane: Rejection of anionic dye and selective adsorption of cationic dye. *Chemosphere, 259*(1), 127390.
83. Yunusa, U., Usman, B., & Ibrahim, M. B., (2021). Cationic dyes removal from wastewater by adsorptive method: A systematic in-depth review. *Algerian Journal of Chemical Engineering AJCE, 1*(2), 6–40.
84. Zhou, M., Ji, Y., Li, F., Hu, H., Wang, P., Jia, K., & Liu, X., (2018). Fe^{3+} mediated self-assembling of polyarylene ether nitrile block copolymer into cationic dye adsorptive sub-micrometer spheres. *Materials Letters, 222*(1), 183–186.
85. Zucca, P., Cocco, G., Sollai, F., & Sanjust, E., (2016). Fungal laccases as tools for biodegradation of industrial dyes. *Biocatalysis, 1*(1), 82–108.

CHAPTER 5

Photochemistry of Dyes

Contents

Unit Introduction 146

5.1. Photochemical Reactions of Dyes in Solution 146

5.2. Photochemistry of Dyes on Substrates 176

5.3. Summary 190

References 191

UNIT INTRODUCTION

The reality of the colorful world can be attributed to the dyes and their intriguing interaction with light. Dyes have gained significant importance in various applications, spanning from dyeing textiles to biological imaging, owing to their remarkable capacity to take in and give off light. The field of photochemistry is a captivating area of study that explores the complex mechanisms underlying the behavior of dyes upon exposure to light. Comprehending the photochemical properties of dyes is of utmost importance not only for enhancing their efficacy but also for elucidating the underlying principles that dictate their chromaticity, durability, and responsiveness (Egerton & Morgan, 1970).

5.1. PHOTOCHEMICAL REACTIONS OF DYES IN SOLUTION

5.1.1. Photofading Model Systems

The quest for techniques to mitigate the deleterious effects of light on colored substances and dyed materials. Since Bancroft's initial precise inquiries into the photo rendering for dyed materials nearly 150 years earlier, numerous significant findings have been garnered regarding the circumstances of fading, the impact of substrate nature, the reliance on dye structure, and other related factors. It is important to note that the intricate nature of the systems as well as the various factors that impact the method of fading have presented numerous challenges in fully understanding the mechanisms of photochemical changes in dyes (Confortin et al., 2010).

The precise examination of various photoreactions can only be accomplished through the simplification of conditions. Consequently, a significant portion of our comprehension regarding the photochemical characteristics of dyes along with dyed materials has been attained through the examination of photoreactions in basic model systems. Simple dyes, which include benzophenone and simple anthraquinones, are commonly selected as models for intricate dyestuffs in various systems, including vat dyes (Allen, 1994). In addition, the aforementioned dyes are solubilized in representative synthetic solvents, which exhibit chemical characteristics that are akin to those associated with fiber. The utilization of model substrates can mitigate the challenges associated with investigating solid-state reactions. Undoubtedly, certain instances exhibit distinct behavior of dyes in designs of organic solvents as compared to their behavior on a substrate that is commonly utilized. The results of measurements indicate that there is no precise similarity in the photochemical actions of aminoanthraquinone substances when they are in solution within ethyl acetate or when they are dyed upon cellulose acetate films. This has led to the inquiry of whether ethanol can truly serve as a substitute for cellulose as model substances within photochemical systems that contain vat dyes.

However, it is imperative to highlight that the examination of the photochemistry for dyes within a solution has significantly enhanced our comprehension of the photochemical conduct of dyes. The aforementioned studies have yielded significant insights into the process of fading in dyed textiles, as well as the relationship between dye structure along with photo tendering behavior. The data obtained from these studies is of great value (Li et al., 2022).

During their research on the light bleaching of dyes within the solution, Oster and colleagues obtained compelling evidence indicating that the chemical reacting state in numerous photochemical processes of dyes is a highly persistent triplet state. Currently, there exists no uncertainty regarding the validity of this assertion, as the revolutionary flash photolysis method developed by Norrish & Porter (1961) has provided empirical proof of the formation of dye triplet conditions in solution through the detection of triplet-triplet absorptions. Porter & Windsor (1957) proposed the formation of triplet states as a widespread phenomenon in a diverse range of compounds based on their flash photolytic investigations of organic molecules within typical fluid solutions. Subsequently, a number of studies have demonstrated the crucial role of the triplet state, which has a lifespan of approximately 0.1 milliseconds, in numerous photochemical reactions involving organic substances as well as dyes in solutions. Photoflash studies have revealed the presence of triplets associated with chlorophyll, fluorescein, eosin, lumiflavine, along with retinene in substances of aromatic hydrocarbons. Numerous experiments have demonstrated that photochemical processes, including the photoreduction for riboflavin or just the photoreactions within ketones along with quinones solutions, occur through triplet states.

Did You Know?

Egyptians, Greeks, and Romans used natural dyes extracted from plants and minerals to color fabrics and other materials.

It is widely accepted that the outcomes of photochemical reactions of dyes within solutions may be extrapolated to more intricate systems. The chemical reactions that occur upon absorption of light by the dye are expected to exhibit similarities in both hypothetical systems along with dyed fibers, in numerous instances. Hence, the outcomes of model systems may offer significant insights into approaches for mitigating photofading as well as photo rendering. An instance of the mitigation of photofading by specific compounds can be illustrated by the reduction of oxygen-deprived photobleaching associated with riboflavin within aqueous solutions through the utilization of KI. This outcome is attributed to the proficient quenching of the three-dimensional intermediate. In certain instances, the process of photo fading may be decelerated through the creation of a charge transfer intricate among a dye as well as an additional compound. Alternatively, the photo-tendering effect of a dye may be inhibited through the strategic introduction of substituents that alter the positioning of reactive states, such as $3(\pi,\pi^*)$ as well as $3(n, \pi^*)$ states. In instances where the processes of light bleaching or just photo-tendering are established, it is plausible that systematic investigations may yield techniques for altering the trajectory associated with a photoreaction as well as mitigating the effects of photofading. It is crucial to note that comprehending the mechanisms for photochemical reactions primarily stems from investigations into the photochemistry of dyes within a solution (Batchelor et al., 2003).

It is noteworthy that there was a limited amount of research conducted on the topic, as evidenced by Venkataraman and Rabinovitch's (1945) research till Oster (1950) along with his colleagues initiated comprehensive investigations into the photochemistry for dyes within solution during the early 1950s. Thus far, these investigations have provided significant understanding regarding the primary mechanisms involved in the photoreactions of dyes. Furthermore, during the process of these investigations, significant issues have garnered notice. In addition to inquiries pertaining to biological phenomena, the utilization of dye-sensitized photo polymers as well as the potential for generating 3-D topographical representations in plastic materials through the photochemical responses of dyes have also been acknowledged (Franck & Livingston, 1941).

5.1.2. Types of Photochemical Reactions of Dyestuffs

A sequence of fundamental photochemical reactions can result in the degradation of a dye. In practical scenarios, it is common for dye molecules to come into close proximity to other compounds. Therefore, it is important to consider not just the photobleaching process for the dye on itself yet the potential for reactions with additional compounds that exist within the system. Reactions caused by chemicals can encompass various processes such as reduction, oxidation, and the generation of free radicals. The initial states considered in this context are the singlet along with triplet states that are excited by the first n, π*, or π electronic transitions. Within this particular context, it is noteworthy to observe that Bancroft (1814) had already deduced from experimental evidence that photofading can be attributed to oxidation or reduction reactions. In contemporary times, the phenomenon of photofading necessitates consideration of not just light intake from the dye yet the photoexcitation for other substances within the system, which can result in the photodecomposition for dyestuffs. Henceforth, it is imperative to deliberate upon the ensuing categories of photoreactions (Winters et al., 1974).

5.1.2.1. Intramolecular Photoreaction

In this particular sort of reaction, Eq.

$D + h\nu\ D^*; \quad D^*$ Products

The photo process does not require interactions with other substances. The photofading for dyestuffs retained on fibers rarely occurs by photodecomposition resulting from a fluctuation associated with the excited pigment molecules. The photoprocessing associated with diazo substances is notably intense and has been extensively utilized as the fundamental principle of image-forming systems. Upon absorption of light, diazonium salts undergo a process whereby the yellow diazo hue fades, nitrogen is released, and colorless disintegration products are generated. These products were thoroughly examined in various solutions. The rate of reducing the concentration within the diazo substance follows a first-order kinetic law. Eggert's (1928) measurement for quantum yield indicates that it is concentration-independent for the diazo substance, with a range of values between 0.2 and 1. When confronted with moisture, irradiation results in the formation of a phenolic disintegration product, as per the available information (Kim et al., 1999; Tsunooka et al., 1984; Rettig et al., 1992).

$$RC_6H_4N = NCI + H_{20}\ \ RC_6H_4OH + N_2 + HC_1$$

The process of image formation is a consequence of the interaction between the phenolic or an additional phenolic element and the unaltered diazo substance.

Photochemical cis-trans isomeric formation is a representative instance of intramolecular processes of a general nature. The salient feature of this molecular reorganization, which implicates the triplet state, is the relaxation of a π-bond within the molecule as well as a rotation about the σ-bond while maintaining continuity within the electron dissemination. Notable instances include the low-temperature cis-trans isomeric formation of indigo, azo substances, and cyanin dyes. The transition from the cis conformation to the trans conformation induced by visible light holds significant importance in the natural world. The process of activating the light-sensitive visions chromophore molecules 11-cis-retinal (also known as vitamin A aldehyde) involves its combination with fat-laden proteins (opsin) to create the visual pigments rhodopsin or just iodopsin. This activation leads to a cis-trans isomerization, as depicted in Figure 5.1, resulting in the formation of all-trans-retinal. Consequently, rhodopsin is transformed into prelumirhodopsin, which contains an all-trans chromophore. Subsequent to the primary process, a sequence of reactions ensues that culminates in the dissociation that separates the chromophore via the protein (Ushakov & Gromov, 2015; Heelis, 2018; Rettig & Lapouyade, 1994).

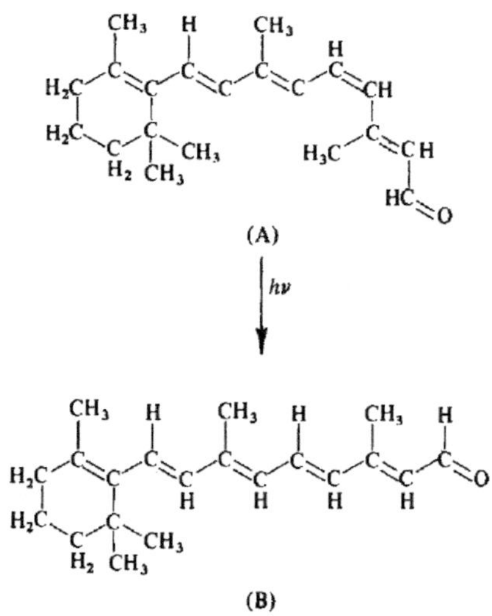

Figure 5.1. *Illustration of cis-trans isomerization of 11-cis-retinal (A); to all-£rans-retinal (B).*

Source: Rowan, Creative Commons License.

It is noteworthy to mention that in the realm of plant photoperiodism, a comparable intramolecular reaction holds significant importance. The phytochrome chromophore, which bears resemblance to the greenish yellow dyes present in humans' bile, undergoes a transformation upon taking in red light (λ_{max} = 634 nm) towards a molecule that absorbs far-red light (λ_{max} = 724 nm). Hendricks (1968) has proposed that the reversed far-red phytochrome development is a consequence of a photochemical processes shift of a pair of hydrogen atoms, as illustrated in Figure 5.2. It is important to acknowledge that in addition to the reversed tautomerization, alternative mechanisms such as cis-trans isomeric formation of the chromophore or just reversible oxidation-reduction reactions have been proposed. The plausibility of an inner oxidation-reduction procedure is doubtful due to the minimal variances observed in the wavelengths of absorption (Walz, 1979). The discussion of cis-trans isomerization, akin to those of rhodopsin mechanisms, necessitates an exploration of various potentialities. The results of kinetic experiments suggest formal analogies involving the phytochrome as well as rhodopsin mechanisms. It is evident that the transformation of phytochrome corresponds to changes in the structure of the molecules within the protein, which are able to be altered upon absorbing far-red light.

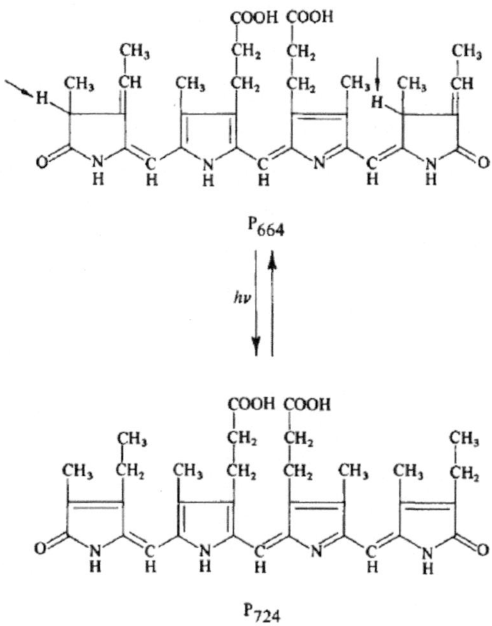

Figure 5.2. *Representation of possible photochemical reaction of the photoperiodism chromophore phytochrome.*

Source: Hendricks, Creative Commons License.

The phytochrome photoprocessing initiates several significant photomorphogenesis outcomes in plants, such as flowering as well as seed germination, which are enzymatically regulated via the far-red phytochrome. Haupt has suggested that the phobophototaxis for Mougeotia chloroplasts within dim light is attributed to the functioning of phytochrome.

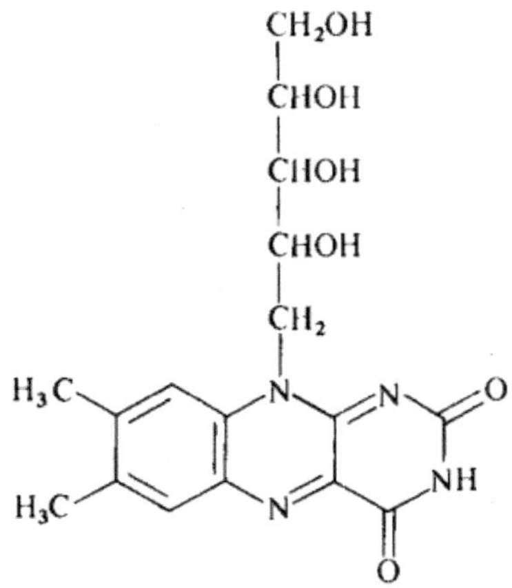

(I)

The activity spectra for phototactic as well as phototropic impacts were measured, revealing that riboflavin (I) may act as both a phototactic as well as phototropic pigment. The vitamin that has biological significance, also known as lactoflavin along with vitamin B2, exhibits reversible reduction as well as reoxidation along with intriguing photochemical reactions. In oxygen-deprived aqueous solutions, the vitamin undergoes fading due to a decrease within the isoalloxazine nucleus upon irradiation, lacking the involvement of any additional agents. Upon aeration, the yellow hue is reinstated, signifying a re-oxidation by the dihydro isoalloxazine moiety. This photoreaction prompts inquiry into whether the production of dihydro riboflavin, formed by the equivalent of a pair of hydrogen atoms, stems from the dissociation of water. Holmstrom along with Oster presented a counterargument to this mechanism based on energetic considerations. They proposed that the discoloration of riboflavin is caused by the photooxidation of the respective ribityl side chain. Thus far, a substantial body of evidence supporting intramolecular photobleaching has been presented through a range of kinetic energy, polarographic, as well as flash photolytic investigations. These investigations have yielded valuable insights into the process of cleavage processes within the side chains.

Song & Metzler (1967) have suggested that the cleavage can be interpreted as an instance of photoelimination, a distinctive reaction observed in aldehydes as well as ketones that possess a y-hydrogen atomic unit, as well as in amino acids (Norrish Type 2 Split). The aforementioned chemical transformation can also take place via a ring with six members intermediate (II), leading to the formation of an enol (1-deoxyribose) (IV) as well as Lumi chrome (III). The substitution of deuterium for the pair of hydrogen atoms located at the 2'-position for the side chain resulted in a significant decrease in the photolysis rate, which is consistent with the proposed mechanism.

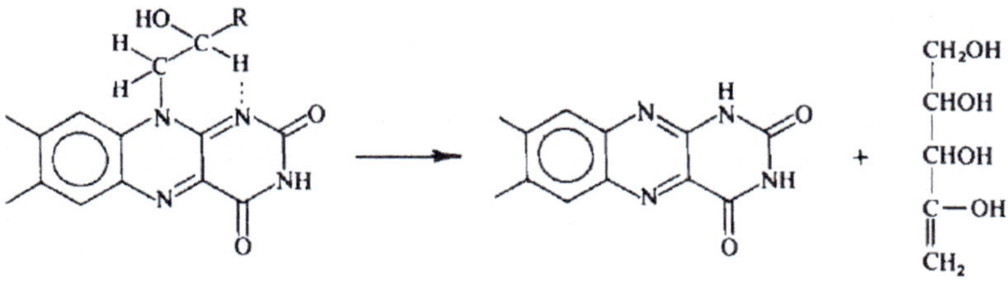

(II) (III) (IV)

Additionally, through the utilization of quenching experiments involving KI and flash work, it was demonstrated that the extended lifespan of the least triplet when excited serves as a photoreactive entity in the anaerobic light bleaching process of riboflavin as well as Flavin mononucleotide. The diminution in fluorescence through KI along with phenol is comparatively minor in relation to the reduction in the pace for aerobic bleaching. Song et al. (1967) has suggested that the singlet form of riboflavin might be involved in photolysis according to specific conditions, based on the comparable fractional suppression of photobleaching instances and fluorescence within ethanol solutions. This implies the possibility of two distinct pathways of photofading, involving a singlet state serving as an intermediary within ethanol as well as a three-dimensional state serving as an intermediate within the water. Additional research is required to address this issue.

Did You Know?

Photochromitc dyes, which undergo reversible color changes upon exposure to light, were first synthesized in the late 19th century. Notable examples include the spiropyran compounds developed by Otto Lehman in 1907.

The reversed photobleaching for methylene blue within phosphate as well as borate-buffered solutions, despite the lack of reduction agents, appears to indicate a photodecomposition response that is akin to the one observed via riboflavin. The electron donation in the process may be attributed to the tertiary amino groups, which can be likened to the ribityl category (Atkinson, 2011).

The inclusion of phototropic color modifications in coloring agents would be a valuable addition to the responses that were previously discussed in this section. The phenomenon referred to as phototropism or just photochromism is marked by a reversible alteration in the pigmentation of a dye, caused by exposure to light or ultraviolet (UV) radiation. Luck declares the existence of the photochromism process is contingent upon the reversed transitions induced by light, which result in states possessing quantum-mechanical balance but thermodynamics instability. The process, as represented by the equation below, is depicted in a schematic manner in Figure 5.3.

$$D \overset{h\nu}{\rightleftarrows} S(\rightleftarrows T) \rightleftarrows X \rightarrow Y$$

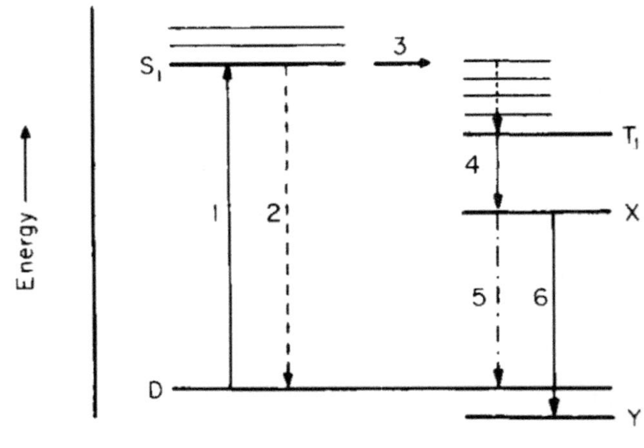

Figure 5.3. *The present study aims to depict photochromic transitions through a simple state diagram.*

Source: Kunkely, Creative Commons License.

Frequently, the reversible procedure is accompanied by a non-reversible response that yields a thermodynamically inert product Y. The lightfastness associated with a dye alongside photochromic color change is compromised due to the sluggishness of the reverse procedure. Conversely, when the photochromic kind X reacts swiftly with D, there is a potential for an improvement in lightfastness. This phenomenon is particularly evident in compounds where the excited single-molecule or triplet state undergoes rapid deactivation, transitioning quickly to the phototropic kinds. As a result, the likelihood of initiating a different photochemical response via an elevated state is low. The phenomenon of photochromism, initially discovered by Marckwald, can manifest in both solution and solid states (Newcomb et al., 1992). It may arise from intramolecular alterations such as tautomerism, ring openings, cis-trans isomerization, as well as the creation of free radicals, plus stereoisomeric adjustments, aggregation development, and other chemical reactions.

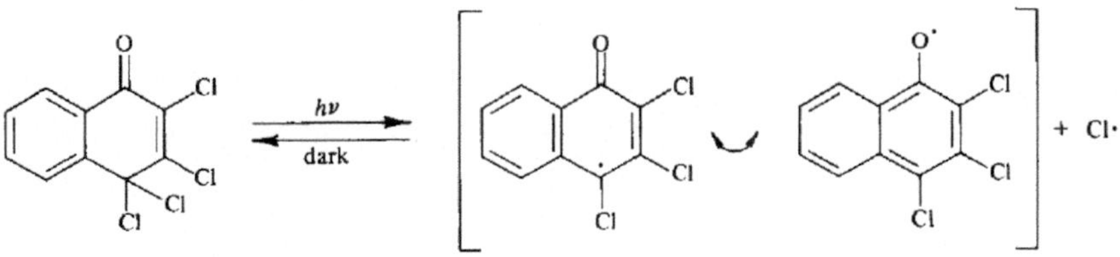

(V)

Two instances of a free radical's creation are the photochromism regarding solid forms for tetrachloro-1 xodihydronaphthalene (V) that have been cooled, and the formation of free radicals in methanolic substances of chlorophyll.

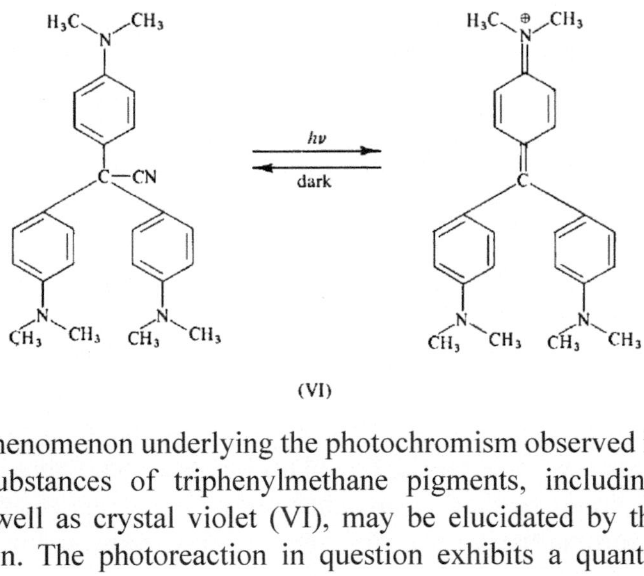

(VI)

The phenomenon underlying the photochromism observed in methanol-ethanol substances of triphenylmethane pigments, including malachite green as well as crystal violet (VI), may be elucidated by the following explanation. The photoreaction in question exhibits a quantum yield of approximate unity, suggesting the absence of any possible radiative or just lack of radiation deactivation pathways between the states of D along with X.

The leucosulfite regarding para fuchsine within aqueous solutions exhibits a photochromic reaction that has potential utility as a UV dosimeter for quantifying the UV rays of the skylight. It is noteworthy for mentioning the photoisomerism associated with the quinoid 1, 2-napthoquinone-2-diphenylhydrazone (VII) due to its formal similarity to phytochrome photochemistry.

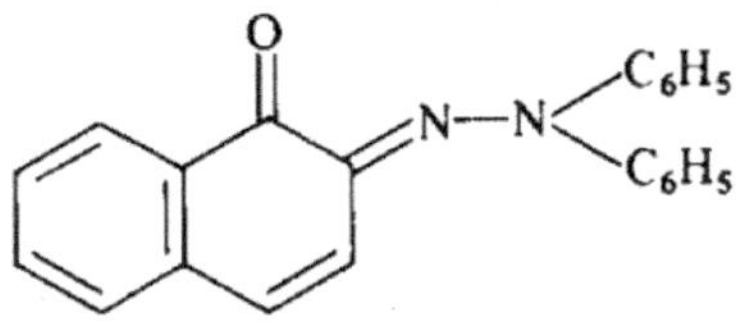

(VII)

The isomeric status exhibiting a band of absorption at a wavelength of 480 nm can be converted to a status that has a 550 nm absorbance peak through irradiation in either polar (ethanol) or just nonpolar solvents. Under low light conditions, the photo equilibrium undergoes a shift towards the standard dark equilibrium state, with a respective 98% as well as 84% distribution of D480 within ethanol and methylcyclohexane isohexane (a nonpolar solvent). The transformation of form D550 towards D480 occurs through irradiation at a wavelength of 550 nm, which bears similarities to the significant phytochrome conversion discussed earlier:

$$D_{480} \underset{\substack{550\text{ nm} \\ \curvearrowleft \\ \text{dark}}}{\overset{480\text{ nm}}{\rightleftarrows}} D_{550}; \qquad P_{664} \underset{\substack{724\text{ nm} \\ \curvearrowleft \\ \text{dark}}}{\overset{664\text{ nm}}{\rightleftarrows}} P_{724}$$

The phenomenon of reversible color changes induced by irradiation within the absorption spectrum associated with the photochromic item species is referred to as added photochromism. The reversible alteration of color in photochromic materials can prevent a decrease in the lightfastness of pigments on textiles. Conversely, it is plausible that added photochromism could lead to an augmentation of lightfastness through exposure to light within the absorbing band associated with the photochromic kinds. Due to photoreaction, there is a resultant outcome.

$$D_1 \underset{h\nu_2}{\overset{h\nu_1}{\rightleftarrows}} D_2$$

No further photochemical processes that are irreversible can occur. These instances demonstrate the essentiality of conducting experiments pertaining to this topic. These studies hold significant value due to the practical applications of phototropic dyes in the development of photochemical fast memory with removable images, as well as mechanical variable-density bright filters.

5.1.2.2. Photo Oxidation

- Mechanism:

i. **Earlier Mechanisms:** The phenomenon of photobleaching, which involves the oxidization of dyes, was effectively demonstrated by Weigert, Hibbert, as well as other researchers during the 1930s. The discovery of oxidation compounds from dyes on dissipated cotton, such as isatin which is an oxidization outcome of the color indigo has provided evidence that the dye is susceptible to photooxidation. The study demonstrated the essentiality of oxygen in facilitating photooxidation, as well as the plausibility of co-occurring photooxidation as well as photoreduction regarding a given dye in a solvent (Wagner et al., 2003). Nevertheless, the role of oxygen as a primary oxidizing agent for the excited dyes wasn't hypothesized. Hillson & Rideal (1953) have discussed the division of fluid into molecules of hydrogen as well as hydroxyl radicals within solutions that are aqueous. The exuberant dye is expected to undergo a reaction with water, while the liberated hydrogen atoms endeavor to mitigate the oxidization of the dye. This process may result in the generation of HO_{2*} radicals through a reaction via oxygen. The potential for the generation of H_2O_2, which can facilitate the oxidation of the dye in both illuminated and non-illuminated conditions, can be inferred from the observed reactions with these radicals. As an instance, it is imperative to note that azo dyes ought to undergo photodecomposition to yield azooxy compounds that lack color, through various mechanisms.

$$D + hv = D^*, \quad D^* + HOH = DOH \cdot + H \cdot$$

Blum along with Spealman conducted a study on photooxidative bleaching regarding xanthene pigments by oxygen. They proposed that the degradation of the dye occurs through an act of • OH as well as HO_{2*} radicals, which are generated through the reactions outlined below:

$$D^* + O_2 \rightarrow D + O_2^* \quad O_2^* + H_2O \rightarrow \cdot OH + HO_2 \cdot$$

Based on this hypothesis, there appears to be no direct correlation between the generation of hydrogen peroxide and the disintegration of the dye.

ii **Important Intermediate Species:** Empirical investigations utilizing flash photolysis have provided compelling indications that the oxidization of colored substances is causally linked to the generation of a partially oxidized form that contains the dye, which arises from transmission of electrons event involving the excited pigment and a substance that facilitates oxidation (Chambon et al., 2009; Matsushita et al., 2007). The inverse relationship between photooxidation and photoreduction reactions can be inferred. Thus, the procedure can be succinctly delineated by:

$$D + h\nu = D^*, \quad D^* + O = D_{ox} + O_{red}$$

The photooxidation procedure involves the utilization of the dye within its minimal triplet state just like a significant intermediary. Within this particular state, dyes undergo oxidation via appropriate oxidizing agents like eosin as well as erythrosine, along with potassium ferricyanide, resulting in their conversion into semi-oxidized eosin (IX) as well as semi-oxidized erythrosine, correspondingly.

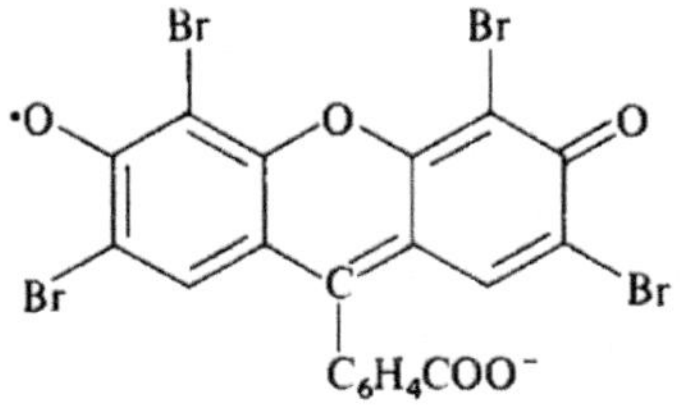

The empirical examination of the assault on triplet forms of dyes through oxidants may serve as compelling corroboration for the plausibility of an immediate response between oxygen as well as the three-dimensional state (Wang et al., 2020; Pattison et al., 2012).

As per the research conducted by Weiss, Imamura, along with other scholars, it can be inferred that photooxidative bleaching frequently occurs by the initial generation of a partially oxidized pigment radical along with a partially diminished oxygen molecule.

$$^{3}D + O_2 \rightarrow D\cdot^{+} + O_{2\cdot}^{-}$$

iii. **Switchover from D-0 to D-D Mechanism:** When considering the aforementioned reaction, it is important to note that the semi-oxidized dye, such as fluorescein, undergoes a transformation into enduring products. Imamura's research suggests that the oxygen attack on the triplet dye is likely to occur only under conditions of low dye levels compared to the oxygen level. Specifically, this phenomenon is observed at dye concentrations of approximately ≤10–5M, given that the level of oxygen within air-saturated substances is approximately $[O_2] \approx 2.5 \times 10$–4M. The photobleaching efficiency of xanthene pigments exhibits a linear relationship with the absorbing capacity of light up to a threshold of 0.2 for the $[D]/[O_2]$ ratio. This is evidenced by the linearity observed within the $\ln\{\exp(\alpha[D]d) - 1\}$ vs. time. Nevertheless, when the ratio of $[D]/[O_2]$ is increased, the aforementioned straightforward correlation becomes inadequate, and the rate of bleaching is intensified. The amount of utilization of oxygen exhibits an upward trend with the escalation of dye concentration until it reaches a threshold of approximately 10–4 M, beyond which it stabilizes (Song et al., 1995). At elevated levels of dye concentration, a greater ratio of oxygen consumption to dye bleaching was observed in comparison to lower concentrations. Additionally, an equivalent amount of hydrogen peroxide generation was noted in relation to the bleached dye. The detection of hydrogen peroxide generation was not feasible at low ratios of [D] / [O2]. Koizumi et al. drew a conclusion that a transition from a D-O process to an alternative mechanism occurs as the amount of the dye increases based on the obtained results. According to Koizumi's postulation, it is plausible to regard the mechanism at an elevated dye concentration as a D-D mechanism, wherein the transfer of electrons within the three-state state, as well as the ground-state pigment, assume a crucial role (Li et al., 2022). Lindqvist has demonstrated through flash photolytic experiments that the semi-reduced as well as semi-oxidized pigment can be produced through an electron transfer event. An instance of electron disintegration responses leads to the emergence of semi-reduced eosin through an optical peak at 405 nm as well as semi-oxidized eosin using a peak at four hundred sixty-two nm.

$$^{3}D+{}^{3}D \rightarrow D^{+}+D^{-} \qquad {}^{3}D+D \rightarrow D^{+}+D^{-}$$

The study showed the recombination for radicals, which possess a lifespan that is approximately ten times longer compared to that found in the three-note state (Ma et al., 2017; Knowles & Gurnani, 1972). Based on the aforementioned findings. Koizumi proposed a series of measures for the process of photooxidative bleaching:

$$^{3}D+D \rightarrow D^{+}\cdots\cdot D^{-}$$

$$D^{+}\cdots\cdot D^{-}+O_{2}+H^{+} \rightarrow HO_{2}\cdot+D^{+}+D^{-}$$

and alternatively

$$^{3}D + D \rightarrow D^{+} + D^{-}$$
$$D^{-} + H^{+} + O_{2} \rightarrow HO_{2}\cdot + D$$

iv. **The Production of Hydrogen Peroxide:** Quantum yields of light bleaching may exhibit significant variations due to the impact of the laboratory conditions as well as the characteristics of the dye upon the lifespan of the reactant intermediate concentrations D+ .... D-, as well as on the generation regarding free radicals D^{+} as well as D-. Nevertheless, the mechanism postulates that the light bleaching caused by the dye cannot be attributed to the production of hydrogen peroxide. Initially, it is feasible to convert D^{+} into a colorless state through various means, such as reacting with oxygen or undergoing dimerization (Korytowski et al., 1987). Additionally, through the generation of HO_2, which has the potential to decrease D^{+} to the initial dye, hydrogen peroxide seems autonomously synthesized via the chemical reaction.

$$2HO_{2}\cdot \rightarrow H_{2}O_{2}+O_{2}$$

When examining the photooxidative bleaching process, it is important to consider the potential for H_2O_2 generation via the semi-reduced oxygen molecules in both the D-0 as well as D-D processes.

$$O_{2\cdot}^{-}+H^{+} \rightarrow HO_{2} \;\; and \;\; 2HO_{2}\cdot \rightarrow H_{2}O_{2}+O_{2}$$

The production of the O_2-~ radical is expected to occur during the initial D-0 step (Sharma et al., 2016). There is existing information regarding the potential occurrence of said process, as demonstrated by photolytic research indicating that oxygen may serve as both a chemical as well as a physical quencher.

$$^{3}D + O_2 \rightarrow D + O_2$$

$$^{3}D + O_2 \rightarrow D + O_2 \cdot^{-}$$

The constant that determines the reaction rate for the formation of the D+… O_2- kind resulting from the 3D + O_2 response is estimated to be within the range of $2.5–3.1\times10^{8}$ M^{-1}. sec^{-1}. The generation of hydrogen peroxide through the involvement of superoxide anion was noted in instances where dyes (such as acriflavine, eosin, as well as vat dyes) along with white pigments (such as zinc oxide as well as zinc sulfide) were excited while being exposed to both oxygens as well as water (Korytowski & Sarna, 1990; Kirkbright et al., 1972).

v. **Formation of [D • • • • O_2] Complexes:** In addition to the generation of semi-oxidized pigment radical as well as semi-reduced oxygen molecules, resulting in the generation of hydrogen peroxide, it's imperative to consider the emergence associated with a metastable in nature intermediary photo peroxide having a fleeting lifespan of $\tau < 10–5$ seconds as a crucial intermediate within the photooxidation mechanism of dyes (Iliev & Ileva, 1995; Tracewell et al., 2001; Bideau et al., 1991; Tso et al., 1985; Demas et al., 1977). Schenck has proposed that the formation associated with a metastable photo peroxide [D .... O_2] within the pigment triplet as well as oxygen occurs through interaction, in charge-transfer, or just a radical complex mechanism. This proposal has been further examined through various kinetic and experimental analyses. It is noteworthy to point out that Havemann et al. (1964) discovered that the process regarding photooxidative bleaching regarding thionine entails the creation of a compound among the dye as well as oxygen, namely [Thionine H+O_2].

To summarize, it has been observed that during the process of photooxidation of dyes, various intermediates are generated, including semi-oxidized pigment radicals, metastable in nature photo peroxides, O_2-, along with additional species. The structural diversity of dyes and variations in laboratory conditions can lead to significant disparities in oxidative photobleaching outcomes. Nevertheless, it is widely accepted that the triplet form represents the active form that contains the dye during photo-oxidation. Furthermore, it is imperative to consider that the dye, when exposed to an oxidation agent, can function as a potent oxidizing agent in various instances. The process by which the oxidized pigment induces oxidation in other compounds involves the transferring of triggered oxygen via the peroxide intermediary or the removal of electrons, resulting in the restoration of the pigment to its initial state.

Quantum Yields and Dye Classification: Typically, the quantum yield associated with the process of pigment photobleaching via dye oxidation can be defined by a notably low value. The quantum generates ΦB associated with the degradation of the yellow hue of acriflavine is approximately 4×10^{-4}. The measurement of ΦB over cyanines falls within the range of 10^{-2} to 10^{-3}. Specifically, thionine exhibits a value of 8×10^{-4}, while eosin displays a value of approximately 3×10^{-4}. The examination of photobleaching mechanisms reveals that dyes within the triplet form tend to undergo photoreduction when exposed to a reducing agent, while they undergo photooxidation with the addition of an oxidizer. Disparities in the responsiveness of dyes for photoreduction as well as photooxidation may be observed through variations in quantum yields among dye photoreduction as well as dye-photosensitized oxidization. Bourdon along with Schnurige (1967) have proposed a classification system for dyes based on their photochemical reactivity towards oxidizing as well as reducing substances, which divides them into three distinct categories (Foote, 1968; Yi et al., 2010).

Class I dyes exhibit a greater affinity towards oxygen as compared to reducing agents. The group of fluorescent dyes that exhibit photo reducibility and photosensitization properties comprises acridines, xanthene-type pigments, thiazines, certain porphyrins, as well as riboflavin. The prevailing characteristic among these dyes appears to indicate a trio of electrons (π,π^*) to be the ground state of lowest energy (Oster et al., 1959; Salari et al., 2009). When considering the main method of photooxidation of pigments, the quantum yield regarding photosensitized oxidization can serve as a useful indicator. Based on this criterion, it is evident that these dyes exhibit a greater propensity for reactivity with oxygen as opposed to reducing agents. Quantum yield values for the photosensitized oxidization of p-toluene diamine via chlorophyllin as well as for photoreduction alongside ascorbic acid are approximately 0.2–0.7 and 0.01–0.001, respectively.

Class II dyes are distinguished by their elevated reactivity towards reducing agents, specifically hydrogen donors. The aforementioned dyes primarily pertain to vat dyes, which are derivative products of anthraquinone, and indanthrene, as well as comparable compounds. These dyes possess a triplet (n, π*) state as their least excited state. Their photo-tendering activity is significantly elevated.

Similar to Class II pigments, Class III pigments are categorized as vat dyes. Nonetheless, it is noteworthy that their least excited state possesses a triplet nature (π,π*). Consequently, their photochemical conduct is akin to that of Class I dyes, wherein they exhibit minimal photo-tendering action but a discernible reaction towards oxygen (Iliev, 2002; Kramer & Mathis, 1980).

The potential for categorizing dyes based on their structure highlights the significance of the dye's framework in determining its photo reactivity, despite the impact of various external factors upon photobleaching reactions. Van Beek & Heertjes (1966) have reported that dyes containing the molecular structures depicted in Figure 5.4(A) exhibit oxidizing properties, as they can undergo photoreduction in the occurrence of lessening agents (Thrall et al., 2013). Dyes that possess the structures depicted in Figure 5.4(B) exhibit characteristics of reducing agents. Consequently, when exposed to oxygen, the dye can undergo photooxidation, as elaborated upon in this part.

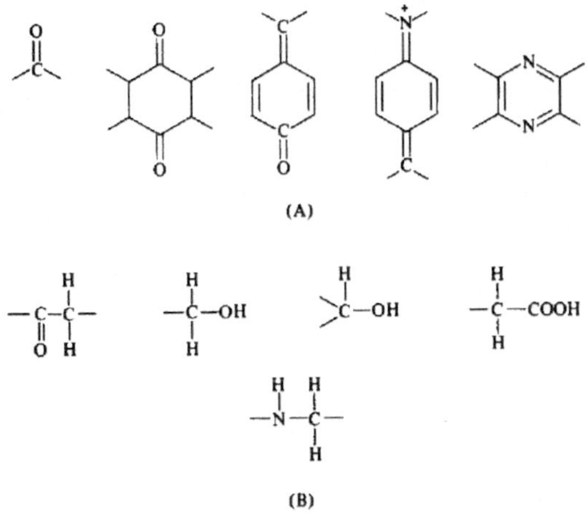

Figure 5.4. *Illustration of (A) structures that give dyes a photooxidizing reactivity against other compounds (corresponding to Class II). And (B) structures which give dyes a photoreducing reactivity, and which can lead to a photooxidation of the dye in the presence of oxygen with subsequent sensitized photooxidation of other compounds.*

Source: P. M. Heertjes, Creative Commons License.

5.1.2.3. Photodecomposition of Dyes by Excitation of Other Substances

1. Principle of the Reaction: The process of photodecomposition for a dye can be facilitated through the excitation of substances that coexist within the same system as the dye molecules (Chen et al., 2010). In photochemical procedures, foreign molecules (F) undergo reactions in an excited state alongside dye molecules, following a specific scheme:

$$F + h\nu = F^*, \quad F^* + D \rightarrow F' + Products$$

The compounds that exhibit a propensity to interact with pigments in this fashion are primarily those that function just like hydrogen donors, including primary as well as secondary alcohols, ketone bodies, carboxylic acid, amine-based compounds as well as similar substances. Additionally, these compounds are also known to diminish excited pigments. The compounds under consideration exhibit a significant increase in their reducing activity upon excitation. This phenomenon enables the movement of hydrogen atoms from methyl, methylene, or just methine categories, which become activated via adjacent carboxyl, hydroxyl molecules, amino aryl categories, etc., to dyes. The photodecomposition of pigment in said reactions is dictated by the wavelength of absorption generated by the diminishing substance, rather than that of the pigment itself, thereby determining its action spectrum (Konstantinou & Albanis, 2004; Dong et al., 2013; Chen et al., 2002).

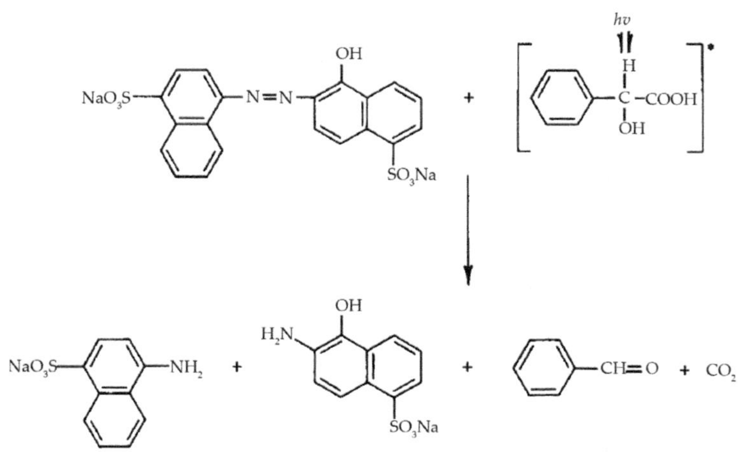

(XII)

The present study provides an illustration of the transformation of azo colorants to their respective amines through the utilization of mandelic acid, as investigated by van Beek along with Heertjes (1963), as evidenced by the notation (XII).

2. **Wavelength Dependence of Photochemical Reactions:** Fibrous materials have the ability to undergo excitation upon exposure to UV light due to their structural similarity to simple compounds. In numerous instances, varying reactivity towards distinct wavelengths may be attributed to this phenomenon (Chen et al., 2002; Liu et al., 1999). As an illustration, basic azo dyes that exhibited minimal photo-fading tendencies on the textiles made of cotton under visible light exposure demonstrated heightened sensitivity when subjected to UV radiation (λ~360 nm). The photoreduction for 2,2-Diphenyl-l-picrylhydrazyl within cumene solution was observed under UV light (λ = 313 nm), while no reduction was observed under visible light (λ = 520 nm). This can be attributed to the engaged reducing agent property possessed by the keen hydrocarbon's isopropyl category. The likelihood of photodecomposition of fast dyes, which exhibit sensitivity solely to UV radiation below 400 nm, is high upon intake of light via a fibrous substrate (Zhao et al., 2002; Yan et al., 2006; Plaetzer et al., 2009).

When examining these reactions, it is important to consider the potential existence of alternative mechanisms. The photoreduction of methylene blue, typically facilitated by reducing agents and visible light, can occur in water-based solutions lacking the inclusion of such agents when exposed to far UV radiation (λ < 250 nm). However, this photoreduction process is not observed when the UV region is excluded. It is possible that the dye, which is in a singlet state of higher excitation, may undergo a reaction with water acting as a reducing agent when exposed to far UV radiation. It has been observed that Lumiflavine has the capability to undergo photoreduction by water, leading to the formation of semiquinone, however, this process is only feasible when exposed to wavelengths below 340 nm, and not when subjected to visible light. Variations in photochemical behavior may arise due to the distinct reactivity of distinct electronic components. Oster noted that triphenylmethane pigments, including malachite green, exhibit photoreduction, and fluorescence solely when anchored to polymers. Upon exiting the long-wavelength group, a distinct effect is observed in contrast to exciting the short-wavelength group. The absorption peaks within methylene blue, chlorophyll, as well as other colorants, are attributed to two subsequent singlet levels (S1,S2) that are excited. However, in triphenylmethane pigments where the 3 groups of phenyls aren't equally replaced, the two absorbing bands (known as x and y bands) are considered independent electronic structures that exhibit distinctive behavior. Upon stimulation within the dichroic direction using red light of longer wavelengths, the extended peak vanishes and is replaced by the emergence of an enduring intermediate that absorbs at 380 nm. The reversible reduction of both absorption peaks as well as the generation of species having a long lifespan was observed upon exiting the short-wavelength dichroic direction with blue light. This phenomenon has been previously reported in purines as well as pyrimidines. The observation that the generation of metastable species, characterized by triplet forms with the long-term sustainability of 10^4 seconds, remains unaffected by elevated dye concentration is noteworthy. This is in contrast to xanthene or just thiazine pigments and is likely attributed to the interaction among the singlet excited pigment as well as an inactive pigment, as expressed by the response $1D + D^{-} \rightarrow 3D + D^{0}$. Furthermore, the impact of oxygen upon the quenching of restricted triphenylmethane colorants is minimal (Pham & Clarke, 2008; Moan et al., 1987).

Did You Know?

During World War I, the photochemical properties of dyes were utilized for camouflage purposes.

5.1.2.4. Photocatalytic Processes

Photocatalytic processes involve the reaction of an intrigued photocatalyst using a colorant or another component within the system, leading to the disintegration of the dye as well as the continuous recuperation of the catalyst (Van et al., 2007). The breakdown of sunlight for a dye exhibits an action spectrum that aligns with the absorbing spectrum produced by the photocatalyst. Typically, the photoreduction of the activated catalyst occurs in the initial photochemical stage through the use of an appropriate reducing agent, followed by subsequent reoxidation via an oxidizing substance in these procedures. The utilization of colorants as oxidizing agents within the photocatalytic procedure may result in an unchangeable reduction of said dyes. The catalyst undergoes continuous regeneration due to reoxidation (Hou et al., 2020; Lasek et al., 2013; Parent et al., 1996).

Figure 5.5 illustrates a scheme depicting the photocatalytic elimination of thionine through ethanol as well as the catalyst ($FeCl_4$). The conversion of thionine to leucothionine as well as ethanol to acetaldehyde occurs without any alteration of the $(FeCl_4)^-$ ion. The catalytic elimination of azo dyes through the utilization of mandelic acid may happen with the involvement of sodium 9,10-anthraquinone 2-sulfonate just like a catalyst.

5.1.3. Relation Between Photo-Reactivity of Dyes and Nature of Excited States

5.1.3.1. General Aspects

The photochemical processes of dyes that result from the photobleaching of fibrous substances are contingent upon various structural factors. These factors include the nature of the state of excitement whereby the process can proceed, the placement of this condition when compared to various other states, the duration of the condition, and the likelihood of its formation as well as elimination by intramolecular as well as intermolecular effects (Demchenko et al., 2013; Arias-Rotondo et al., 2016).

The utilization of flash photolysis and other experimental techniques has revealed that a considerable number of photochemical responses of dyes occur through the formation of the least excited three-note state. However, it is important to note that not every one of these reactions follows this pathway. The production of the least excited three-note state from the least excited individual state through intersystem crossing can occur with significant efficacy under specific conditions. The progression of photochemical processes is contingent upon not solely the singlet or just triplet state that constitutes the least excited states, but additionally its n, π*; π, π*; or CT attributes. The relative arrangement of such states exhibits variability across dye categories and can be altered by substituents as well as solvents, thereby inducing significant variations in the photochemical properties of dyes belonging to a particular class. Upon initial observation, it appears that the chromaticity of the dye produced from conversions to the singlet form is not correlated with photoreactions that occur through the triplet state. The possibility of establishing a correlation between color as well as photochemistry arises due to the association between the likelihood of three-note state development along with singlet-triplet division. In addition to this correlation, spectrum absorption can be influenced by n, π*, as well as CT states (Damrauer et al., 1997; Watanabe et al., 1977).

5.1.3.2. Results on Model Systems

5.1.3.2.1. Vat (Anthraquinoid) Dyes

The utilization of the conceptual system method, initially introduced by Bolland & Cooper (1954), has provided noteworthy comprehension regarding the mechanism underlying the fading as well as photo-tendering induced by vat dyes. Prior to the implementation of this methodology, a sequence of patterns had been noticed, yet an adequate explanation was unattainable. These patterns include the requirement for the existence of oxygen, the decrease in activity due to negative substitutes (basic electron donors), the heightened susceptibility of anthraquinonoid colorants through substitution within the 2-position as opposed to the 1-position, the definitive action of un-substituted quinones (anthraquinones, naphthoquinones, dibenzpyrenequinones, as well as pyran-throne), and especially, the significant photo-tendering action of yellow along with orange vat colorants (which otherwise exhibit favorable qualities) with intake below nm, and lack of activity (in most cases) of green as well as blue colorants with an intake in the long-wavelength area ($\lambda > 550$ nm).

The model structures employed in vat dyeing involve the substitution of commercial colorants with ketones (such as benzophenone) as well as quinones (such as simple anthraquinones), as well as the replacement of textile materials with suitable organic solvents (such as ethanol and isopropyl alcohol) (Bridge, 1960). Through the utilization of various techniques including flash photolysis as well as quantum yield evaluations, researchers were able to provide explanations for previous observations regarding the sunlight reduction of ketones. As a result, a theoretical mechanism known as the "photo-tendering mechanism" was proposed. As per this method, a cyclic process operates effectively in water-based solutions that are non-alkaline. In this method, hydrogen is removed via the carbon atom adjacent to the hydroxyl compound in ethanol via the excited ketones or just quinone. The photoreaction results in the formation of a semiquinone radical which exhibits a degree of stability in irradiated substances that lack oxygen (Xu et al., 2004; Shen et al., 2001; Balakrishna & Devi, 2005). However, in a solution full of oxygen, the resulting radical undergoes rapid reversion to the genuine quinone, thereby enabling renewed excitation as well as hydrogen conceptualization from the substrate. Stated differently, the process can be regarded in terms of the photosensitized oxidization of alcohols through oxygen with the quinone's existence. The process of hydrogen abstraction leading to the development of a semiquinone is feasible only when the bonding energies among R and H within the substrate are below 100 kcal/mol. Consequently, a highly reactive reaction can occur between benzophenone as well as alcohols or just toluene ($E = 75$ kcal/mol), yet not using benzene.

The overarching framework is:

$$D + h\nu \rightarrow D^{*} \qquad D^{*} + RH \rightarrow DH\cdot + R\cdot$$

$$DH\cdot + O_2 \rightarrow D + HO_2\cdot \qquad R\cdot + O_2 \rightarrow \text{Phototendering products}$$

When analyzing this particular scheme, it is important to consider the issue that arises when contrasting the photosensitized oxidization of ethanol to that of textile fibers. The phenomenon of cellulose deterioration desensitization caused by anthraquinone colorants in the far UV region appears to contradict the occurrence of vigorous hydrogen transfer among ethanol as well as aminoanthraquinone within the near UV, far UV, and visible regions. Furthermore, there is uncertainty regarding the potential for adequate contact between the reactive component of the pigment as well as an abstractable water atom present in the polymer or just textile fiber for particular instances. However, as elucidated in this part, the process of photo fading from a dye may take place through various mechanisms. In cases where direct hydrogen abstraction is not feasible in dyed substances, the deterioration of textile fibers through photosensitization may occur within dry atmospheres that include oxygen. This degradation could be caused by the fibers being attacked by O^{2-} radicals, via electronic means of excited oxygen. Consequently, photo-tendering can take place in a way akin to photoreaction regarding acriflavine or eosine, as mentioned previously. The degradation by sunlight of fiber through hydrogen peroxide can occur in a setting of water vapor.

Through a range of investigations, such as flash photolysis, intake spectroscopy, and fluorescence evaluations, the through photosynthesis reactive state within quinone-sensitized responses has been elucidated, along with the impact of substituents as well as solvents upon the response of this state. These findings have been documented in various papers as well as reviews by scholars such as Bridge along with Dorr (1965). Currently, it is widely acknowledged that the n,λ* three-note state constitutes the active form of ketones along with quinones (such as anthraquinone dyes) during the procedure of hydrogen abstractions. The singlet state responsiveness of duroquinone (tetramethyl-p-benzoquinone) has been observed, and it is possible that this phenomenon is due to an intramolecular response. The demonstration of the responsiveness of the n,λ* three-note state within carbonyl compounds involves the encouragement by nonbonding n electrons to the anti-bonding λ (i.e., λ*) orbital, accompanied by quick intersystem crossing. This can be achieved by comparing the neutral state as well as n,λ* excited state, as depicted in Figure 5.5.

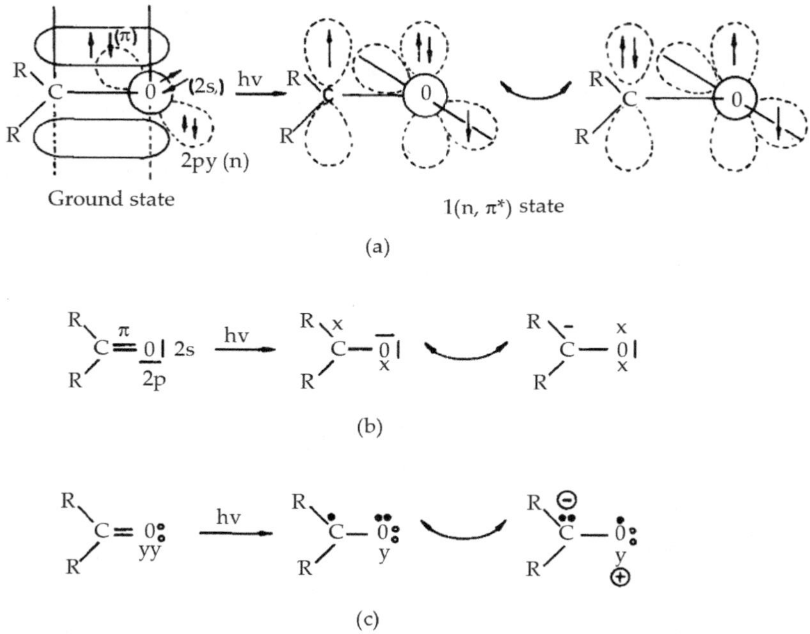

Figure 5.5. *Illustration of n–ϖ* excitation process: (A) simple atomic orbital representation; (B) usual valence bond notation; and (C) notation according to Zimmerman (λ-electrons are indicated as solid dots, s electrons as circular dots, and the* p_y *electrons as small y's).*

Source: El-Sayed, Creative Commons License.

The state denoted as n, π* can be regarded as an impact hybrid of two distinct structures, wherein the electron that has been promoted is present either in the oxygen p_x circular or within the carbon px orbitals. Along with the augmentation of the overall electron concentration in the π-system within the excited state due to the incorporation of a single p_y electron, there's additionally a decrease or inversion of the magnetic field of the neutral state $C^{\delta+} = O^{\delta-}$ owing to the negative as well as positive charges upon oxygen as well as carbon, accordingly, in a single of the resonance forms. Upon removing the n electrons from the atomic oxygen, an electrophilic center is generated at carbonyl oxygen due to its electron deficiency. Additionally, the n orbital exhibits a radicalized character owing to its limited overlap with different orbitals of molecules. The n, π* orbital of carbonyl compounds exhibits a heightened level of reactivity in terms of its ability to separate an atom of water or just an electron via other substances.

The proclivity for hydrogen abstractness could conceivably exhibit similarity between 1(n, π*) and 3(n, π-*) in principle. A more favorable prospect for response arises from the comparatively protracted duration of the triplet state in contrast to the singlet state. Due to the narrow energy gap between the singlet as well as triplet states, a swift intersystem crossing occurs from 1(n, π*) to 3(n, π*), leading to the likelihood of 3(n, π*) being the predominant reactive state within carbonyl compounds.

The minimal and therefore the most reactive state within a sequence of anthrax dyes resembling benzophenone belongs to the n, π* state. The greatest wavelength of absorption within compounds of this nature is attributed to the weak n-π* change (ϵmax ~ 20–100) towards the 1(n, π*) state, which is succeeded by a highly intense π-π* absorption. The detection of n, π* changes is often limited to the polarized luminescence spectrum in numerous instances.

The excitability energy of carbonyl substances, specifically the n-π* excitement energy, has been found to be equivalent to the energy that exists in the CO π-bond in formaldehyde. This energy is estimated to be approximately eighty kcal/mol and includes both the significant energy levels of the neutral state (E_g) as well as the excited state (E_e). These energies are influenced because of the resonance among the group that contains carbon and any substituting categories. The equation hv(n–»π*) = 80 + E_g – E_e (kcal/mol-) is used to describe this relationship, where E_e is greater than E_g. The n, π* absorbance bands of various carbonyl compounds, such as ketones as well as quinones, are typically located in the UV region and towards the shorter end of the visible spectrum. It is imperative to emphasize in this particular context which direct excitation of n-π* is unnecessary. Upon irradiation within the strong π-π* intake band, the first-created (π,π*) state undergoes a swift transformation into an n, π* state via internal transformation, as previously expounded.

The distinct responsiveness of anthrax vat dyes (and additional carbonyl compounds) is determined by the following crucial factors (Vatansever et al., 2013):

- The positioning of n, π* states is minimally impacted by the introduction of electron-donating categories, whereas conjugation of an aromatic nucleus can significantly decrease the energy of π, π* states. Typically, the less intense n-π* bands, which are situated towards the longer end of the electromagnetic spectrum, are obscured by the more prominent π-π* bands with the longest wavelength.

- Porter along with Suppan (1965) has suggested that within compounds containing firmly electron-donating categories, including hydroxy or just amino acid groups, it is feasible for the lowest contends to be comprised of fierce intramolecular CT suggests.
- The relative positions for n, π*, π, π*, as well as CT states in substances may vary depending on the solvent used, despite their identical positions. In polar solutions such as alcohol, it is possible for the fewest states to belong to the π, π* state, as well as the CT state. This is due to the hypsochromic move of n, π* contends as well as the bathochromic move of π, π*, and CT states, accordingly, when contrasted with nonpolar solutions like hydrocarbons.
- When considering π, π* triplet, as well as specifically when the CT status is the least excited state, an electrical shift towards the oxygen that surrounds the group of carbons is observed, which is opposing to the shift observed during the n, π* state. Within the n, π * state, the oxygen atom of the carbonyl exhibits a positive charge within one of its resonance networks. As the system transitions through the (π, π *) status to the (CT) state, there is a progressive increase in the negative charge on the oxygen atom. The hydrogen along with the electron abstraction responsiveness of energized molecules is contingent upon the character of the least excited state. Specifically, the n, π * state elicits a pronounced abstraction effect, whereas the π, π * state yields a negligible or imperceptible effect. Lastly, the CT state does not permit any abstraction.

The aforementioned regulations have been substantiated through a sequence of experiments concerning the responsiveness of carbonyl molecules to photoreduction in connection with substituents and surroundings.

The compound p-amino benzophenone exhibits the ability to remove hydrogen via cyclohexane, yet not via isopropanol. This is due to the fact that the lowest energy state in cyclohexane is 3 (n, π *), while in the solution with alcohol, it is three (CT).

Primarily, the provided concept facilitates an explanation for the unique photo-tendering capability exhibited by vat dyes. Figure 5.6 illustrates the energy state scheme of anthraquinone along with replaced anthraquinone dyes.

The diagram indicates that the colorants within yellow and orange hues are distinguished by a 3 (n, π *) status that has the least energy level.

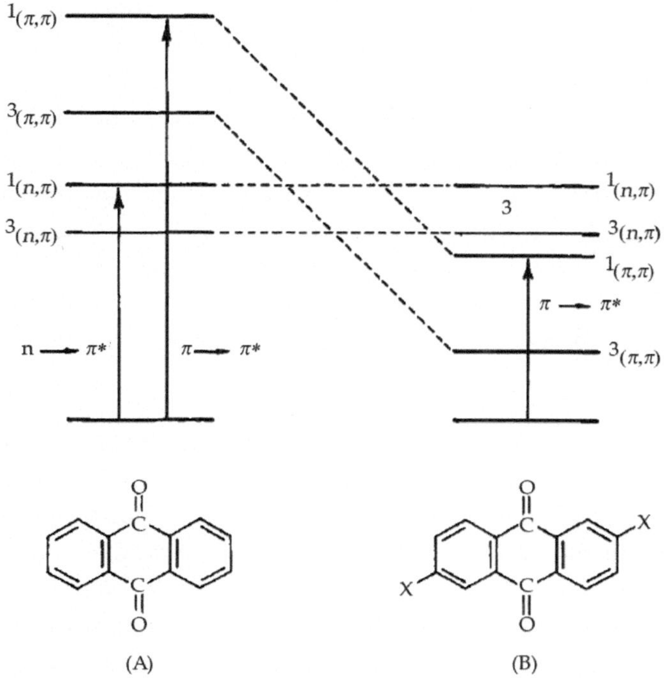

Figure 5.6. *Dependence of the position of 3(n, ϖ *) and 3(ϖ, ϖ *) in anthraquinone on electron-donating substituents X: (A) high photoreactivity towards H abstraction; and (B) no photoreactivity towards H abstraction.*

Source: Aveline, Creative Commons License.

Moreover, it is imperative to consider the factors that affect the n-π* and π - π* change while addressing photoreactivity. An instance of a bond made up of hydrogen which involves the n electrons upon the carbonyl oxygen atom or the attachment of a proton with it can impede the n- π * change. Consequently, the photochemical action of a colorant may be reduced. It is noteworthy to consider Venkataraman's debate regarding the hydrogen bonds of vat dyes within solid (cellulose) polymers within this setting. This is due to the potential influence of such bonding upon the n- π* change. Additionally, it is imperative to consider that dyes with low sensitization capabilities may potentially undergo a primary photooxidation reaction with their substrate in an environment of oxygen, through the 3(n, π *) status of the colorant.

When considering the fading as well as photo-tendering effects caused by anthraquinone pigments, it is important to acknowledge the correlation between the redox capacity and photoactivity for a given dye. An increase in redox potential corresponds to an increase in reactivity. However, certain deviations arise due to the uptake of electrons coming from the unbound π-orbital (π*-orbital) during electrochemical lessening, leading to the formation of semiquinone radicals. Conversely, the optical n- π* change involves the absorption of an electron in the initial stage due to the nonbonding oxygen orbitals. It is worth noting that the equality of unrestrained energy in all colorants is not a necessary condition for both reactions.

5.1.3.2.2. Azo Dyes

Limited research has been conducted on the degradation of azo colorants in solution. The research conducted by van Beek along with Heertjes (1966) pertaining to the photochemical activity of water-based solutions for azo colorants in an environment of various soluble substrates yielded significant findings. The phenomenon of reduction-induced fading of azo dyes can be demonstrated through the relationship between molecular substrates in an excited state as well as pigment molecules that are within an unexcited or just excited state. The decay kinetics of the reactions exhibit a first-order dependence on the amount of each of the substrate particles in their excited state. The reactivity of decreased pigment radicals through oxygen has been observed to impede the discoloration of dye. Amines are the resultant species of decrease in solutions devoid of oxygen (Aleboyeh et al., 2008; Nam et al., 2001; Cho et al., 2007).

In addition to these findings, it is noteworthy to mention the research conducted by Hillson & Rideal (1953) pertaining to the quantification of the Becquerel effect. Their experimentation yielded noteworthy observations of both the oxidization and decrease of azo dyes. The discoloration system of azo colorants remains incompletely understood, as evidenced by research on the hue changes of azo substances generated through exposure to UV rays. There exists a significant level of interest in comprehending the process of fading, as well as its structural impacts that can potentially enhance lightfastness, particularly in the context of dye photographing (De Long et al., 1986).

The examination of the lightfastness regarding azo colorants involves a discourse on the emission characteristics of uncomplicated models of azo colorants. These models refer to aromatic compounds that comprise two adjacent nitrogen atoms, including 9,10-diazaphenanthrene (XIII) and are highly significant in this context. Based on the research carried out as well as comparisons regarding cis- and trans-azobenzene, it can be inferred that a prompt radiationless deactivation occurs through the 3(n, π*) state in cases where the lowest states are a combination of the 1(n, π*) as well as 3(n, π*) states, which can happen due to the nonplanarity for the molecules. No phosphorescence was detected within the 3(n, π*) state, leading to a proposed actual lifetime of fewer than $10–^9$ seconds for these three-note states. The high lightfastness of azo molecules with the smallest 3(n, π*) state is a significant outcome, which distinguishes them from anthraquinone dyes. Conversely, azo compounds exhibiting the least energetic 1(π,π*) state demonstrate fluorescence. This observation suggests that azo dyes possessing the lowest π, π* states exhibit an extended excited state lifelong and enhanced photoreactivity.

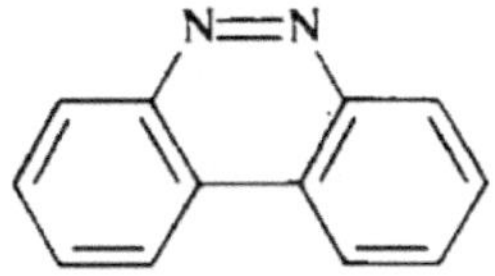

(XIII)

Figure 5.7 depicts the extreme scenarios of the n, π* and n, π* states with respect to their photoactivity and inactivity. When analyzing this particular plan, it is important to acknowledge the fact that the n, π* states may be achieved through either adsorbing on fibers or just through the attachment of protons with the n electrons. The phenomenon of pigment fluorescence along with azo substance photoreactivity can happen when the energy level of the non-bonding to π* electronic transition surpasses that of the π to π* transition. An instance of this is observed in the reaction of cfs-azobenzene according to acidic conditions, where 9,10-diazaphenanthrene along with hydrazobenzene (XIV) are produced upon exposure to X = 405 nm. Additional research is required to further investigate these relationships.

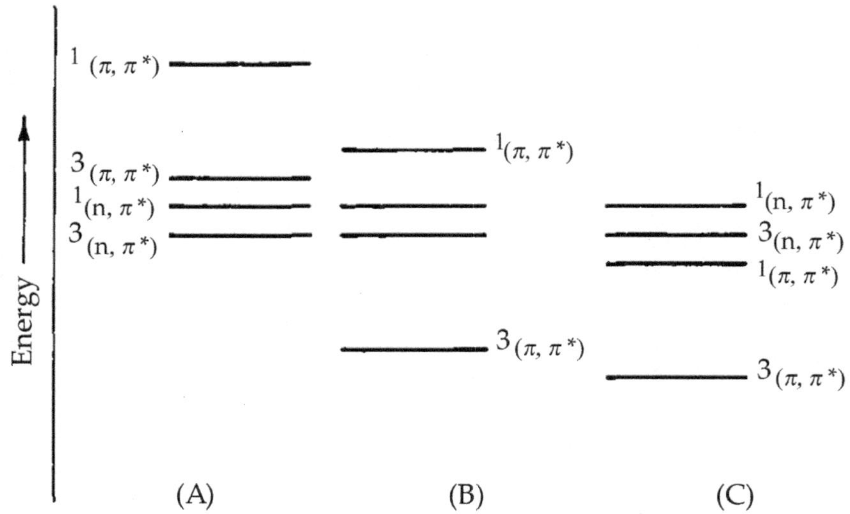

Figure 5.7. *Lightfastness of azo dyes: (A) and (B) high lightfastness; and (C) small lightfastness. [Note: (B) corresponds to cis-azobenzol].*

Source: R. M. Acheson, Creative Commons License.

5.2. PHOTOCHEMISTRY OF DYES ON SUBSTRATES

5.2.1. Testing Methods

5.2.1.1. Measurements of Lightfastness

For several years, it was successfully established that the rate of fading is contingent upon the substrate, atmospheric humidity, temperature, and spectral distribution of radiation. Despite the intricate nature of various fading factors, it is widely accepted that the fundamental photochemical procedure for dyed materials can be compared to one of the extensively researched mechanisms in modeling systems. These mechanisms have demonstrated the significance of the dye framework, as well as the impact of reduction or oxidization agents, oxygen, as well as water (Cristea & Vilarem, 2006).

The determination of lightfastness for a dye is subject to the impact of various factors, and therefore, can only be precisely established within a specific medium and under particular circumstances. The determination of the rate of fading can be accomplished through measures of the spectral density of colorants within the solution or retained in robust films, or through the reflectance just like an indicator of the period of illumination.

These measurements can be conducted using a spectrophotometer as well as a Spectro reflectometer, accordingly. A method that enables a swift differentiation within dyes of varying lightfastness on the substrate is imperative for pragmatic applications.

In the United States, a technique is employed to ascertain the duration required for a specific degree of fading through exposure to a standardized light source. Sunlight as well as artificial light are commonly employed as conventional sources of illumination. However, due to variations in the proportion of actinic radiation to total radiation, such as variations in the UV amount in daylight, along with other drawbacks, the encased carbon arc, commonly referred to as the Fade-Ometer, has been established as a conventional light source. Consequently, the parameter commonly utilized to measure lightfastness in the United States is the Normal Fade-Ohmmeter hour. The composition of carbon curve rays and daylight differs, leading to discussions regarding the adoption of xenon arc as a standard source (Giles & McKay, 1963).

The British Standard (BS) procedure circumvents the challenges associated with standard-source techniques by determining the rate of fading through contrast with a set of conventional dyeings that are concurrently exposed. The present method bears resemblance to the one employed by Dufay in 1729 during his methodical investigations on the durability of colors under light exposure. Considerable effort has been required to produce proficient dyes that meet the lightfastness criteria, commonly referred to as actinometers. In the past, the BA technique utilized a single standard that was substituted with a new standard following a significant reduction in specimen intensity, equivalent to one standard fading period (SFP). Presently, the SDC Blue dimension (ISO The suggestion No. 117, part 11) employs eight standards. The standards in question, commonly utilized in Europe, consist of woolen fragments that have undergone dyeing with varying shades of blue. Upon being exposed to bright summertime light equivalent to 37 Langleys [cal/cm2] (≈37 × 1,200 lux hours), the lightfastness level 1 undergoes noticeable fading within a duration of one hour. The remaining standards are selected in a manner that results in Standard 1 exhibiting a decay rate that is approximately twice the rate as Standard 2, requiring 82 Langleys, and continuing in this pattern up to Standard 6. Standard 7 exhibits a speed that is fourfold greater than that of Standard 6, while Standard 8 demonstrates a twofold increase in speed relative to Standard 7. Specifically, the irradiation of 5,400 Langleys is required for Standard 8. Standard 1 (XVI) pertaining to lightfastness incorporates reactive methylene groups, which may lead to fading due to either an intramolecular or just intermolecular (D-D) oxidative-reducing process or a photoreaction using oxygen (Duxbury, 1993).

In the United States, a distinct blue scale, as suggested by AATCC, is employed. This scale utilizes a rapid discoloration dye, Mordant Blue 1 (CI 43830), which is blended in varying proportions with a highly fast colorant, Solubilized Vat Blue 8 (CI 73801), and applied to cotton swatches. The resulting grades are approximately one grade higher than those provided by the European ISO scales.

During the process of evaluating lightfastness, the colored sample is subjected to exposure with the established standards until a noticeable level of fading becomes apparent. The classification of dyes is based on the standard number that indicates the degree of fading exhibited by the dye. In order to achieve a precise evaluation of lightfastness, the "first break" is not utilized. Instead, the alteration in color is compared to the typical gray scale (GS) that outlines the differentiation between two regions of color (DIN 54001). It is feasible to ascertain the lightfastness for a dyed sample using a standard deviation around 0.5 grade.

It is important to acknowledge that the prescribed technique for measuring lightfastness (BS 1006: 1953) had been limited to daylight exposures due to significant variations observed within fading-lamp as well as daylight fastness values within this particular framework. McLaren's findings indicate that inconsistencies arise from various factors, including variations in the spectral makeup of fading luminaries and sunlight, as well as the moisture that covers the surface throughout contact, which is influenced by the exterior temperature as well as the humidity level of the air around it. Currently, there exists a prevailing inclination towards the utilization of light generated by a xenon curve lamp for the purpose of irradiating materials. This approach, when coupled with suitable filters, affords the optimal potential for replicating outdoors conditions. Additionally, the Xenotest equipment (Quarzlampen GmbH, Hanau, Germany) offers the potential for effective adjustment of exposure parameters to simulate natural exposure circumstances, such as humidity and black-panel temperatures (Baxter et al., 1955).

5.2.1.2. Measurement of Photo-Tendering

As elaborated in Section 5.1, there is a potential occurrence of the catalytic offering of dyed fibers, wherein the fabric undergoes disintegration without exhibiting any discernible discoloration of the pigment. This phenomenon is commonly observed in practical applications, particularly with specific types of rapid and transient vat dyes. Experimental results indicate that sitting in daylight over a duration of 4 weeks resulted in a significant reduction in the breaking strength of cotton. Specifically, the observed reduction was around forty percent for CI Vat Yellow 2, CI Vat Orange 9, as well as other Vat colorants (Chacon et al., 1988; Hihara et al., 2004).

The assessment of photo-tendering action cannot solely rely on the discoloration of the pigment as an accurate parameter. Therefore, it is imperative to ascertain the reduction in the tear strength of the yarn. The calculation of viscosities or just cuprammonium fluidities (the reciprocal for fluidity) is considered the most precise method for determining photochemical degeneration. Furthermore, it is possible to consider an elevation in copper quantity or acidity levels (estimated just like uronic acid percentage of carbon dioxide) by referring to relevant sources. It is noteworthy in this particular context that obtaining precise outcomes can be challenging due to the non-uniformity of the inflicted damage. Despite a severe attack on the caps of the displayed threads, it is possible for the interior components to remain unaffected in numerous instances (Hihara et al., 2002; Griffiths & Hawkins, 1973).

5.2.1.3. Testing of the Fading Products

Analytical research on substances from degradation can provide valuable insights into their nature. Spectrophotometric techniques or chromatographic examination can enable comparisons with oxidative products that are chemically synthesized. The Weiss ring-oven method, previously utilized for the partially quantitative analysis of small quantities of dyes, may have the potential for application in this context (Cumming et al., 1956).

5.2.1.4. Measurements of the Rate of Fading

Valuable insights into the physical condition of dyes within fibers can be gleaned from measures of the degree of discoloration of painted materials. Additionally, the study of enhancements regarding the lightfastness of dyed textiles through special processes is possible. Various plot types are being utilized to analyze fastness data (Cristea & Vilarem, 2006).

1. **Fading-Rate Curves:** Fading-rate arcs can be categorized into multiple groups based on the remaining dye content (or its logarithm) plotted against time.

- In instances of discoloration, the rate exhibits an exponential decline over time in accordance using the first-order or just pseudo-first-order principle (as depicted in Figure 5.8(A)). Specifically, the proportion of unaltered dye present over time conforms to a standard exponential curve (Li et al., 2008). The phenomenon of first-order fading can be demonstrated through the use of molecular scattered dyes such as merocyanine colorants in collodion as well as dissolved in water surface-active azo colorants within methyl ethyl cellulose, as evidenced by the observation of true first-order discoloration arcs.

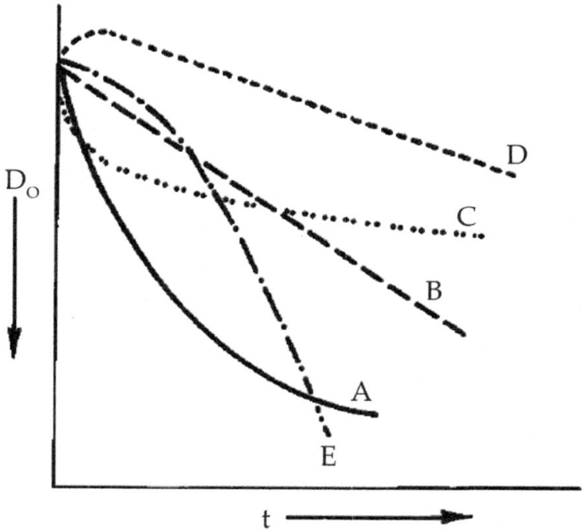

Figure 5.8. *Illustration of types of fading-rate curves.*

Source: Criestea, Creative Commons License.

- The detection of fading in accordance with the zero-order may be achieved through the observation of a constant discoloration rate. This is characterized by a linear correlation between the quantity of dye visible and the duration of time, as depicted in Figure 5.8(B). The phenomenon of fading rate is observed in materials that have been dyed with dyestuffs having relatively large particles. More specifically, pigments as well as rapid dyes exhibit this type of fading kinetics (Crews, 1987).
- The fading process of a combination of monomeric along with aggregated pigment forms may be described by a first-order equation that represents the fading associated with the monomers

initially, followed by a gradual zero-order discoloration of the unreacted larger particles, as depicted in Figure 5.8(C). This phenomenon of fading is commonly observed in practical applications (Auclair et al., 2003).

- Negative first discoloration, characterized by an acceleration of the discoloration rate over time, may manifest in certain circumstances where significant dye particles disintegrate due to the heat generated by illumination. This phenomenon is illustrated in Figure 5.8, panel D. The initial fading-rate curve surpasses the subsequent decrease due to the accelerated fading of fewer colored particles, owing to the heightened optical density associated with a specific weight of dye caused by reduced-size particles. The reaggregation of recently formed particles is observed in low light conditions, resulting in a perceived discoloration of the colorant. Conversely, illumination leads to the repair of the pigment's color. A diagram outlining the various fading pathways during dye lighting is presented in Figure 5.9.

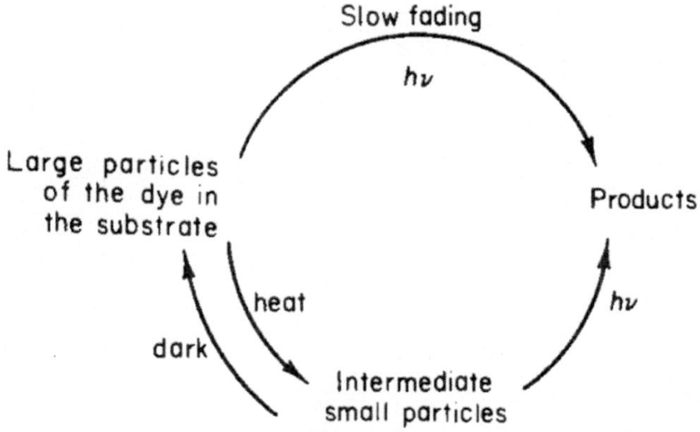

Figure 5.9. *Illustration of explaining the abnormal fading behavior of a dyed material.*

Source: Radin, Creative Commons License.

v. In certain dyed materials, including insoluble azo-colored substances in cellulose, the process of fading may be expedited over time due to a persistent degradation of dye particulates, as illustrated in Figure 5.8(E). The brief discourse presented to explicate fading-rate arcs illustrates that such curves can be associated with distinct fading patterns exhibited by diverse dyed substances. This study addresses the challenges associated with measuring the opacity of dyed content and identifying the various types of discoloration.

2. **Characteristic Fastness-Grade (CFG) Curves:** It is a well-established phenomenon that shows the lightfastness associated with a pigment on fabric fibers often exhibits an upward trend with a rise in depth. This trend can be leveraged to create CFG curves, which provide significant insights into the physical properties of the dyes within the fibers. Giles has proposed that the displaying of the mean lightfastness levels of the pigment classes toward their respective depth, specifically the logarithm regarding the initial dye content in the material, results in the acquisition of straight lines that constitute the CFG arcs (Kamel et al., 1979).

Baxter & Giles et al. (1955) demonstrated the feasibility of generating CFG curves through spectrophotometric means by verifying the linear correlation among log [D0] as well as log t over a specific degree of fading, in accordance with the empirical guidelines established by Cunliff along with Lambert (1929).

Giles conducted a comprehensive investigation and discovered that in numerous instances, the slants of CFG arcs exhibited variations either among dye categories or among fibers. Two constraints for gradients have been proposed by relating the inclination of the CFG arc along with the rate during which the assortment of dimensions of pigment particles shifts with a rise in the amount of pigment in the yarn.

Systems exhibiting uniformity in pigment particle sizes across different concentrations are commonly associated with low slopes. The lightfastness does not exhibit a significant alteration upon an escalation in dye concentration.

This scenario is plausible when the dye generates minute particles, resulting in low lightfastness, or when it produces large particles, leading to high lightfastness (Giles et al., 1961; Shakra et al., 1981).

Systems exhibiting nonuniform pigment particles display steep slopes, which are attributed to the growth of said particles due to a rise in pigment concentration. It is pertinent to note in this particular context that the explication of CFG slopes (*refer* Figure 5.10), which has been corroborated by empirical evidence, is founded on the fundamental principle that the degree of attenuation is contingent upon the area of the dye particulates. The rate of fading is influenced by the size of particles, with larger particles exhibiting a slower rate of fading. This can be attributed to the fact that smaller particles possess a larger surface area, which facilitates their interaction with light, air, humidity, as well as fiber material.

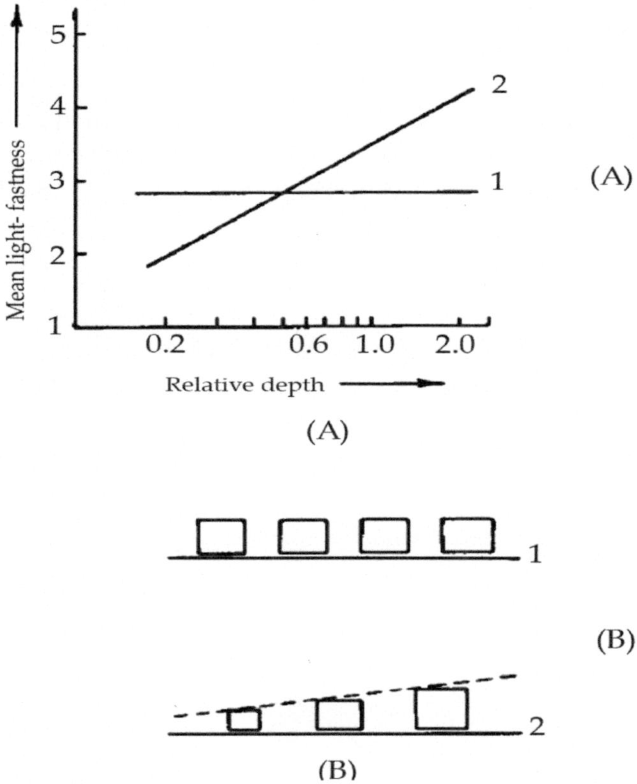

Figure 5.10. *Illustration of (A) example of CFG curves. (B) Scheme of particle-growth possibility with increasing dye concentration: 1 – the number of particles increased (see curve 1 of (A)); 2 – the range of sizes of particles increased (see curve 2 of (A)).*

Source: Rossow, Creative Commons License.

Giles et al. (1961) presented a number of significant relationships based on the provided hypothesis regarding CFG curves. The data suggests that the CFG curve exhibits a greater slope on fibers with larger pore sizes. Additionally, it has been observed that the gradients are more pronounced for insoluble colorants such as azoic as well as sulfur, as compared to soluble pigment categories. This phenomenon can be attributed to the obstruction of dye connection or growth through solvating water atoms. Furthermore, based on the data obtained from CFG, it can be inferred that determining the lightfastness associated with a dye in isolation is not feasible. Instead, it is imperative to define the lightfastness of specific systems, including the combination of dye, fiber, and other components. Additionally, it should be noted that in numerous instances, the fastness solely applies to a certain amount or depth of the pigment on a specific fiber (Giles et al., 1977).

Characteristic Fading (CF) Curves: The characteristic discoloration curves, also known as CF or CFO arcs, are generated through the logarithmic plotting of the period of discoloration tF for a specific proportion of the initial dye concentration [D0], such as a ten percent loss, towards the logarithmic value of [D0]. The CF curves exhibit dissimilarities from their CFG arcs by substituting the fastness grade with the logarithmic tF value (Tera et al., 1985).

The equivalence between fastness ranks as well as log tF amounts has been established. It is feasible to generate CF curves by utilizing the data obtained from basic fading-rate arcs, which record the decline in the density of light or reflectance resulting from exposure to different initial dye levels (*see* Figure. 5.11).

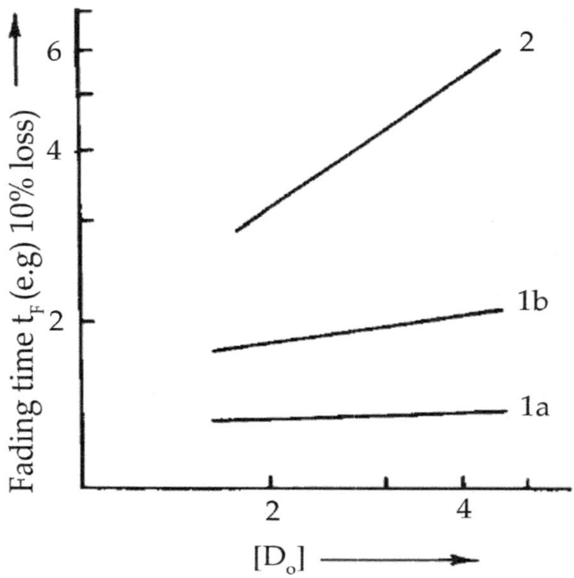

Figure 5.11. *Scheme of CF curves: 1 – low slope (fastness of lb higher than that of la); and 2 – high slope (fastness higher than dye-substrate system 1).*

Source: Vlachopoulos, Creative Commons License.

Similar to the CFG arcs, the CF arcs depicted in Figure 5.11 typically exhibit linearity and can provide insights into the dye's chemical makeup within the fiber. In structures where the condition of the pattern of particle sizes remains constant across every concentration level of the pigment, i.e., with a consistent surface area/weight ratio, the CF curve exhibits a significantly low slope. Mechanisms that consist of an entirely monodisperse pigment or an evenly distributed pigment exhibit fading behavior that can be described by a first-order response. Although rare, monodisperse structures that possess poor lightfastness do exist (Baxter et al., 1957). Vat dyes, certain fast-to-light influence fiber dyes, as well as metal intricate dyes that exhibit neutral-dyeing properties, are known to form particulates of similar dimensions with minimal CF slopes. Conversely, in cases where dyes generate particles of varying sizes that expand as the number of dyes increases, the gradient of the correlation function curves is elevated. The colorfastness of certain dyes, including azoic as well as sulfuric acid dyes, exhibits a decrease in performance when used at lower concentrations, or depths of shade, as compared to higher concentrations.

Giles et al. (1977) recommends that in order to accurately assess the lightfastness properties of a dye, it is necessary to plot the two: fading rates as well as CF (or CFG) curves. An example of how valuable insights into the generation of exceptional lightfastness can be obtained is through the examination of the CF arc of a dye-fiber framework following a treatment. Furthermore, significant principles have been derived by analyzing the CF curves for various dye categories and examining dyes on yarn with varying regain, which corresponds to distinct pore sizes. It has been demonstrated that the rate of dye fastness augmentation is higher within hydrophilic filaments that have elevated porosity (i.e., regain) as compared to fibers with low porosity, as the size of the yarn increases. Additionally, the correlation between the increase in fastness and an upsurge in regain for hydrophilic material was established based on the analysis of CF arcs (Campbell et al., 1958; Khamis & Arafah, 2020).

- **Comparison Method:** Giles suggested a comparative approach to gain an understanding of the distinct photochemical behavior exhibited by different fibers. This method involves plotting the rapidity of particular dyes belonging to the same coloring class on a single fiber against their fastness on an additional fiber. The equivalent line depicted in Figure 5.12 can only be achieved if the rapidity of each color on both fibers is identical. Any variation in the fastness of the dyes on the fibers may be observed through the divergence of the observed line compared to the equivalence line. In instances where dyes on a particular fiber exhibit greater fastness compared to the other, the measured line will be positioned above the equivalence line. This is determined by utilizing fastness level as the ordinate. It is noteworthy that an intersection between the equivalents line as well as the calculated protein/nonprotein fastness-grade line of various dye series reveals a distinct photochemical reaction between protein as well as nonprotein material (Auclair et al., 2003; Yin et al., 2006; Whitmore et al., 1999).

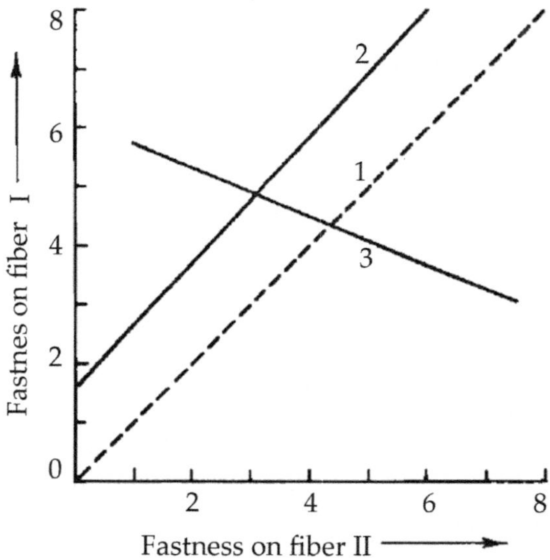

Figure 5.12. *Illustration of comparison method: 1 – equivalence line; no difference in effect of fastness; 2 – dyes of fiber I have higher fastness than on II; and 3 – difference in chemical nature of fading.*

Source: Pandey, Creative Commons License.

5.2.2. External Influences on Light Fading

The fading speed of dyed content is significantly impacted by a range of outside variables.

5.2.2.1. Oxygen

The process of dye fading is observed to be accelerated by atmospheric oxygen in numerous instances. This phenomenon has been documented since Lasareff's (1907) investigations into the fading characteristics of cyanine colorants. The indispensability of oxygen within the photodegradation process of colored textile fibers such as cotton, rayon, and linen appear to be unequivocal. The process of photo rendering, specifically in relation to vat dyes, can be characterized as a dye-sensitized mechanism whereby the reduced form of the pigment is subsequently regenerated to its original state through the involvement of oxygen (Fujibuchi et al., 2006). Furthermore, there exists the potential for fiber vulnerability to hydrogen peroxide in an environment of water, as well as oxygen may exhibit reactivity through association with Class I or III $3(\pi,\pi^*)$ excited colorants. The experiments conducted by Egerton are noteworthy in this particular context. These experiments involved the tendering of fibers that split from the pigments (namely, xanthene as well as acridine dyes) by gaps of millimetric order. The likelihood of producing H_2O_2 as an oxidizing agent is attributed to the requirement for an adequate amount of water. Egerton postulated that under conditions of reduced oxygen pressure as well as limited distance, the charged singlet oxygen molecules or 02.- could potentially undergo oxidation, thereby supporting Kautsky's hypothesis. This proposition has been reviewed by (Bourdon & Schnuriger, 1968).

Conversely, oxygen may exhibit inhibitory properties toward fading in certain instances. The fastness regarding Diaz cyanine blue upon cellulose or just protein fibers were observed to be greater in oxygen compared to nitrogen. This phenomenon is comparable to the discoloration outcomes of azo colorants within solution according to UV light as well as other materials. Due to the intricate nature of fading reactions, it is not feasible to provide a straightforward explanation of these phenomena. It may be necessary to consider the impact of oxygen's quenching impact on excited photochemical reactions' active states. Van Beek & Heertjes (1966) have posited that the degree of inhibition is contingent upon the comparative velocities of the potential mechanisms, whereby the stimulated substrate either transfers hydrogen into the dye or undergoes oxidation via the creation of peroxides within an environment of oxygen (Blaznik et al., 2022).

5.2.2.2. Humidity

Through a series of experiments, (McLaren et al., 1952) were able to demonstrate that the inclusion of vaporized water accelerates the loss of color rate. A number of vat as well as azoic colorants exhibit a reduction in their fastness properties upon exposure to increased levels of humidity. It is noteworthy to mention that the comprehension of the impact of water and oxygen resulted in a technique, developed in 1893, for safeguarding paintings (Chen et al., 2018; Liu et al., 2022). This method involves enclosing them in transparent containers under a vacuum or enveloping dyed materials with an inert gas like nitrogen or argon. The dyeing process of wool exhibits a relatively lower sensitivity towards fluctuations in humidity levels as compared to cotton. Hence, in terms of rapidity benchmarks, woolen articles are favored. Elucidating the impact of humidity is a complex task. It is highly probable that the generation of H_2O_2 in an environment full of oxygen may contribute to the phenomenon of humidity impacts. The intricate nature of photochemical reactions resulting in fading and tendering often entails the differential impact of humidity across different reaction stages. This viewpoint is substantiated by empirical evidence regarding the impact of copper upon the susceptibility of vat dyes to tendering as a result of changes in relative humidity. Additionally, the influence of kinetic impacts can be examined, such as in the instance of diffusion-limited bimolecular reactions. At reduced levels of humidity, the movement of reactants towards each other is impeded, resulting in an elevated likelihood of energy dissipation. The phenomenon of photodimerization within fiber materials may additionally be attributed to humidity (Aydemir & Yenidoğan, 2018; Cristea et al., 2006).

5.2.3. Internal Physical Factors

5.2.3.1. Physical State of the Dye

The fading of colored materials is significantly influenced by the physical condition of both the dye as well as the surrounding environment. (Giles et al., 1957) conducted a series of both scientific and practical studies that demonstrated the significance of the correlation between discoloration and the placement of dye molecules within the fiber. The fading of dye molecules is influenced by the presence of water and oxygen, causing accessible molecules to fade at a faster rate compared to those that are not exposed to moisture as well as oxygen. It is noteworthy in this particular context that significant regulations pertaining to the impact of physical variables are deduced from the curves of fading rate and the consequences induced by additives (Weissbein & Coven, 1960; Mojbafan et al., 2020).

As evidenced by kinetic data, there is a positive correlation between the typical size of gathered dye particles and the degree of lightfastness. This regulation is a consequence of the limitation that exclusively dye molecules located on the outermost layer of the fragments are capable of undergoing a reaction. It is widely accepted that the stability of dye is greatly influenced by the physical condition of both the dye and the fiber. Primarily, any form of treatment that alters the size for dye aggregates has the potential to either decrease or enhance the fading characteristics of a dyed substrate. Several techniques have been developed to modify the lightfastness of dyes, and numerous established phenomena are rooted in these physical conditions' impacts. Several points should be noted:

The correlation between fastness grades and a logarithm for dye concentration exhibits a linear trend, which can be attributed to the enlargement of dye particles. Consequently, deeper shades tend to exhibit a slower rate of fading compared to lighter shades. Dyes possessing high solubility and finer dispersion in substrates exhibit a greater tendency towards rapid discoloration, as compared to dyes with lower solubility. Changes within the condition of aggregation resulting from modifications in surface activity may lead to alterations in lightfastness. The surface activity of a dye can be enhanced by the incorporation of lengthy paraffin chains. This results in the dye spreading into a monolayer with limited light stability. Conversely, dyes containing short alkyl chains typically develop into micelles, which exhibit a greater degree of lightfastness compared to the monolayer state. The surface action of flat dyes has been observed to be significantly elevated when sulfonate groups are present at the tip as well as one side within the molecule. This can be attributed to the fact that the remaining portion of the molecules is recognized as a substantial hydrophobic group. The rapidity of these dyes is comparatively lower than that of colorants with sulfonate categories of low surface action, which are uniformly distributed around the molecule. This observation is supported by the research conducted by Giles et al. (1977) on various series of colorants belonging to the sulfonated anthraquinone, sulfonated as well as unsulfonated azo, azomethine, as well as merocyanine classes (Cao et al., 2022).

5.2.3.2. Physical State of the Fiber

The permeability of fibers plays a crucial role in governing the formation of pigment particles as well as the transfer of water through the aggregates. Consequently, a correlation exists regarding the lightfastness property as well as the physical condition of the fiber. The enhancement of light stability within hydrophilic fibers is positively correlated with an augmentation in porosity due to the potential formation of substantial aggregates within this milieu. Conversely, the inclusion of solid materials such as titanium dioxide for delustering or resins for crease rebellion can diminish the lightfastness provided by the fiber due to a decrease in porosity (Wang et al., 2021). Fibers that are hydrophobic and possess limited porosity impede the spread of moisture toward the dye fragments that are integrated within them, thereby minimizing the occurrence of fading. When examining the impact of supplementary agents, it is imperative to consider the potential for subsequent chemical reactions of said substances. The impact of methylal amide agents employed in the finalization of dyed cotton fabric to enhance wrinkle obstruction as well as wash-and-wear characteristics appears to stem from the agent's direct involvement in the photochemical process that expedites light-fading. The emergence of cyanines through the utilization of said resins can be ascribed to either the creation of an oxygen-dense intermediate via photooxidation or a reaction caused by sunlight.

5.3. SUMMARY

Photochemistry of dyes refers to the study of the chemical processes and reactions that occur when dyes interact with light. Dyes are colored substances that can absorb light of certain wavelength and reflect or transmit light of other wavelengths, giving them their characteristic color. Photochemical reactions in dyes involve the absorption of photons and subsequent molecular transformations. It is a complex and diverse field of study with applications in various areas such as photography, color science, sensor and optoelectronics.

REFERENCES

1. Abraham, D. P., Knuth, J. L., Dees, D. W., Bloom, I., & Christophersen, J. P., (2007). Performance degradation of high-power lithium-ion cells—Electrochemistry of harvested electrodes. *Journal of Power Sources, 170*(2), 465–475.
2. Alaton, I. A., & Balcioglu, I. A., (2001). Photochemical and heterogeneous photocatalytic degradation of waste vinylsulphone dyes: a case study with hydrolyzed Reactive Black 5. *Journal of Photochemistry and Photobiology A: Chemistry, 141*(2–3), 247–254.
3. Aleboyeh, A., Kasiri, M. B., Olya, M. E., & Aleboyeh, H., (2008). Prediction of azo dye decolorization by UV/H2O2 using artificial neural networks. *Dyes and Pigments, 77*(2), 288–294.
4. Allen, N. S., (1994). Photofading and light stability of dyed and pigmented polymers. *Polymer Degradation and Stability, 44*(3), 357–374.
5. Arias-Rotondo, D. M., & McCusker, J. K., (2016). The photophysics of photoredox catalysis: a roadmap for catalyst design. *Chemical Society Reviews, 45*(21), 5803–5820.
6. Atkinson, M. B. J., (2011). *Fundamentals and Applications of Co-Crystal Methodologies: Reactivity, Structure Determination, and Mechanochemistry*. Doctoral dissertation, University of Iowa.
7. Auclair, M., Lamothe, M., & Huot, S., (2003). Measurement of anomalous fading for feldspar IRSL using SAR. *Radiation Measurements, 37*(4–5), 487–492.
8. Aveline, B. M., Matsugo, S., & Redmond, R. W., (1997). Photochemical mechanisms responsible for the versatile application of naphthalimides and naphthaldiimides in biological systems. *Journal of the American Chemical Society, 119*(49), 11785–11795.
9. Aydemir, C., & Yenidoğan, S., (2018). Light fastness of printing inks: A review. *Journal of Graphic Engineering and Design, 9*(1), 37–43.
10. Ba, C., Fu, Y., Niu, F., Wang, M., Jin, B., Li, Z., & Li, X., (2020). Effects of environmental stresses on physiochemical stability of β-carotene in zein-carboxymethyl chitosan-tea polyphenols ternary delivery system. *Food Chemistry, 311*, 125878.
11. Badour, Y., Jubera, V., Andron, I., Frayret, C., & Gaudon, M., (2021). Photochromism in inorganic crystallised compounds. *Optical Materials: X, 12*, 100110.

12. Balakrishna, R. G., & Devi, L. G., (2005). A study of photocatalytic oxidation of indanthrene red LGG an anthraquinone vat dye on TiO_2. *Polish Journal of Chemistry, 79*(5), 919–931.
13. Bancroft, E., (1813). *Experimental Researches Concerning the Philosophy of Permanent Colors; and the Best Means of Producing Them by Dying (sic), Callico (sic) Printing, etc.* (Vol. 2, PP.4-9).
14. Batchelor, S. N., Carr, D., Coleman, C. E., Fairclough, L., & Jarvis, A., (2003). The photofading mechanism of commercial reactive dyes on cotton. *Dyes and Pigments, 59*(3), 269–275.
15. Baxter, G., Giles, C. H., & Lewington, W. J., (1957). Relation between physical state and rate of fading of dyes. *Journal of the Society of Dyers and Colorists, 73*(8), 386–392.
16. Baxter, G., Giles, C. H., McKee, M. M., & Macaulay, N., (1955). The influence of the physical state of dyes upon their light fastness. *Journal of the Society of Dyers and Colorists, 71*(5), 218–235.
17. Beek, H. V., & Heertjes, P. M., (1966). Fading by light of organic dyes on textiles and other materials. *Studies in Conservation, 11*(3), 123–132.
18. Bideau, M., Claudel, B., Faure, L., & Kazouan, H., (1991). The photo-oxidation of acetic acid by oxygen in the presence of titanium dioxide and dissolved copper ions. *Journal of Photochemistry and Photobiology A: Chemistry, 61*(2), 269–280.
19. Blaznik, B., Kovač, F., Bizjak, G., & Bračko, S., (2022). Fastness of dye-based ink-jet printing inks in aqueous solution in the presence and absence of oxygen. *Color Research & Application, 47*(5), 1193–1199.
20. Bolland, J. L., & Cooper, H. R., (1954). The photo-sensitized oxidation of ethanol. Proceedings of the Royal Society of London. *Series A. Mathematical and Physical Sciences, 225*(1162), 405–426.
21. Bridge, N. K., (1960). The effect of environment and substituents on the photochemical activity of anthraquinonoid vat dyes and the Rôle of n-π* transitions. *Journal of the Society of Dyers and Colorists, 76*(8), 484–489.
22. Brimblecombe, P., & Shooter, D., (1986). Photo-oxidation of dimethylsulphide in aqueous solution. *Marine Chemistry, 19*(4), 343–353.
23. Brunell, M., Tignor, S., Pierson, E., & Walsh, P. L., (2022). Photo-oxidation mechanisms in liquid pharmaceutical formulations:

The overlooked role of singlet oxygen presented as a case study. *Pharmaceutical Research, 39*(10), 2529–2540.

24. Buylaert, J. P., Vandenberghe, D., Murray, A. S., Huot, S., De Corte, F., & Van den Haute, P., (2007). Luminescence dating of old (> 70 ka) Chinese loess: A comparison of single-aliquot OSL and IRSL techniques. *Quaternary Geochronology, 2*(1–4), 9–14.
25. Campbell, D. S. E., & Giles, C. H., (1958). A study of the physical state of methylene blue (CI basic blue 9) in dyed films and its relation to light fading rates. *Journal of the Society of Dyers and Colorists, 74*(3), 164–168.
26. Cao, Z. J., Tang, A., Wang, J., Zhang, Y. L., Yang, G., & Zhang, X. Q., (2022). Improvement in dyeing and physical properties of wool fabrics through pretreatment based on the bacterial culture of Stenotrophomonas maltophilia DHHJ. *Textile Research Journal, 92*(23–24), 4596–4605.
27. Carroccio, S., Puglisi, C., & Montaudo, G., (2003). Photo-oxidation products of polyetherimide ULTEM determined by MALDI-TOF-MS. Kinetics and mechanisms. *Polymer Degradation and Stability, 80*(3), 459–476.
28. Chacon, J. N., McLearie, J., & Sinclair, R. S., (1988). Singlet oxygen yields and radical contributions in the dye-sensitised photo-oxidation in methanol of esters of polyunsaturated fatty acids (oleic, linoleic, linolenic and arachidonic). *Photochemistry and Photobiology, 47*(5), 647–656.
29. Chambon, S., Rivaton, A., Gardette, J. L., & Firon, M., (2009). Reactive intermediates in the initiation step of the photo-oxidation of MDMO-PPV. *Journal of Polymer Science Part A: Polymer Chemistry, 47*(22), 6044–6052.
30. Chen, C., Li, X., Ma, W., Zhao, J., Hidaka, H., & Serpone, N., (2002). Effect of transition metal ions on the TiO_2-assisted photodegradation of dyes under visible irradiation: a probe for the interfacial electron transfer process and reaction mechanism. *The Journal of Physical Chemistry B, 106*(2), 318–324.
31. Chen, C., Ma, W., & Zhao, J., (2010). Semiconductor-mediated photodegradation of pollutants under visible-light irradiation. *Chemical Society Reviews, 39*(11), 4206–4219.
32. Chen, X., Yang, H., & Zhang, W., (2018). Simulation-based approach

to optimize passively designed buildings: A case study on a typical architectural form in hot and humid climates. *Renewable and Sustainable Energy Reviews, 82*, 1712–1725.

33. Cho, I. H., & Zoh, K. D., (2007). Photocatalytic degradation of azo dye (Reactive Red 120) in TiO2/UV system: Optimization and modeling using a response surface methodology (RSM) based on the central composite design. *Dyes and Pigments, 75*(3), 533–543.
34. Confortin, D., Neevel, H., Brustolon, M., Franco, L., Kettelarij, A. J., Williams, R. M., & van Bommel, M. R., (2010). Crystal violet: study of the photo-fading of an early synthetic dye in aqueous solution and on paper with HPLC-PDA, LC-MS and FORS. In: *Journal of Physics: Conference Series* (Vol. 231, No. 1, p. 012011). IOP Publishing.
35. Crews, P. C., (1987). The fading rates of some natural dyes. *Studies in Conservation, 32*(2), 65–72.
36. Cristea, D., & Vilarem, G., (2006). Improving light fastness of natural dyes on cotton yarn. *Dyes and Pigments, 70*(3), 238–245.
37. Cumming, J. W., Giles, C. H., & McEachran, A. E., (1956). A study of the photochemistry of dyes on proteins and other substrates. *Journal of the Society of Dyers and Colorists, 72*(8), 373–381.
38. Cumpston, B. H., & Jensen, K. F., (1995). Photo-oxidation of polymers used in electroluminescent devices. *Synthetic Metals, 73*(3), 195–199.
39. Dalsgaard, T. K., Otzen, D., Nielsen, J. H., & Larsen, L. B., (2007). Changes in structures of milk proteins upon photo-oxidation. *Journal of Agricultural and Food Chemistry, 55*(26), 10968–10976.
40. Damrauer, N. H., Cerullo, G., Yeh, A., Boussie, T. R., Shank, C. V., & McCusker, J. K., (1997). Femtosecond dynamics of excited-state evolution in [Ru (bpy) 3] 2+. *Science, 275*(5296), 54–57.
41. Davies, M. J., & Truscott, R. J., (2001). Photo-oxidation of proteins and its role in cataractogenesis. *Journal of Photochemistry and Photobiology B: Biology, 63*(1–3), 114–125.
42. De Long, M. J., Prochaska, H. J., & Talalay, P., (1986). Induction of NAD (P) H: quinone reductase in murine hepatoma cells by phenolic antioxidants, azo dyes, and other chemoprotectors: a model system for the study of anticarcinogens. *Proceedings of the National Academy of Sciences, 83*(3), 787–791.
43. Demas, J. N., Harris, E. W., & McBride, R. P., (1977). Energy transfer from luminescent transition metal complexes to oxygen. *Journal of the*

American Chemical Society, 99(11), 3547–3551.

44. Demchenko, A. P., Tang, K. C., & Chou, P. T., (2013). Excited-state proton coupled charge transfer modulated by molecular structure and media polarization. *Chemical Society Reviews, 42*(3), 1379–1408.
45. Dominoni, D. M., Carmona-Wagner, E. O., Hofmann, M., Kranstauber, B., & Partecke, J., (2014). Individual-based measurements of light intensity provide new insights into the effects of artificial light at night on daily rhythms of urban-dwelling songbirds. *Journal of Animal Ecology, 83*(3), 681–692.
46. Dong, H., Chen, G., Sun, J., Li, C., Yu, Y., & Chen, D., (2013). A novel high-efficiency visible-light sensitive Ag_2CO_3 photocatalyst with universal photodegradation performances: simple synthesis, reaction mechanism and first-principles study. *Applied Catalysis B: Environmental, 134*, 46–54.
47. Duxbury, D. F., (1993). The photochemistry and photophysics of triphenylmethane dyes in solid and liquid media. *Chemical Reviews, 93*(1), 381–433.
48. Egerton, G. S., & Morgan, A. G., (1970). The photochemistry of dyes II—some aspects of the fading process. *Journal of the Society of Dyers and Colorists, 86*(6), 242–249.
49. Eggert, (1928). Zur Quantenausbeute bei der photochemischen Zersetzung lichtempfindlicher Diazoverbindungen. *Zeitschrift für Elektrochemie und angewandte physikalische Chemie, 34*(9), 602–605.
50. El-Sayed, M. A., (1963). Spin—orbit coupling and the radiationless processes in nitrogen heterocyclics. *The Journal of Chemical Physics, 38*(12), 2834–2838.
51. Feldman, D., (2002). Polymer weathering: photo-oxidation. *Journal of Polymers and the Environment, 10*, 163–173.
52. Foote, C. S., (1968). Mechanisms of photosensitized oxidation: there are several different types of photosensitized oxidation which may be important in biological systems. *Science, 162*(3857), 963–970.
53. Franck, J., & Livingston, R., (1941). Remarks on the fluorescence, phosphorescence and photochemistry of dyestuffs. *The Journal of Chemical Physics, 9*(2), 184–190.
54. Fryer, M. J., Oxborough, K., Mullineaux, P. M., & Baker, N. R., (2002). Imaging of photo-oxidative stress responses in leaves. *Journal of Experimental Botany, 53*(372), 1249–1254.

55. Fujibuchi, T., Funabashi, N., Hashimoto, M., Abe, Y., Iimori, T., Matsubayashi, F., & Komuro, I., (2006). Examination of beam profile measurement using an imaging plate by the light exposure fading method for quality assurance of external radiation therapy. *Japanese Journal of Radiological Technology, 62*(12), 1697–1706.
56. Gardette, J. L., Mailhot, B., & Lemaire, J., (1995). Photooxidation mechanisms of styrenic polymers. *Polymer Degradation and Stability, 48*(3), 457–470.
57. Geng, H., Wang, Z., Liang, Y., Cheng, Y., & Alsaadi, F. E., (2017). Tobit Kalman filter with fading measurements. *Signal Processing, 140*, 60–68.
58. Giles, C. H., & McKay, R. B., (1963). The lightfastness of dyes: A review. *Textile Research Journal, 33*(7), 528–577.
59. Giles, C. H., & Rahman, S. M., (1961). Studies on high fastness to light in coloring matters in hydrophobic substrates: disperse dyes in cellulose acetates and polyesters. *Textile Research Journal, 31*(12), 1012–1019.
60. Giles, C. H., (1964). The fading of coloring matters. *Studies in Conservation, 9*(1), 8–26.
61. Giles, C. H., (1965). The fading of coloring matters. *Journal of Applied Chemistry, 15*(12), 541–550.
62. Giles, C. H., Baxter, G., & Rahman, S. M., (1961). Studies of high fastness to light in coloring matters in hydrophilic substrates. *Textile Research Journal, 31*(10), 831–844.
63. Giles, C. H., Walsh, D. J., & Sinclair, R. S., (1977). The relation between light fastness of colorants and their particle size. *Journal of the Society of Dyers and Colorists, 93*(9), 348–352.
64. Gómez, M. L., Previtali, C. M., Montejano, H. A., & Bertolotti, S. G., (2007). Photoreaction and photopolymerization studies on phenoxazin dyes/diphenyliodonium chloride salt. *Journal of Photochemistry and Photobiology A: Chemistry, 188*(1), 83–89.
65. Gouvea, C. A., Wypych, F., Moraes, S. G., Duran, N., Nagata, N., & Peralta-Zamora, P., (2000). Semiconductor-assisted photocatalytic degradation of reactive dyes in aqueous solution. *Chemosphere, 40*(4), 433–440.
66. Griffiths, J., & Hawkins, C., (1973). II—Evidence for the formation of singlet oxygen by phototendering vat dyes. *Journal of the Society of*

Dyers and Colorists, 89(5), 173–177.

67. Griffiths, J., (1972). II. Photochemistry of azobenzene and its derivatives. *Chemical Society Reviews, 1*(4), 481–493.
68. He, Y., Zhou, W., Wu, F., Li, M., & Wang, E., (2004). Photoreaction and photopolymerization studies on squaraine dyes/iodonium salts combination. *Journal of Photochemistry and Photobiology A: Chemistry, 162*(2–3), 463–471.
69. Heelis, P. F., (2018). The photochemistry of flavins. In: *Chemistry and Biochemistry of Flavoenzymes* (pp. 171–193). CRC Press.
70. Hendricks, S. B., & Siegelman, H. W., (1967). Phytochrome and photoperiodism in plants. In: *Comprehensive Biochemistry* (Vol. 27, pp. 211–235). Elsevier.
71. Hendricks, S. B., (1968). How light interacts with living matter. *Scientific American, 219*(3), 174–189.
72. Hihara, T., Okada, Y., & Morita, Z., (2002). Photo-oxidation and-reduction of vat dyes on water-swollen cellulose and their lightfastness on dry cellulose. *Dyes and Pigments, 53*(2), 153–177.
73. Hihara, T., Okada, Y., & Morita, Z., (2004). Relationship between photochemical properties and colorfastness due to light-related effects on monoazo reactive dyes derived from H-acid, γ-acid, and related naphthalene sulfonic acids. *Dyes and Pigments, 60*(1), 23–48.
74. Hillson, P. J., & Rideal, E. K., (1953). The Becquerel effect in the presence of dyestuffs and the action of light on dyes. *Proceedings of the Royal Society of London. Series A. Mathematical and Physical Sciences, 216*(1127), 458–476.
75. Hillson, P. J., & Rideal, E. K., (1953). The Becquerel effect in the presence of dyestuffs and the action of light on dyes. *Proceedings of the Royal Society of London. Series A. Mathematical and Physical Sciences, 216*(1127), 458–476.
76. Hou, H., Zeng, X., & Zhang, X., (2020). Production of hydrogen peroxide by photocatalytic processes. *Angewandte Chemie International Edition, 59*(40), 17356–17376.
77. Iliev, V., & Ileva, A., (1995). Oxidation and photooxidation of sulfur-containing compounds in the presence of water soluble phthalocyanine complexes. *Journal of Molecular Catalysis A: Chemical, 103*(3), 147–153.
78. Iliev, V., (2002). Phthalocyanine-modified titania—catalyst for

photooxidation of phenols by irradiation with visible light. *Journal of photochemistry and photobiology A: Chemistry, 151*(1–3), 195–199.

79. Jain, M., Buylaert, J. P., Thomsen, K. J., & Murray, A. S., (2015). Further investigations on 'non-fading'in K-Feldspar. *Quaternary International, 362*, 3–7.
80. Kamel, M., Amin, S. A., Afifi, M. S., & Tera, F. M., (1979). The relationship between light-fastness and the physical properties of simple mono-azo dyestuffs. *Textile Research Journal, 49*(5), 260–261.
81. Kars, R. H., Wallinga, J., & Cohen, K. M., (2008). A new approach towards anomalous fading correction for feldspar IRSL dating—tests on samples in field saturation. *Radiation Measurements, 43*(2–6), 786–790.
82. Kasimova, K. R., Sadasivam, M., Landi, G., Sarna, T., & Hamblin, M. R., (2014). Potentiation of photoinactivation of gram-positive and gram-negative bacteria mediated by six phenothiazinium dyes by addition of azide ion. *Photochemical & Photobiological Sciences, 13*, 1541–1548.
83. Khamis, F. M., & Arafah, D. E., (2020). Improved thermoluminescence properties of natural NaCl salt extracted from Mediterranean sea water relevant to radiation dosimetry. *European Journal of Applied Physics, 2*(3).
84. Kim, D., Ingram, M. A., & Smith, W. W., (2003). Measurements of small-scale fading and path loss for long range RF tags. *IEEE Transactions on Antennas and Propagation, 51*(8), 1740–1749.
85. Kim, J. H., Tani, Y., Matsuoka, M., & Fukunishi, K., (1999). Topochemical photoreaction of self-assembled styryldicyanopyrazine dyes and their solid-state spectral properties. *Dyes and Pigments, 43*(1), 7–14.
86. Kirkbright, G. F., Mayne, P. J., & West, T. S., (1972). Photo-oxidation of thallium (I) with the production of hydrogen peroxide. *Journal of the Chemical Society, Dalton Transactions, 17*, 1918–1920.
87. Knowles, A., & Gurnani, S., (1972). A study of the methylene blue-sensitized oxidation of amino acids. *Photochemistry and Photobiology, 16*(2), 95–108.
88. Kolb, T., Sontag-González, M., & Fuchs, M., (2022). Testing the potential of a standardized growth curve approach for improving the applicability and performance of fading correction. *Quaternary*

Geochronology, 73, 101375.

89. Konstantinou, I. K., & Albanis, T. A., (2004). TiO_2-assisted photocatalytic degradation of azo dyes in aqueous solution: Kinetic and mechanistic investigations: A review. *Applied Catalysis B: Environmental, 49*(1), 1–14.
90. Korytowski, W., & Sarna, T., (1990). Bleaching of melanin pigments. Role of copper ions and hydrogen peroxide in autooxidation and photooxidation of synthetic dopa-melanin. *Journal of Biological Chemistry, 265*(21), 12410–12416.
91. Korytowski, W., Pilas, B., Sarna, T., & Kalyanaraman, B., (1987). Photoinduced generation of hydrogen peroxide and hydroxyl radicals in melanins. *Photochemistry and Photobiology, 45*(2), 185–190.
92. Kramer, H., & Mathis, P., (1980). Quantum yield and rate of formation of the carotenoid triplet state in photosynthetic structures. *Biochimica et Biophysica Acta (BBA)-Bioenergetics, 593*(2), 319–329.
93. Lamothe, M., Auclair, M., Hamzaoui, C., & Huot, S., (2003). Towards a prediction of long-term anomalous fading of feldspar IRSL. *Radiation Measurements, 37*(4–5), 493–498.
94. Lasareff, P., (1907). Über das Ausbleichen von Farbstoffen im sichtbaren Spektrum. *Annalen der Physik, 329*(14), 661–671.
95. Lasek, J., Yu, Y. H., & Wu, J. C., (2013). Removal of NOx by photocatalytic processes. *Journal of Photochemistry and Photobiology C: Photochemistry Reviews, 14*, 29–52.
96. Lee, R. F., (2003). Photo-oxidation and photo-toxicity of crude and refined oils. *Spill Science & Technology Bulletin, 8*(2), 157–162.
97. Li, B., & Li, S. H., (2008). Investigations of the dose-dependent anomalous fading rate of feldspar from sediments. *Journal of Physics D: Applied Physics, 41*(22), 225502.
98. Li, Q., Mei, H., Zhang, Z., Jiang, H., & Zhang, W., (2022). Valorization of flavonoid-rich invasive weed-Eupatorium adenophorum in the cleaner production of coloristic and functional wool fabric with a research focus on photofading mechanism. *Industrial Crops and Products, 189*, 115835.
99. Li, Z., Kang, Z., & Zhu, Z., (2022). A photo-switch for enzymatic biofuel cells based on the photo-oxidization of electron acceptor in cathode by C-dots nanozyme. *Chemical Engineering Journal, 428*, 131258.

100. Liu, G., Wu, T., Zhao, J., Hidaka, H., & Serpone, N., (1999). Photo-assisted degradation of dye pollutants. 8. Irreversible degradation of alizarin red under visible light radiation in air-equilibrated aqueous TiO_2 dispersions. *Environmental Science & Technology, 33*(12), 2081–2087.

101. Liu, Y., Fearn, T., & Strlič, M., (2022). Photodegradation of iron gall ink affected by oxygen, humidity and visible radiation. *Dyes and Pigments, 198*, 109947.

102. Ma, J., Lv, C., Xu, M., Hao, P., Wang, Y., Shen, W., & Lv, C., (2017). Analysis of chlorophyll a fluorescence and proteomic differences of rice leaves in response to photooxidation. *Acta Physiologiae Plantarum, 39*, 1–14.

103. MacCallum, J. R., (1985). Kinetics and mechanisms of photo-oxidation processes. *Developments in Polymer Degradation, 6*, 191–205.

104. Maleki, A., Mahvi, A. H., Ebrahimi, R., & Zandsalimi, Y., (2010). Study of photochemical and sonochemical processes efficiency for degradation of dyes in aqueous solution. *Korean Journal of Chemical Engineering, 27*, 1805–1810.

105. Matsushita, M., Tran, T. H., Nosaka, A. Y., & Nosaka, Y., (2007). Photo-oxidation mechanism of l-alanine in TiO_2 photocatalytic systems as studied by proton NMR spectroscopy. *Catalysis Today, 120*(2), 240–244.

106. Memming, R., (1972). Photochemical and electrochemical processes of excited dyes at semiconductor and metal electrodes. *Photochemistry and Photobiology, 16*(4), 325–333.

107. Min, P., (2021). Design and implementation of LED single lamp dimming and color temperature smart control system for tunnel lighting. In: *2021 IEEE Conference on Telecommunications, Optics and Computer Science (TOCS)* (pp. 822–825). IEEE.

108. Mojbafan, M., Zanjanchi, F., & Taherkhani, M., (2020). The influence of physical factors on the halochromic behavior of the pH-sensitive sulfonphthaleine dyes: A DFT study. *Chemical Papers, 74*(12), 4355–4363.

109. Nam, S., Renganathan, V., & Tratnyek, P. G., (2001). Substituent effects on azo dye oxidation by the FeIII–EDTA–$H2O_2$ system. *Chemosphere, 45*(1), 59–65.

110. Newcomb, M., Johnson, C. C., Manek, M. B., & Varick, T. R., (1992).

Picosecond radical kinetics. Ring openings of phenyl-substituted cyclopropylcarbinyl radicals. *Journal of the American Chemical Society, 114*(27), 10915–10921.

111. Oster, G., Bellin, J. S., Kimball, R. W., & Schrader, M. E., (1959). Dye sensitized photooxidation. *Journal of the American Chemical Society, 81*(19), 5095–5099.
112. Pandey, A., Saini, B. S., Singh, B., & Sood, N., (2017). An integrated approach using chaotic map & sample value difference method for electrocardiogram steganography and OFDM based secured patient information transmission. *Journal of Medical Systems, 41*, 1–20.
113. Pattison, D. I., Rahmanto, A. S., & Davies, M. J., (2012). Photo-oxidation of proteins. *Photochemical & Photobiological Sciences, 11*(1), 38–53.
114. Pham, T. N., & Clarke, R. J., (2008). Solvent dependence of the photochemistry of the styrylpyridinium dye RH421. *The Journal of Physical Chemistry B, 112*(20), 6513–6520.
115. Philippart, J. L., & Gardette, J. L., (2001). Thermo-oxidation of isotactic polypropylene in 32O2–36O2: comparison of the mechanisms of thermo-and photo-oxidation. *Polymer Degradation and Stability, 73*(1), 185–187.
116. Plaetzer, K., Krammer, B., Berlanda, J., Berr, F., & Kiesslich, T., (2009). Photophysics and photochemistry of photodynamic therapy: fundamental aspects. *Lasers in Medical Science, 24*, 259–268.
117. Porter, G., & Suppan, P., (1965). Primary photochemical processes in aromatic molecules. Part 12.—Excited states of benzophenone derivatives. *Transactions of the Faraday Society, 61*, 1664–1673.
118. Radin, M. D., Vinckeviciute, J., Seshadri, R., & Van der Ven, A., (2019). Manganese oxidation as the origin of the anomalous capacity of Mn-containing Li-excess cathode materials. *Nature Energy, 4*(8), 639–646.
119. Rånby, B., (1989). Photodegradation and photo-oxidation of synthetic polymers. *Journal of Analytical and Applied Pyrolysis, 15*, 237–247.
120. Rettig, W., Majenz, W., Lapouyade, R., & Vogel, M., (1992). Adiabatic photochemistry with luminescent products. *Journal of Photochemistry and Photobiology A: Chemistry, 65*(1–2), 95–110.
121. Roger, A., Sallet, D., & Lemaire, J., (1986). Photochemistry of aliphatic polyamides. 4. Mechanisms of photooxidation of polyamides 6, 11, and 12 at long wavelengths. *Macromolecules, 19*(3), 579–584.

122. Rossow, W. B., Del Genio, A. D., Limaye, S. S., Travis, L. D., & Stone, P. H., (1980). Cloud morphology and motions from Pioneer Venus images. *Journal of Geophysical Research: Space Physics, 85*(A13), 8107–8128.

123. Rowan, R. I., & Sykes, B. D., (1974). Carbon-13 nuclear magnetic resonance study of the visual chromophores and model compounds. *Journal of the American Chemical Society, 96*(22), 7000–7008.

124. Sai, N., Leung, K., Zádor, J., & Henkelman, G., (2014). First principles study of photo-oxidation degradation mechanisms in P3HT for organic solar cells. *Physical Chemistry Chemical Physics, 16*(17), 8092–8099.

125. Schnuriger, B., Bourdon, J., & Bedu, M. J., (1968). Photosensitized oxidation through stearate monomolecular films. *Photochemistry and Photobiology, 8*(5), 361–368.

126. Scoponi, M., & Ghiglione, C., (1997). Photo-oxidation mechanisms of poly(2, 6-dimethyl-1, 4-phenylene oxide). Die Angewandte Makromolekulare Chemie: *Applied Macromolecular Chemistry and Physics, 252*(1), 237–256.

127. Shen, Z., Wang, W., Jia, J., Ye, J., Feng, X., & Peng, A., (2001). Degradation of dye solution by an activated carbon fiber electrode electrolysis system. *Journal of Hazardous Materials, 84*(1), 107–116.

128. Song, L., Hennink, E. J., Young, I. T., & Tanke, H. J., (1995). Photobleaching kinetics of fluorescein in quantitative fluorescence microscopy. *Biophysical Journal, 68*(6), 2588–2600.

129. Song, P. S., & Metzler, D. E., (1967). Photochemical degradation of flavins-IV. Studies of the anaerobic photolysis of riboflavin. *Photochemistry and Photobiology, 6*(10), 691–709.

130. Tera, F. M., Abdou, L. A., Michael, M. N., & Hebeish, A., (1985). Fading characteristics of some monoazo dyes on cellulose diacetate and polyamide films: Part II. *Polymer Photochemistry, 6*(5), 375–383.

131. Thrall, E. S., Preska Steinberg, A., Wu, X., & Brus, L. E., (2013). The role of photon energy and semiconductor substrate in the plasmon-mediated photooxidation of citrate by silver nanoparticles. *The Journal of Physical Chemistry C, 117*(49), 26238–26247.

132. Tracewell, C. A., Vrettos, J. S., Bautista, J. A., Frank, H. A., & Brudvig, G. W., (2001). Carotenoid photooxidation in photosystem II. *Archives of Biochemistry and Biophysics, 385*(1), 61–69.

133. Tsunooka, M., Cha, Y., Tanaka, M., & Konishi, F., (1984). Photochemical reactions of polymers containing thiocyanatoacetyl groups and dyeing the resulting polymer films. *Journal of Polymer Science: Polymer Chemistry Edition, 22*(3), 739–747.

134. Ushakov, E. N., & Gromov, S. P., (2015). Supramolecular methods for controlling intermolecular photocycloaddition reactions of unsaturated compounds in solutions. *Russian Chemical Reviews, 84*(8), 787.

135. Valdés, Á., Qu, Z. W., Kroes, G. J., Rossmeisl, J., & Nørskov, J. K., (2008). Oxidation and photo-oxidation of water on TiO_2 surface. *The Journal of Physical Chemistry C, 112*(26), 9872–9879.

136. Van Beer, H. C. A., & Heertjes, P. M., (1963). Photochemical reactions of azo dyes in solution with different substrates. *Journal of the Society of Dyers and Colorists, 79*(12), 661–670

137. Van Gerven, T., Mul, G., Moulijn, J., & Stankiewicz, A., (2007). A review of intensification of photocatalytic processes. *Chemical Engineering and Processing: Process Intensification, 46*(9), 781–789.

138. Visocekas, R., (1985). Tunnelling radiative recombination in labradorite: its association with anomalous fading of thermoluminescence. *Nuclear Tracks and Radiation Measurements (1982), 10*(4–6), 521–529.

139. Vlachopoulos, N., Zhang, J., & Hagfeldt, A., (2015). Dye-sensitized solar cells: new approaches with organic solid-state hole conductors. *Chimia, 69*(1–2), 41–41.

140. Vogler, A., & Kunkely, H., (1993). Charge transfer excitation of coordination compounds. Generation of reactive intermediates. *Photosensitization and Photocatalysis Using Inorganic and Organometallic Compounds*, 71–111.

141. Walz, D., (1979). Thermodynamics of oxidation-reduction reactions and its application to bioenergetics. *Biochimica et Biophysica Acta (BBA)-Reviews on Bioenergetics, 505*(3–4), 279–353.

142. Wang, L., Wang, Z., Shen, B., & Wei, G., (2020). Recursive filtering with measurement fading: A multiple description coding scheme. *IEEE Transactions on Automatic Control, 66*(11), 5144–5159.

143. Wang, S., Newland, M. J., Deng, W., Rickard, A. R., Hamilton, J. F., Muñoz, A., & Wang, X., (2020). Aromatic photo-oxidation, a new source of atmospheric acidity. *Environmental Science & Technology, 54*(13), 7798–7806.

144. Weissbein, L., & Coven, G. E., (1960). The physical state of direct dyes in viscose and its influence on lightfastness: Part I: a method of examining the physical state of direct dyes in viscose. *Textile Research Journal, 30*(1), 58–62.
145. Whitmore, P. M., Pan, X., & Bailie, C., (1999). Predicting the fading of objects: identification of fugitive colorants through direct nondestructive lightfastness measurements. *Journal of the American Institute for Conservation, 38*(3), 395–409.
146. Whitten, D. G., (1974). Photochemical reactions in organized monolayer assemblies. I. Cis-trans isomerization of thioindigo dyes. *Journal of the American Chemical Society, 96*(2), 594–596.
147. Williamson, C. L., (2015). *The Analysis of Ballpoint Inks with APCI-MS after Fading with Light, Hydrogen Peroxide and Sodium Hypochlorite Bleach*. Doctoral dissertation, University of Central Lancashire *2*(1), 1–13.
148. Winters, B. H., Mandelberg, H. I., & Mohr, W. B., (1974). Photochemical products in coumarin laser dyes. *Applied Physics Letters, 25*(12), 723–725.
149. Xu, X. R., Li, H. B., Wang, W. H., & Gu, J. D., (2004). Degradation of dyes in aqueous solutions by the Fenton process. *Chemosphere, 57*(7), 595–600.
150. Yan, X., Ohno, T., Nishijima, K., Abe, R., & Ohtani, B., (2006). Is methylene blue an appropriate substrate for a photocatalytic activity test? A study with visible-light responsive titania. *Chemical Physics Letters, 429*(4–6), 606–610.
151. Yin, J., Holland, G., Elbatt, T., Bai, F., & Krishnan, H., (2006). DSRC channel fading analysis from empirical measurement. In: *2006 First International Conference on Communications and Networking in China* (pp. 1–5). IEEE.
152. Zguris, G. C., (1996). A review of physical properties of separators for valve-regulated lead/acid batteries. *Journal of Power Sources, 59*(1–2), 131–135.
153. Zhao, W., Chen, C., Li, X., Zhao, J., Hidaka, H., & Serpone, N., (2002). Photodegradation of sulforhodamine-B dye in platinized titania dispersions under visible light irradiation: influence of platinum as a functional co-catalyst. *The Journal of Physical Chemistry B, 106*(19), 5022–5028.

CHAPTER 6

Dyeing of Textile Fiber Blends

Contents

Unit Introduction 206

6.1. Key Fiber Blends 206

6.2. Dyeing of Textile Blends 211

6.3. Main Dyeing Methods 218

6.4. Future Trends 221

6.5. Summary 226

References 227

UNIT INTRODUCTION

Blending textile fibers is performed to achieve visually pleasing color effects, reducing overall production expenses and enhancing the functional characteristics of the resultant textile goods. The selection of appropriate dyes and dyeing purposes conditions for achieving superior combined dyeing outcomes is influenced by various factors (Davidson & Taylor, 1965). These factors comprise the reaction behavior of blend constituents and their respective proportions toward various dyes and coloration conditions, the efficacy of dyebath substances, suitable dyeing tools, the growing need for innovative effects attained through blending, subsequent concluding procedures, the economy, and ecological considerations.

The present chapter pertains to dyeing mixes and is structured into four sections. These sections encompass the topics: of prominent fiber blends, significant challenges encountered in dyeing purposes blends, primary techniques employed for dyeing particular blends, and recent advancements and potential prospects concerning blends and their respective positions in dyeing properties (Egan & Salmon, 2022). Moreover, supplementary sources for data are provided.

The initial section of the text deals with textile fibers, blends of fibers, the benefits of blending, prevalent textile blends, and their possible applications. The subsequent section comprehensively summarizes pollution mitigation strategies in pre-treatment processes, categorizing dyes, dyeing additives, dyeing phases, dye-fiber interactions, chromatic impacts, and prominent dye amalgamations (Cui & Xue, 2021). The third section pertains to the various stages of the dyeing process, encompassing dye selection and appropriate techniques for dyeing some of its most significant fiber blends. The fourth section of the study examines contemporary advancements and prospective tendencies in the dyeing of blends, encompassing the evolution of fibers and fiber blends, as well as alternatives for augmenting the dyeing of blends. Ultimately, the chapter concludes by providing a list of references for readers to access additional information.

6.1. KEY FIBER BLENDS

6.1.1. Textile Fibers

The categorization of textile fibers is generally divided into two main groups: natural and synthetic fibers. The paramount natural fibers comprise cellulosic fibers, such as fibers made from cotton, flax, jute, or sisal; protein fibers, including wool, silk, and many more; and mineral fiber, exemplified by asbestos (Ugbolue, 1990). Conversely, fibers that are produced by human intervention can be categorized into two groups: regenerated fibers, which can be either unaltered (such as rayon and lyocell) or altered (such as acetate of cellulose along with triacetate), and synthetic fibers, which include polyester fibers polyamide, polyacrylic, polyurethane paint and others. There are also refractory and manufacturing fibers, such as ceramic, carbon, and glass. The chemical composition of the primary textile fibers is illustrated in Figure 6.1.

6.1.2. Fiber Blending

The primary objective of fiber blending is to create a uniform and cohesive single yarn that combines two or more fiber types, whether similar or dissimilar. This process is intended to address the individual fiber components' limitations while offering a wide range of options that would otherwise be unattainable (Friele, 1952). The result is a diverse array of products that enrich our daily experiences.

The degree of enhancement in the characteristics of blended yarn is determined by various factors such as the nature of the fiber, the proportion of blending, and the method of blending. These methods may include utilizing distinct filaments or stable fibers within the same yarn, combining different yarns, incorporating them as warp and weft within the same material, or utilizing core spun yarns for knitting.

6.1.3. Advantages of Blending

The textile industry (Numata et al., 2001). The rising need for fiber blending has been observed to affect the textile sector favorably indicated that:

- One potential strategy for attaining economic benefits is amalgamating high-cost fibers with lower-cost alternatives.
- The efficacy, as well as the characteristics, can be improved by synergistically combining the desirable characteristics of both fiber components.
- The creation of innovative textile products to fulfill market needs.
- One potential approach to enhancing the longevity of textile products is integrating a more resilient fiber component.

6.1.4. Most Common Textile Blends and Potential Uses

Providing comprehensive information regarding the diverse textile blends and their potential permutations within a single chapter would be unfeasible. The examples provided in Table 6.1, which outline the key textile blends and their potential applications, are inherently limited in scope (Patti & Acierno, 2022). Alternatively, the appropriate selection of fiber blend constituents, blending proportion, and blending methodology can guarantee the creation of binary form or ternary textile mixes with improved properties and extended applications, such as for sports, recreational activities, style, apparel, safety gear, households textiles, and so on.

Table 6.1. Major Textile Blends and Their Suitable Uses

Blend	Current Uses
(A) Binary Blends	
The topic under consideration pertains to the comparison between materials that are of human origin and those that occur naturally.	
The fabric in question is a woven blend of polyester and cotton, with a composition ratio of 67% polyester.	The three categories of clothing items under consideration are shirting, outerwear, and workwear.
The fabric in question is a woven blend of polyester and linen (Ibrahim, 2011).	Fashion merchandise, dining utensils, and bedding linens.
The fabric in question blends polyester and cotton, specifically in a knitted form.	The items mentioned included sportswear, T-shirts, and dress wear.
The fabric composition under consideration is a woven blend of acrylic and cotton fibers, with a ratio of 80/20 or 50/50.	Attire is characterized by its low weight, including suits, sports clothing, and formal wear.
The textile option referred to is modacrylic/cotton or viscose blends with varying ratios of 80/20 or 50/50 and a knitted form.	Intimate apparel, legwear, and casual attire.

The present study investigates the properties of cellulose viscose/cotton acetate/cellulosic blends with varying ratios of (30/70) and (35/65, 50/50).	Attire comprises of suits, skirts, and dress wear. The items above include poplin shirts, tops, dresses, knitwear, and terry towels.
Towels are nylon and cotton with staple lengths ranging from 10 to 50 for nylon and 90 to 50 for cotton.	Leisure shirts are commonly worn as dress attire.
The materials under consideration are nylon filament bags and cotton piles, which belong to the categories of man-made artificial natural (protein) materials, respectively (Harmsen et al., 2021).	The textile product in question is a type of toweling produced using a weft knitting.
The fabric composition under consideration is a blend of polyester and wool, specifically in the ratios of 80/20 and 55/45 (40/60)	Fabrics suitable for suits. High-end automotive textiles.
The fabric composition in question is a blend of nylon and wool, with a ratio of 20% nylon.	Woven carpets are a type of textile flooring produced by interlacing yarns or fibers.
A blend of acrylic and fine wool within a woven fabric with a composition ratio of either 50/50 or 80/20.	Garments
The material in question is a blend of acrylic and coarser wool (Kalliala & Talvenmaa, 2000).	A blanket is a type of textile material commonly used as a covering for warmth or comfort, while a floor covering refers to any material used to cover the floor.
Viscose/wool	Attire that is lightweight, such as suits, dress wear, and knitwear.
Lyocell/wool	Summer attire that is suitable for sports and leisure activities.

The fabric composition in question is a blend of polyester and viscose, with a ratio of 50:50.	Textile fabrics are used for making clothing.
Polyester/polynosic	Tubular-knitted textiles and clothing items.
Polyester/vinyl	Rainwear
Acrylic/viscose	Carpets that are not prone to crushing.
The composition of the material in question ranges from 85% polyester and 15% acrylic to 70% polyester and 30% acrylic.	Textile fabrics that are either woven or knitted are commonly used as upholstery and sportswear materials.
The fabric under consideration is composed of a blend of natural cotton and wool fibers in an 80/20 ratio, with the cotton and wool.	Garments designed to be laundered and reused multiple times and are commonly referred to as washable apparel. The items mentioned earlier include blazer garments, rain-resistant attire, tops, and sleepwear.
Jute/cotton	The textile providing and apparel business sectors.
A blend of wool and ramie in the proportion of 70:30.	Dress fabrics
Wool/silk	Apparel of superior quality.
Silk/cellulose	The three categories of fabrics under consideration are upholstery, furnishing, and shirting fabrics.
B. Ternary Blends	
The material composition of the fabric in question includes a blend of nylon, the polyurethane (10–20%)	Knitted undergarments.
Nylon/wool/viscose	Carpets

The fabric composition in question is a woven blend of triacetate, nylon, and viscose, with a ratio of either 60/20/20 or 55 (Valera et al., 2006).	The attire in question consists of a shirt and a pleated dress.
The blended staple yarn is composed of polyester, wool, and acrylic.	The attire is comprised of overcoats and suits.
The fabric in question is composed of a polyester/cotton blend for the warp (65/35) and an acrylic/cotton blend for the weft (80/20).	The subject matter at hand pertains to household textiles.
The composition of the material in question is a blend of acetate of, cellulose, and polyester, which is nylon in a ratio of 25:50.	Fabric is suitable for dress wear.
The composition of the material in question is a blend of cellulose triacetate, polyester yarn, and viscose, with a ratio of either 65/15/20 or 30/40/30 (Erbil et al., 2018).	Fabrics are suitable for dress wear.

Source: D. M. Levis, Creative Commons License.

6.2. DYEING OF TEXTILE BLENDS

6.2.1. Pre-Treatment Processes

The production of textile goods involves several key phases, including the creation of yarn through spinning, the creation of fabric through weaving or the craft of knitting, the process of wet processing, which encompasses pretreatments, dyeing purposes, and/or the printing press, functioning finishing, and textiles fabrication through cutting and stitching (Chen et al., 2008). To achieve superior dyeing results, it is imperative to uniformly eliminate all impurities, including those that are organic, inherent, introduced for an improved spin or weave capacity, and any accidental contaminants. This process must also eradicate any residual coloring agents to prevent complications and enhance the responsiveness of already treated substrates to afterward wet-treatment techniques, including dyeing, printing, and chemical-based finishing. The production of textile goods involves several key phases, including the creation of yarn through spinning, the creation of fabric through loom weaving or the craft of knitting, dry processing, which encompasses pretreatments, dyeing purposes, and/or the printing press, functioning finishing, and textiles fabrication through cutting and stitching. To achieve superior dyeing results, it is imperative to uniformly eliminate all impurities, including those that are organic, inherent, introduced for an improved spin or weave capacity, and any accidental contaminants (Ujhelyiova et al., 2007). This process must also eradicate any residual coloring agents to prevent complications and enhance the responsiveness of already treated substrates to afterward wet-treatment techniques, including dyeing, printing, and chemical-based finishing.

The effective elimination of pollutants and impurities in textile fibers is crucial in preventing common dyeing issues, including insufficient shading, blotchy dyeing, poor fastness properties, washing off, filtering in package dyeing, and unlevel dyeing. To achieve this, it is imperative to utilize high-quality water for processing and select appropriate textile auxiliaries. Without a doubt, a superior pretreatment process is a crucial prerequisite for achieving uniform dyeing results, regardless of the type of material being dyed. In contrast to natural and renewed textile fabrics, artificial fiber-containing fabrics necessitate a distinct preparatory procedure, namely heat setting, to confer a significant level of dimensional rigidity while preventing the formation of folds and creases during afterward wet processing (Ujhelyiova et al., 2007). The efficacy of stain, grease, and designing agent removal while performing the scouring procedure may be reduced by implementing heat-establishing scouring. An inadequate heat setting may result in suboptimal fabric stability, whereas a high-temperature establishment may lead to discoloration, stiffness, and/or reduced elasticity. Regarding environmental concerns, the focus on preventing pollution opportunities about pretreatments is emphasized:

- Effective management of household tasks.
- The process of ensuring the high standard of incoming materials that are raw, such as fibers, yarns, and fabrics.
- The practice of conducting preliminary assessments on chemicals before their utilization.
- The rationalization, optimization, and alteration of processes.
- One potential strategy for reducing the negative environmental impact of chemical substances involves substituting dangerous

chemicals with more environmentally friendly alternatives. For example, peroxide of hydrogen could be used in place of sodium hydroxide; biodegradable compounds could replace not recyclable ones, and specific enzymes such as α-amylase, an enzyme known as cellulases, lipases, and proteases could be utilized as substitutes for harsh chemicals (Said et al., 2020).

- On-site recovery methods such as size, cola, wool wax, and heat recovery are utilized.
- The seventh option for sustainable water management is on-site reuse, which involves utilizing clean washing water following H_2O_2 bleaching and reusing dye baths.
- Implementing on-site recycling practices for both reworkable and non-reworkable materials, including hard wastes, is recommended.
- The selection of machinery is a crucial aspect of various industrial processes. Examples of such machinery include opposition washing, inadequate material-to-liquor proportion systems, and ongoing ranges.
- Training of employees with a focus on practices aimed at preventing pollution.

6.2.2. Dye Classification

Chemical composition and method of application:

- The chemical structures of various dyes, such as azo dyes, xanthene dyes, thioindigo dyes, anthrax dyes, and indigo, are of interest in this context (Ajmal et al., 2014).
- The method of dye application can vary and may include reactive, regulated, acid, basic, vat, and dissipate dyes.

6.2.3. Dyeing Auxiliaries

To achieve dyeing of superior quality, it is imperative to employ dyeing auxiliaries with precision. The possible roles of dyeing auxiliary products, such as surfactants, electrolytes, oxidants, reductants, buffers, defoamers, storing and soluble agents, leveling agents, delaying agents, and anti-precipitants, encompass the following (Aleboyeh et al., 2012).

- Enhancing the efficacy of pre-treatment procedures for afterward dyeing.
- Augmenting the degree of dye absorption and fixation.
- The maintenance of dyeing shower stability and integration of its constituents is imperative.
- Altering the textile substrates and preventing their impairment.
- Enhancing the fastness of real estate of the resultant dyeings.
- Mitigating the ecological consequences of wet processing within the textile industry, with a particular emphasis on dyeing (Gao et al., 2014).

6.2.4. Dyeing Stages

The dyeing textile substrates have four distinct stages (Lara et al., 2022).

- The process of conveying dye molecules via a solution to the surface of a substrate.
- The phenomenon of dye molecules adhering to the surface of a substrate.
- The pigment diffusion or permeation from the outermost layer of the substrate into the inside of the fiber occurs via its amorphous regions.
- The dye is affixed throughout the substrate via various mechanisms, including ion exchange, covalent bonds, hydrogen bonds, van der Waals forces, or dissolution of the pre-dissolved pigment within the fiber.

6.2.5. Dye Fiber Bonds

The degree of dye fixation and the durability of the resulting dyeings are influenced by how the dye molecules interact with the textile fibers, which may occur through any one of the subsequent mechanisms:

- Ionic bonding is observed in various instances, such as the interaction between acid and basic dyes with protein fibers (Gorenšek, 1999).
- The topic of discussion pertains to covalent bonding, specifically in the context of reactive dyes and their interaction with cellulosic fibers.

- One example of a type of bonding that occurs between direct dyes and cellulosic fibers is hydrogen bonding.
- The process of insolubilization from pre-dissolved pigment is after their penetration into fibers, such as vat and sulfate dyes in cellulosic fibers.
- The absorption of dye vapors, specifically volatile dispersed dyes used for polyester, is being considered.

The dyeing process is influenced by various factors such as the type of fiber and shape of the material, choice of dyestuff, the amount of textile auxiliaries, dyeing situations, and the equipment used for dyeing (Luttringer, 1993). The efficacy of the dyeing procedure is contingent upon attaining the appropriate hue, uniform dispersion of dye molecules across the substrate, superior resistance to fading, and the absence of any harm to the textile material.

6.2.6. Color Effects Obtained by Dyeing Binary Blends

The process of dyeing textile blends results in the emergence of four primary categories of color effects:

- The solid effect is observed when both fiber components exhibit identical coloring.
- The reserved effect can be achieved by selectively dyeing a single element while attempting to maintain the other fiber in its natural white state (Samanta et al., 2009; Van Tran et al., 2009).
- The shadow effect refers to a visual phenomenon where both components exhibit identical hue and brightness yet differ in the intensity of their shades.
- The contrast effect refers to using contrasting hues to dye both fiber components.

This text discusses the primary dye combinations utilized in binary and ternary blends. Table 6.2 illustrates the categorization of binary combines amenable to dyeing with the principal dye combinations to achieve complete and rapid dyeing.

Table 6.2. Recommended Dye Combinations for Proper Dyeing of Most Important Fiber

Binary Blends	Dye Type Combinations
AA blends, such as… The wool/silk fiber is a textile material composed of wool and silk fibers (Körbahti et al., 2011). The combination of nylon and wool fibers is a commonly used in various applications. The wool/polyurethane fiber is a composite material of wool and polyurethane. The combination of nylon and polyurethane fibers.	Acid dyes
The combination of AB blends, such as wool and basic-dyeable polyester yarn, is being considered.	Acid dye/basic dye
The combination of nylon and acrylic fibers. The textile material in question is a blend of silk and acrylic fibers. Polyurethane/acrylic fiber is a composite material that combines the properties of polyurethane and acrylic fibers.	Acid dye/basic dye
AC blends, such as. The combination of wool and cellulosic fibers. Acid and cellulosic dyes are two dyes commonly used in the textile industry (Iqbal et al., 2021). The combination of silk and cellulosic fiber is an interest in academic research. The combination of nylon and cellulosic fibers is commonly used in the textile industry. Bers composites have been extensively studied for their potential applications in various fields.	The utilization of acid/cellulosic dye on wool/cellulosic fiber is a subject of interest in the academic community.
Using CB blends, such as the combination of cellulosic/acrylic and modacrylic fiber, is a prevalent practice in the textile industry.	The topic of interest pertains to the differences between cellulosic and basic dyes.
The CC blends, such as cellulosic/cellulosic fiber, are a noteworthy topic in the field.	Cellulosic dye refers to a type of dye that is specifically designed to be used on cellulosic fibers.

The combination of cellulose acetate and protein fiber is an example of a DA blend. The combination of polyester and protein fibers is a subject of interest in textile engineering.	Disperse dye/acid dye
DB blends, such as those composed of polyester/acrylic and modacrylic fibers (Iqbal et al., 2011). The present study focuses on cellulose acetate's properties and characteristics and triacetate/acrylic fiber.	Disperse dye/basic dye
DC blends, such as cellulose acetate and triacetate/cellulosic fiber. The combination of polyester and cellulosic fibers is a commonly utilized textile blend.	Disperse dye/cellulosic dye
The utilization of DD: DD Blenre used lose acetae/tacetate fusedtion of acetates of cellulose or triacetate with polyester is being considered.	Disperse dyes

Source: J. Bone, Creative Commons License.

Furthermore, the careful choice of fibers and their respective proportions in three-component mixtures, meticulously choosing dyes and dye arrangements, and adherence to optimal dyeing conditions are crucial factors in achieving innovative outcomes (Azbar et al., 2004; Hai et al., 2007). Table 6.3 compiles valuable ternary blends alongside their corresponding dye combinations.

Table 6.3. Some Useful Ternary Blends and Their Proper Dye Combination

Binary Blends	**Dye Type Combinations**
The AAA blends consist of a combination of nylon, wool, and polyurethane materials.	Acid dyes
Blends of AAB, Nylon/Wool/Acrylic, and Polyurethane/Wool/Basic-Dye Able Polyester are commonly used in textile manufacturing (Yang et al., 2013; Harillo-Baños et al., 2020).	Acid dye/basic dye
The AAC blends comprise various combinations such as nylon/wool/cotton as well as viscose and nylon/polyurethane/cotton.	Acid/acid/cellulosic dyes
The CBA blends refer to the combination of cotton, acrylic, and nylon fibers.	Cellulosic/basic/acid dyes
The DAA blends comprise cellulosic acetate or triacetate, nylon, and wool. Another blend consists of polyester, nylon, and wool.	Disperse/acid/acid dyes

The blends as mentioned above are commonly used in the production of DAC fabrics. These blends consist of various combinations of polyester, cellulose acetate, viscose, wool, cotton, nylon, polyurethane, and linen.	Disperse/acid/cellulosic dyes
The DBA blends combine polyester, acrylic, and wool fibers.	Disperse/basic/acid dyes
The DBC blends combine polyester, acrylic, textiles, or viscose materials.	Disperse/basic/cellulosic dyes
The topic of discussion pertains to the blending of DDA (Cabanillas-Gonzalez et al., 2003). The material is composed of cellulose acetate, polyester, and nylon.	Disperse/disperse/acid dyes
The topic of interest pertains to the combination of DDC blends. The utilization of cellulose triacetate, polyester fibers, cotton, or viscose has been observed in various applications.	Disperse/disperse/cellulosic dyes

Source: M. R. Christie, Creative Commons License.

6.3. MAIN DYEING METHODS

The dyeing method can be executed at various stages of textile manufacturing, including but not limited to staple fiber, yarn, fabric, and pieces (Prabu & Sundrarajan, 2002; Van der Schueren & Clerck, 2012). Dyeing machinery can be categorized based on specific criteria.

6.3.1. Type of Textile Material to Be Processed

- A loose-stock coloring machine is utilized for dyeing fibers.
- Using Hank/package/rope dyeing devices for yarns is a prevalent practice in the textile industry.
- The utilization of jigger, winch, jet, beam, and cushioning mangle coloring machines is common in the dyeing process of fabrics.
- Using rotating drums/side paddles for coloring garments is a prevalent technique in the textile industry.

6.3.2. Processing Methods

- The discontinuous method can be carried out in rope form using a hook, jet, or inaccessible width using a jigger (Misnon et al., 2004).

- Semi-continuous systems operate intermittently, with periods of continuous operation interspersed with periods of downtime or maintenance.
- Continuous systems designed for high production rates.

The semi-continuous variation for fabric dyeing is commonly characterized by a set of preferred sequences that have gained popularity among practitioners in the field:

Pad → cold batch → wash

Pad → preheating → hot batch → wash

Pad → dry → jigger development

The prevalent sequence utilized for the ongoing pad-steam fluctuation in the dyeing process is commonly acknowledged:

Pad → dry → pad → steam fixation → wash

In contrast, the thermos dyeing purposes process for fabrics featuring polyester comprises the subsequent stages:

Padding → pre-drying (using IR-pre-heater to avoid migration) → final drying (using steam cylinders) → dye fixation at 200–220°C for 30–60 s → washing.

6.3.3. Operating Principles

- Systems for circulating dyebaths;
- Systems for moving textile materials;
- The circulation of the dyebath and the movement of the textile material are essential systems (Paosangthong et al., 2019; Eyvaz et al., 2017).

6.3.4. Process Conditions

The categorization of systems as either closed (operating under high pressure and high temperatures) or open depends on the material's nature and the intended procedure.

6.3.5. Selection of Dyeing Methods

The selection of dyeing techniques for specific blends to ensure optimal dyeing outcomes is influenced by various factors:

- The objective is to minimize the utilization of natural resources and chemical consumption, thereby reducing the pollution load (Sano et al., 2005).
- The required color effects and level of fastness real estate of the resulting dyeings.
- The durability of the acquired dyeings with regards to subsequent finishing procedures.
- The influence of dye type and coloring equipment, specifically batch, semi-continuous, and/or ongoing ranges, on the outcome, is being examined.
- The economic and ecological expenses associated with the complete dyeing procedure.

6.3.6. Selection of Dyeing Machineries

To ensure the appropriate selection of dyeing equipment, it is imperative to consider various factors:

- The safeguarding of the textile material.
- Achieving consistent permeation, uniform dyeing, increased dye accumulation, and optimal dye fixation (Kumar et al., 2019; Perkins, 1991).
- The Importance of Achieving Correct Dyeing Results on the First Attempt and Ensuring Reproducibility in Dyeing Processes.
- The Ecological and Economic Aspects.

6.3.7. Dye Selection and Dyeing Methods for Some of the Most Important Textile Blends

The procedure of blend dyeing presents a significant challenge owing to variations in the properties of blend constituents, color substantivity, dyeing conditions, types of dyeing auxiliaries employed, and the dyeing process itself, which may be batch, semi-continuous, or continuous. As previously noted, the classification of textile fibers according to their dye substantivity may be divided into four main categories (Baban et al., 2010).

These categories include:

- **Group A:** The utilization of acid and pre-metalized acid-based colors for the purpose of coloring wool, silk, polyester, and polyurethane components is facilitated through the mechanism of ionic interactions, as well as non-polar forces called van der Waals.
- **Group B:** Basic dyes for acrylics and modacrylics, cationic dyeable polyesters via electrostatic (Coulombic) bond formation.
- **Group C:** Various types of cellulosic dyes, such as reactive in nature direct, vat, and sulfur, among others, are utilized to establish both physical and/or covalent links.
- **Group D:** Hydrophobic bonding is utilized to disperse dyes for various synthetic fibers such as polyester, acetate of cellulose, nylon, and polyurethane compounds.

Achieving solid shades is the most challenging task regarding blend dyeing, irrespective of the blend constituents (Shang, 2013; Nunes-Pereira et al., 2015).

6.4. FUTURE TRENDS

Dyeing facilities have been compelled to enhance their dyeing techniques and merchandise to achieve superior dyings with unique attributes, diminish effluent burdens, and consequently mitigate the harmful effects of dyeing. Additionally, these facilities must comply with regulations and maintain a competitive edge in a market with increasingly narrow profit margins. In recent years, significant endeavors have been undertaken to enhance innovation within the dyeing industry (Kalaoğlu et al., 2003). The primary focus of advancements in blend dyeing has been improving fiber blends and auguring their dyeability.

6.4.1. Development of Fiber Blends

- The present study aims to introduce a high-performance fiber that utilizes advanced technology and science (S&T).
- The present study aims to introduce the concept of biodegradable cellulose fibers.
- The present study concerns the implementation of nano-structured fibers with enhanced functionality.

- The research and innovation in synthetic fibers have led to their potential use as a substitute for natural fibers (Ibrahim et al., 2019).
- The present study involves the solid-phase merging method for preparing polymer blends comprising polysaccharide-based polymers that are natural and artificial.
- This text pertains to producing and processing sustainable fibers, including but not limited to lyocell, organic cotton, bamboo, polylactide acid, fiber, and organic wool.

The current state of fiber blending technology necessitates further technical advancements to achieve consistent and refined yarns from diverse fiber types. This is crucial for producing blend fabrics that exhibit superior physiomechanical and dyeing characteristics while also considering the economic or ecological elements and customer demands.

6.4.2. Enhancing Dyeability of Fiber Blends

Efforts are currently underway and anticipated to continue in the realm of study and creation to improve the dyeability of textile blends:

- Clean methods such as plasma and laser therapy have been employed to alter the surface and dyeing characteristics of various textile fibers, including, complicated polyolefin, polyester, polyamides, and flax (Lewis & Mcllroy, 1997; Ujhelyiova et al., 2007).
- The enhancement of dyeability alongside acid dyes and disperse/reactive colorants in one bath has been reported through the surface alteration of silk fibers using chitosan and the prior treatment of polyester/cotton clothes employing chitin biopolymers.
- The present study explores the potential of incorporating specific additives, including α-cyclodextrin, into the dyeing creation of polyamide/wool cellulose combined using acid dyes, along with the use of a high bursting swelling agent, namely polyethylene glycol, within a single-stage dyeing process for polyester/cotton blends alongside dispersed colored pigments (Zhang et al., 2019).
- A recent study has presented a new method for the union disperse dyeing along with ultra-violet protection of cotton/polyester blend. This approach involves alkali-stable disperse dyes and

monochlorotriazine ß-cyclodextrin (MCT-ß-CD), as illustrated in Figure 6.2).

- The coon part of a blend can be permanently modified by utilizing hydrophobic aromatic substances such as cyclohexyl carbodiimide and benzoyl chloride (Egan & Salmon, 2022; Kaliyamoorthi & Thangavelu, 2015). This modification enables the one-bath coloring of polyester/cotton to combine with dispersed or reactive spread dyes.
- Using conventional receptive disperse dyes in the acidic one-step disperse/reactive dyeing purposes process of polyester/cotton blends has resulted in significant reductions in resource consumption, including supplies, water, energy, and manufacturing time. Furthermore, this approach has led to the environmental footprint of the dyeing process.
- The efficacy of enzymatic procedures in augmenting the hydrophilicity and dyeability of textile fibers has been assessed (Bechtold et al., 2003). In contrast, a study has reported the utilization of NADH-dependent reductase via Bacillus subtilis as a substitute for sodium dithionite in the indigo dyeing process of polyamide 6 along with polyamide 6,6. This approach includes mediator 1,8-dihydroxy 9 and 10-anthraquinones as an antioxidant, which is environmentally sustainable. The study also demonstrated the successful reusability of the enzyme, as depicted in Figure 6.1.
- Supercritical atmospheric carbon dioxide dyeing purposes electronic devices have been established as a sustainable and alternative dyeing method to improve the dyeability of synthetic and natural textiles using dispersed reactive dyes (Lawler et al., 2003). Nonetheless, the economic feasibility of this procedure has been diminished due to the exorbitant expenses associated with the necessary equipment.

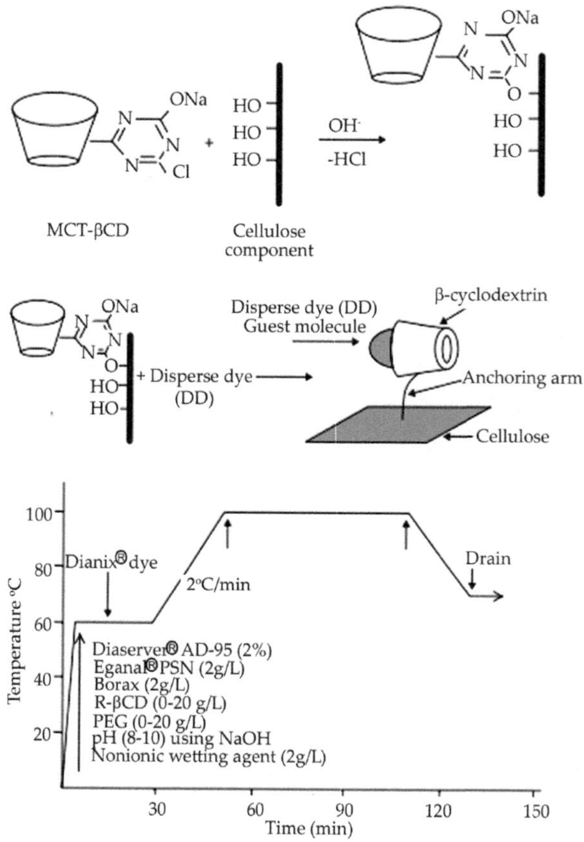

Figure 6.1. *Mechanism of modification of cellulose component and dyeing profile of cotton/polyester blend.*

Source: M. Clark, Creative Commons License.

- A report has been published on an eco-friendly substitute for the dangerous textiles auxiliaries utilized in the dispersed dyeing for polyester and nylon, yarn dyeing, metal intricate dye application, and dyeing of polyester-wool combinations (Amor et al., 2021; Ahmed & An, 2018).
- The optimization of dye exhaustion, fixation, and wash-off processes while minimizing the use of salt and reusing the dyebath has been assessed. The results indicate that these practices do not negatively impact product quality or reproducibility while reducing the environmental impact of dye-house effluent (Ibrahim et al., 2011; Correia et al., 1994).

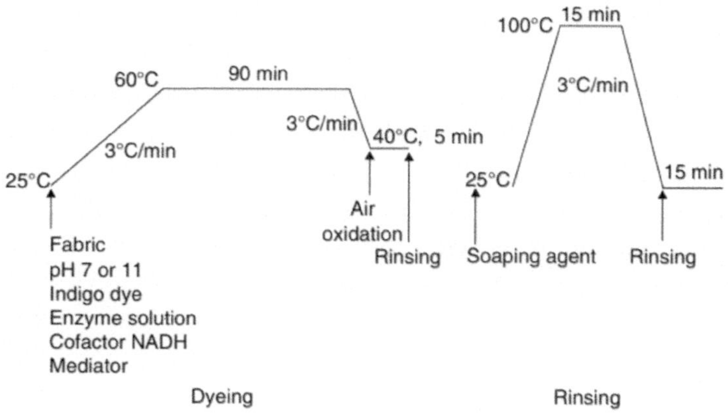

Figure 6.2. *Mediated enzymatic indigo dyeing and soaping processes.*

Source: N. A. Ibrahim, Creative Commons License.

The present study investigates the mechanism of cellulose element alteration and the dyeing persona of a cotton/polyester blend (Li et al., 2014; Božič et al., 2009).

- The aim is to create universal dyes, specifically reactive disperse dyes, which can effectively dye fibers of varying chemical compositions. This would simplify the manufacturing process and facilitate one-bath blend dyeing, such as cotton/acrylic, polyester/cotton, silk/nylon, silk/polyester, and wool/polyester (Božič et al., 2008).
- The preparation of micro-encapsulated disperse dyes, wherein the disperse dyes are utilized to be the center component, and diphenylmethane-4,4-diisocyanate is employed as the wall-former to form the capsules. The resulting capsules were characterized and found to be suitable for dyeing PET during high temperatures lacking the requirement of clear reduction post-dyeing. According to research findings, the dyeings acquired through this technique exhibited similarity to those achieved through the conventional dispersed dyeing approach (Ibrahim, 2011; Abadulla et al., 2000). After the dyeing process, the polyurea microcapsules undergo separation from the dyebath to facilitate the recovery and subsequent reuse of the spent dyebath.
- Nano-disperse dyes were synthesized using oil/water nano-emulsions, specifically sodium lauryl sulfate/caprylic triglyceride,

while subjected to ongoing ultrasound therapy. The study revealed that the microfiber demonstrated elevated K/S values compared to conventional polyester, indicating superior dyeing efficacy on the ultra-microfiber polyester (Arputharaj et al., 2016).

Nonetheless, certain methods mentioned earlier are still undergoing growth due to scientific and economic constraints. The potential outcomes of these options cannot be speculated upon after they have been carried out on a large scale by dye houses (Xu & Salmon, 2008).

6.5. SUMMARY

Deying of textile blends involves the process of coloring or staining fibers that are composed of two or more different types of fibers. Blends are commonly made by combining natural fibers like cotton, wool or silk, or linen eith synthetic fibers such as polyester, nylon or acrylic. Dyeing these blends requires careful considerations of the different fiber types and their individual dyeing properties. It is recommended to refer to specific dyeing guidelines, consult professionals or conduct further research to ensure successful and satisfactory results.

REFERENCES

1. Abadulla, E., Tzanov, T., Costa, S., Robra, K. H., Cavaco-Paulo, A., & Gübitz, G. M., (2000). Decolorization and detoxification of textile dyes with a laccase from *Trametes hirsuta. Applied and Environmental Microbiology*, *66*(8), 3357–3362.
2. Ahmed, M. T., & An, S. K., (2018). Efficient dyeing mechanism of cotton/polyester blend knitted fabric. *Fibers and Polymers*, *19*, 2541–2547.
3. Ajmal, A., Majeed, I., Malik, R. N., Idriss, H., & Nadeem, M. A., (2014). Principles and mechanisms of photocatalytic dye degradation on TiO_2 based photocatalysts: a comparative overview. *Rsc Advances*, *4*(70), 37003–37026.
4. Akrman, J., & Přikryl, J., (1997). Dyeing behavior of polypropylene blend fiber. II. Ionic exchange mechanism of dyeing. *Journal of Applied Polymer Science*, *66*(3), 543–550.
5. Aleboyeh, A., Kasiri, M. B., & Aleboyeh, H., (2012). Influence of dyeing auxiliaries on AB74 dye degradation by UV/H_2O2 process. *Journal of Environmental Management*, *113*, 426–431.
6. Amor, N., Noman, M. T., & Petru, M., (2021). Classification of textile polymer composites: recent trends and challenges. Polymers, 13(16), 2592.
7. An, Y., Miao, J., Fan, J., Li, M., Shao, M., & Shao, J., (2021). High-efficiency dispersant-free polyester dyeing using D5 non-aqueous medium. *Dyes and Pigments*, *190*, 109303.
8. Arputharaj, A., Raja, A. S. M., & Saxena, S., (2016). Developments in sustainable chemical processing of textiles. *Green Fashion: Volume 1*, 217–252.
9. Arroyo-Figueroa, G., Ruiz-Aguilar, G. M., Cuevas-Rodriguez, G., & Sanchez, G. G., (2011). Cotton fabric dyeing with cochineal extract: influence of mordant concentration. *Coloration Technology*, *127*(1), 39–46.
10. Arslan-Alaton, I., & Alaton, I., (2007). Degradation of xenobiotics originating from the textile preparation, dyeing, and finishing industry using ozonation and advanced oxidation. *Ecotoxicology and Environmental Safety*, *68*(1), 98–107.
11. Aspland, J. R., (1992). Disperse dyes and their application to polyester. *Textile Chemist and Colorist*, *24*, 18–18.

12. Azbar, N. U. R. I., Yonar, T., & Kestioglu, K., (2004). Comparison of various advanced oxidation processes and chemical treatment methods for COD and color removal from a polyester and acetate fiber dyeing effluent. *Chemosphere, 55*(1), 35–43.
13. Baban, A., Yediler, A., & Ciliz, N. K., (2010). Integrated water management and CP implementation for wool and textile blend processes. *CLEAN–Soil, Air, Water, 38*(1), 84–90.
14. Bechtold, T., Burtscher, E., & Hung, Y. T., (2004). Treatment of textile wastes. In: *Handbook of Industrial and Hazardous Wastes Treatment,* 1(pp. 412–452). CRC Press.
15. Bechtold, T., Turcanu, A., Ganglberger, E., & Geissler, S., (2003). Natural dyes in modern textile dyehouses—how to combine experiences of two centuries to meet the demands of the future? *Journal of Cleaner Production, 11*(5), 499–509.
16. Božič, M., & Kokol, V., (2008). Ecological alternatives to the reduction and oxidation processes in dyeing with vat and sulur dyes. *Dyes and Pigments, 76*(2), 299–309.
17. Božič, M., Kokol, V., & Guebitz, G. M., (2009). Indigo dyeing of polyamide using enzymes for dye reduction. *Textile Research Journal, 79*(10), 895–907.
18. Bright, L., (1996). Trouble shooting continuous thermosol dyeing of polyester fiber and blends. *American Dyestuff Reporter, 85*(8), 60–61.
19. Cabanillas-Gonzalez, J., Yeates, S., & Bradley, D. D. C., (2003). Solar cells from thermally treated polymer/dye blends with good spectral coverage. *Synthetic Metals, 139*(3), 637–641.
20. Chavan, R. B., (2011). Environmentally friendly dyes. *Handbook of Textile and Industrial Dyeing*, 515–561.
21. Chen, C. H., Wang, F. Y., Mao, C. F., Liao, W. T., & Hsieh, C. D., (2008). Studies of chitosan: II. Preparation and characterization of chitosan/poly (vinyl alcohol)/gelatin ternary blend films. *International Journal of Biological Macromolecules, 43*(1), 37–42.
22. Chen, Q., & Tang, X., (2021). Fiber art and its application in illustration creation. In: *2nd International Conference on Language, Art and Cultural Exchange (ICLACE 2021)* (pp. 538–543). Atlantis Press.
23. Correia, V. M., Stephenson, T., & Judd, S. J., (1994). Characterization of textile wastewaters-a review. *Environmental Technology, 15*(10), 917–929.

24. Cowger, C., & Weisz, R., (2008). Winter wheat blends (mixtures) produce a yield advantage in North Carolina. *Agronomy Journal, 100*(1), 169–177.
25. Crone, H. R., (1952). Fiber blends as carpet surface yarns. *Journal of the Textile Institute Proceedings*, *43*(8), P533-P555.
26. Cui, P., & Xue, Y., (2021). An intelligent weavable dyeing method for the environmentally-friendly dyeing of fabric. *Alexandria Engineering Journal*, *60*(5), 4639–4649.
27. Davidson, H. R., & Taylor, M., (1965). Prediction of the color of fiber blends. *JOSA*, *55*(1), 96–100.
28. Egan, J., & Salmon, S., (2022). Strategies and progress in synthetic textile fiber biodegradability. *SN Applied Sciences*, *4*, 1–36.
29. Emsermann, H., Foppe, R., & Bayer, A. G., (1995). Dyeing processes. *Acrylic Fiber Technology and Applications*, 285–312.
30. Erbil, Y., Babaarslan, O., & Ilhan, İ., (2018). A comparative prediction for tensile properties of ternary blended open-end rotor yarns using regression and neural network models. *The Journal of The Textile Institute*, *109*(4), 560–568.
31. Eyvaz, M., Arslan, S., Gürbulak, E., & Yüksel, E., (2017). Textile materials in liquid filtration practices: current status and perspectives in water and wastewater treatment. *Textiles for Advanced Applications*, 293–320.
32. Friele, L. F. C., (1952). The application of color measurement in relation to fiber-blending. *Journal of the Textile Institute Proceedings*, *43*(8), P604-P611.
33. Furferi, R., & Servi, M., (2023). A machine vision-based algorithm for color classification of recycled wool fabrics. *Applied Sciences*, *13*(4), 2464.
34. Gao, M., Li, N., Lu, W., & Chen, W., (2014). Role of cellulose fibers in enhancing photosensitized oxidation of basic green 1 with massive dyeing auxiliaries. *Applied Catalysis B: Environmental*, *147*, 805–812.
35. Gawish, S. M., Ramadan, A., & Mosleh, S. E. S., (2019). Improvement of polypropylene (PP) dyeing by modification methods. *Egyptian Journal of Chemistry*, *62*(1), 49–62.
36. Gorenšek, M., (1999). Dye–fiber bond stabilities of some reactive dyes on cotton. *Dyes and Pigments*, *40*(2–3), 225–233.
37. Hai, F. I., Yamamoto, K., & Fukushi, K., (2007). Hybrid treatment

systems for dye wastewater. *Critical Reviews in Environmental Science and Technology*, *37*(4), 315–377.

38. Harillo-Baños, A., Rodríguez-Martínez, X., & Campoy-Quiles, M., (2020). Efficient exploration of the composition space in ternary organic solar cells by combining high-throughput material libraries and hyperspectral imaging. *Advanced Energy Materials*, *10*(1), 1902417.
39. Harmsen, P., Scheffer, M., & Bos, H., (2021). Textiles for circular fashion: The logic behind recycling options. *Sustainability*, *13*(17), 9714.
40. Ibrahim, H., El-Zairy, E. M., Emam, E. A. M., & Saad, E. A., (2019). Combined antimicrobial finishing & dyeing properties of cotton, polyester fabrics and their blends with acid and disperse dyes. *Egyptian Journal of Chemistry*, *62*(5), 965–976.
41. Ibrahim, N. A., (2011). Dyeing of textile fiber blends. In: *Handbook of Textile and Industrial Dyeing* (pp. 147–172).
42. Ibrahim, N. A., El-Badry, K., Eid, B. M., & Hassan, T. M., (2011). A new approach for biofinishing of cellulose-containing fabrics using acid cellulases. *Carbohydrate Polymers*, *83*(1), 116–121.
43. Iqbal, M., Khatri, Z., Ahmed, A., Mughal, J., & Ahmed, K., (2012). Prediction of low-sensitivity reactive dye recipe in exhaust dyeing influenced by material to liquor ratio and nature of salt. *Journal of Saudi Chemical Society*, *16*(1), 1–6.
44. Kalaoğlu, F., Önder, E., & Özipek, B., (2003). Influence of varying structural parameters on abrasion characteristics of 50/50 wool/polyester blended fabrics. *Textile Research Journal*, *73*(11), 980–984.
45. Kaliyamoorthi, K., & Thangavelu, R., (2015). Union dyeing of cotton/nylon blended fabric by plasma-nano chitosan treatment. *Fashion and Textiles*, *2*, 1–10.
46. Kalliala, E., & Talvenmaa, P., (2000). Environmental profile of textile wet processing in Finland. *Journal of Cleaner Production*, *8*(2), 143–154.
47. Kamel, M. M., El-Shishtawy, R. M., Hanna, H. L., & Ahmed, N. S. E., (2003). Ultrasonic-assisted dyeing: I. Nylon dyeability with reactive dyes. *Polymer International*, *52*(3), 373–380.
48. Khansari, S., Duzyer, S., Sinha-Ray, S., Hockenberger, A., Yarin, A. L., & Pourdeyhimi, B., (2013). Two-stage desorption-controlled release of fluorescent dye and vitamin from solution-blown and electrospun

nanofiber mats containing porogens. *Molecular Pharmaceutics, 10*(12), 4509–4526.

49. Khatri, A., Peerzada, M. H., Mohsin, M., & White, M., (2015). A review on developments in dyeing cotton fabrics with reactive dyes for reducing effluent pollution. *Journal of Cleaner Production, 87*, 50–57.
50. Körbahti, B. K., Artut, K., Geçgel, C., & Özer, A., (2011). Electrochemical decolorization of textile dyes and removal of metal ions from textile dye and metal ion binary mixtures. *Chemical Engineering Journal, 173*(3), 677–688.
51. Kumar, A., Purtell, C., & Lepola, M., (1994). Enzymatic treatment of man-made cellulosic fabrics. *Textile Chemist and Colorist, 26*(10), 25–28.
52. Kumar, P. S., Joshiba, G. J., Femina, C. C., Varshini, P., Priyadharshini, S., Karthick, M. A., & Jothirani, R., (2019). A critical review on recent developments in the low-cost adsorption of dyes from wastewater. *Desalin. Water Treat, 172*, 395–416.
53. Lara, L., Cabral, I., & Cunha, J., (2022). Ecological approaches to textile dyeing: A review. *Sustainability, 14*(14), 8353.
54. Lawler, J. S., Wilhelm, T., Zampini, D., & Shah, S. P., (2003). Fracture processes of hybrid fiber-reinforced mortar. *Materials and Structures, 36*, 197–208.
55. Lewis, D. M., & Mcllroy, K. A., (1997). The chemical modification of cellulosic fibers to enhance dyeability. *Review of Progress in Coloration and Related Topics, 27*(1), 5–17.
56. Li, M., Zhang, K., & Xie, K., (2014). Grafting printing of cellulose fabric with the reactive disperse dyes containing N-substituted 3-chloro-2-hydroxypropyl group. *Carbohydrate Polymers, 113*, 77–82.
57. Luttringer, J. P., (1993). A New Generation of Reactive Dyes for Cotton. *Textile Chemist & Colorist, 25*(5).
58. Mahbubul Bashar, M., & Khan, M. A., (2013). An overview on surface modification of cotton fiber for apparel use. *Journal of Polymers and the Environment, 21*, 181–190.
59. McConnell, B. L., Graham, L. A., & Swidler, R. A., (1979). A new reactive system for continuous dyeing and printing of cellulose and blends. *Textile Research Journal, 49*(8), 458–467.
60. Misnon, M. I., Islam, M. M., Epaarachchi, J. A., & Lau, K. T., (2014). Potentiality of utilizing natural textile materials for engineering

composites applications. *Materials & Design*, *59*, 359–368.

61. Nguyen, T. A., & Juang, R. S., (2013). Treatment of waters and wastewaters containing sulfur dyes: A review. *Chemical Engineering Journal*, *219*, 109–117.
62. Numata, T., Amemiya, C., Kumeuchi, T., Shirakata, M., & Yonezawa, M., (2001). Advantages of blending $LiNi_0$. $8Co_0$. $2O_2$ into $Li_1^+ xMn_2-xO_4$ cathodes. *Journal of Power Sources*, *97*, 358–360.
63. Nunes-Pereira, J., Costa, C. M., & Lanceros-Méndez, S., (2015). Polymer composites and blends for battery separators: state of the art, challenges and future trends. *Journal of Power Sources*, *281*, 378–398.
64. Paosangthong, W., Torah, R., & Beeby, S., (2019). Recent progress on textile-based triboelectric nanogenerators. *Nano Energy*, *55*, 401–423.
65. Patti, A., & Acierno, D., (2022). Towards the sustainability of the plastic industry through biopolymers: properties and potential applications to the textiles world. *Polymers*, *14*(4), 692.
66. Perkins, W. S., (1991). A Review of Textile Dyeing Processes. *Textile Chemist & Colorist*, *23*(8).
67. Philips-Invernizzi, B., Dupont, D., & Caze, C., (2002). Formulation of colored fiber blends from Friele's theoretical model. *Color Research & Application: Endorsed by Inter-Society Color Council, The Color Group (Great Britain), Canadian Society for Color, Color Science Association of Japan, Dutch Society for the Study of Color, The Swedish Color Centre Foundation, Color Society of Australia, Centre Français de la Couleur*, *27*(3), 191–198.
68. Prabu, H. G., & Sundrarajan*, M., (2002). Effect of the bio-salt trisodium citrate in the dyeing of cotton. *Coloration Technology*, *118*(3), 131–134.
69. Price, J. M., Silbernagel, J., Nixon, K., Swearingen, A., Swaty, R., & Miller, N., (2016). Collaborative scenario modeling reveals potential advantages of blending strategies to achieve conservation goals in a working forest landscape. *Landscape Ecology*, *31*, 1093–1115.
70. Ristić, N., Jovančić, P., Canal, C., & Jocić, D., (2009). One-bath one-dye class dyeing of PES/cotton blends after corona and chitosan treatment. *Fibers and Polymers*, *10*, 466–475.
71. Sahinbaskan, B., Koçak, E. D., Merdan, N., & Akalın, M., (2017). Dyeing of polypropylene blends by using microwave energy. *Journal of Engineered Fibers and Fabrics*, *12*(2), 155892501701200203.

72. Said, B., M'rabet, S., Hsissou, R., & El Harfi, A., (2020). Synthesis of new low-cost organic ultrafiltration membrane made from Polysulfone/Polyetherimide blends and its application for soluble azoic dyes removal. *Journal of Materials Research and Technology*, *9*(3), 4763–4772.
73. Shahid, M., Mohammad, F., Chen, G., Tang, R. C., & Xing, T., (2016). Enzymatic processing of natural fibers: white biotechnology for sustainable development. *Green Chemistry*, *18*(8), 2256–2281.
74. Shang, S. M., (2013). Process control in dyeing of textiles. In: *Process Control in Textile Manufacturing* (pp. 300–338).
75. Shindhal, T., Rakholiya, P., Varjani, S., Pandey, A., Ngo, H. H., Guo, W., & Taherzadeh, M. J., (2021). A critical review on advances in the practices and perspectives for the treatment of dye industry wastewater. *Bioengineered*, *12*(1), 70–87.
76. Singha, K., (2013). Characterization of dyeing P/C blends fabric: a thermodynamic view. *International Journal of Textile Science*, *2*(1), 1–6.
77. Ugbolue, S. C. O., (1990). Structure/property relationships in textile fibers. *Textile Progress*, *20*(4), 1–43.
78. Ujhelyiova, A., Bolhova, E., Oravkinova, J., Tiňo, R., & Marcinčin, A., (2007). Kinetics of dyeing process of blend polypropylene/polyester fibers with disperse dye. *Dyes and Pigments*, *72*(2), 212–216.
79. Valera, T. S., Morita, A. T., & Demarquette, N. R., (2006). Study of morphologies of PMMA/PP/PS ternary blends. *Macromolecules*, *39*(7), 2663–2675.
80. Van der Schueren, L., & De Clerck, K., (2012). Coloration and application of pH-sensitive dyes on textile materials. *Coloration Technology*, *128*(2), 82–90.
81. Van Tran, T. T., Kumar, S. R., & Lue, S. J., (2019). Separation mechanisms of binary dye mixtures using a PVDF ultrafiltration membrane: Donnan effect and intermolecular interaction. *Journal of Membrane Science*, *575*, 38–49.
82. Wallenberger, F. T., Holfeld, W. T., & Turner, G. R., (1981). Energy Demand in Dyeing and Finishing. *Textile Chemist & Colorist*, *13*(8).
83. Weil, E. D., & Levchik, S. V., (2008). Flame retardants in commercial use or development for textiles. *Journal of Fire Sciences*, *26*(3), 243–281.

84. Xu, F., & Salmon, S., (2008). Potential applications of oxidoreductases for the re-oxidation of leuco vat or sulfur dyes in textile dyeing. *Engineering in Life Sciences*, *8*(3), 331–337.

85. Yang, L., Yan, L., & You, W., (2013). Organic solar cells beyond one pair of donor–acceptor: ternary blends and more. *The Journal of Physical Chemistry Letters*, *4*(11), 1802–1810.

CHAPTER 7

Dyes for the Medical Industry

Contents

Unit Introduction 236
7.1. Dyes for Indication 238
7.2. Dyes as Colorants in Medicines 240
7.3. Dyes for Therapy 241
7.4. Medical Photosensitisers 248
7.5. Potential Adverse Effects of Dyes 253
7.6. Summary 255
References 256

UNIT INTRODUCTION

The foundation of contemporary existence is predicated upon the maintenance of optimal health. When experiencing illness, our primary objective is to regain optimal health. However, once this goal is achieved, it is common to overlook the significance of this restored state. Although prevention is often considered more effective than treatment, a significant disparity exists in funding allocation between the two. Individuals residing in affluent nations are privileged to have convenient access to healthcare services throughout their lifespan (Michaels, 1988). Undoubtedly, the continuous progressions in technology have resulted in the integration of healthcare delivery as an omnipresent facet of contemporary life in industrialized countries. The contemporary medical sector encompasses a diverse array of products, including physical items such as medications, equipment, and hospital beds, as well as non-physical services like counseling and other forms of support.

The role of dyes in this context may not be readily apparent. A significant portion of the contemporary pharmaceutical industry is founded on the principles of dyes and their interplay with tissue, which were initially uncovered in the late 1800s by notable researchers, including Koch, Ehrlich, and Gram.

Please consider, as an illustration, a patient who is hospitalized and experiencing discomfort in their throat. In cases where a bacterial etiology is suspected, the overseeing healthcare provider will prescribe a swab test, which involves the examination of cellular material obtained from the posterior pharyngeal wall. A component of the analysis will entail the microscopic inspection of cells that have been subjected to staining with a blend of dyes, commonly triphenylmethane crystal violet and iodine, succeeded by an alcohol rinse and the application of safranine, a phenazine derivative (Stasiak et al., 2014). In the presence of bacterial cells, their coloration typically manifests as either purple or reddish-brown, contingent upon the degree to which crystal violet is retained following the alcohol rinse. The procedure is referred to as the Gram staining technique. Gram-positive bacteria such as streptococci exhibit a characteristic of retaining crystal violet, whereas Gram-negative bacteria such as Hemophilus influenzae do not exhibit this property. Upon detection of a bacterial infection, the healthcare provider will likely prescribe a regimen of antibiotics for the individual. In the field of pediatrics, it is common to administer a suspension of amoxicillin, which is a derivative of penicillin. To enhance the palatability and acceptability of the suspension, it is common practice to incorporate several additives, such as a flavoring agent and a dye, such as quinoline yellow (Jarman & Ballschmiter, 2012). Amoxicillin capsules frequently exhibit dichromaticity, with azoic dyes comprising the two halves in shades of red and yellow (Wainwright, 2021).

Hence, it can be observed that within this comparatively uncomplicated exposition, multiple dyes have been discerned. It is plausible that additional applications of colorants exist within the patient's surroundings, including textile dyes utilized in bed linens and pigments in the paint coating the ward's walls. However, it should be noted that these particular dyes and pigments are not exclusively employed in the healthcare industry. The illustrations in this chapter are specifically pertinent to the medical sector and are largely construed to involve some form of engagement with patients. Figure 7.1 provides an overview of the justification for using dyes in the healthcare industry (Sabnis, 2010).

Did You Know?

Methylene Blue, a vital dye used in medicine was first synthesized in 1876 by the German chemist Heinrich Caro.

Based on the information presented in Figure 7.1, it can be observed that the application of dyes in healthcare is primarily focused on three key areas: pharmaceutical coloration, cell staining, and chemotherapy (Wainwright, 2011). As previously mentioned, the initial two methods were concurrently developed during a comparable timeframe. Conversely, the third approach originated from cell staining and served as the primary source of chemotherapeutic agents before being overshadowed by penicillin and subsequent antibiotics.

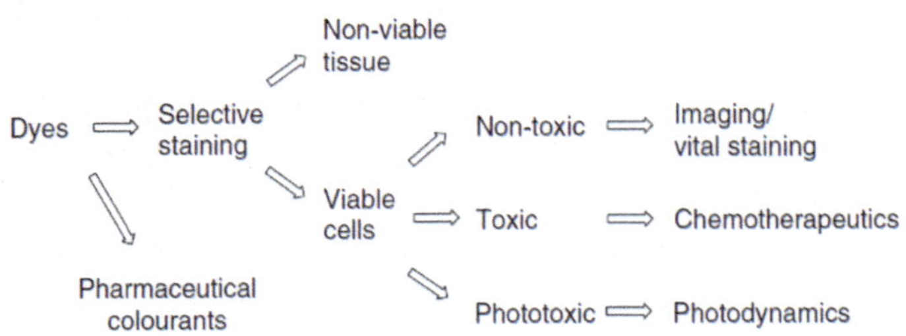

Figure 7.1. *Rationale for dye use in healthcare.*

Source: Matthew Clark, Creative Commons License.

In recent years, the third area of research has experienced a resurgence of interest due to the growing fascination with the potential of the photodynamic approach. This is particularly noteworthy given the prevalence of bacterial penicillin resistance.

7.1. DYES FOR INDICATION

7.1.1. Biological Stains

The scientific principle underlying the process of coloring biological samples bears a resemblance to that of textile dyeing. In both domains, adding dyes to a substance facilitates the attachment of dye molecules, typically onto a polymeric substrate (Descalzo & Rurack, 2009). In the field of textiles, the chemical composition of the substrate is widely recognized, such as cotton/cellulose and silk/protein. This enables a significant degree of regulation over the final color and shade (Price et al., 1978). Conversely, biological samples are typically intricate chemical blends, with the resulting color and distribution contingent upon the relative ratios of various biopolymers. The composition and allocation of these biopolymers are subject to variation contingent upon the developmental stage and chronological age of the cellular entity.

As previously stated, the Gram staining technique is fundamental for prompt bacterial identification within healthcare facilities. Christian Gram, a Danish scientist, developed the process mentioned above in 1884 when the 'Germ Theory' of the disease concept was not widely accepted. This was a pioneering effort by Gram. The inquiry mentioned above was subsequently expanded upon by modern-day scholars such as Robert Koch, Paul Ehrlich, and their peers, ultimately culminating in establishing the scientific discipline of histology (Assaf et al., 2016). Histology encompasses various specializations, one of which is histopathology. This field involves examining disease processes through staining techniques, which aligns with the original intentions of Gram and other researchers. Contemporary scientific officers in medical laboratories perform regular bacterial staining procedures and conduct supplementary biochemical analyses to obtain data on distinct bacterial strains, particularly regarding their susceptibility to drugs.

From the perspective of infection control, biological stains continue to hold significant importance in contemporary medicine. Nevertheless, cell staining encompasses a broader scope beyond the mere discrimination of bacterial and human cells.Typically, medical microbiologists analyze cellular specimens collected from individuals. Typically, the specimens collected for diagnostic purposes comprise blood, sputum, urine, feces, or tissue swabs, contingent upon the manifestation of the ailment (Wilms & Eilers, 2007). The procedure follows a sequence of determining the presence of a pathogenic microorganism in the given context. If such is the case, what precisely is it? Certain pathological conditions may present a challenge in identifying the underlying cause, particularly those involving aberrations in the process of cellular differentiation in humans. It is evident that when differentiation is necessary for a living patient, there are stricter criteria for dye toxicity compared to the analysis of removed tissue specimens.

While it may be an oversimplification, Homo sapiens can be viewed as a complex cell assemblage that ideally functions in unison (Ezati et al., 2019). In instances with abnormal cellular activity, such as erroneous biochemical processes or uncontrolled proliferation, the typical outcome is the manifestation of organic ailments, such as diabetes and cancer.

Diabetes is characterized by insufficient secretion of insulin hormone by the pancreatic Islets of Langerhans. Demonstrating a deficiency in chemical production through microscopic observation using staining techniques may pose a challenge regarding simplicity and speed. Testing the glucose level in blood or urine is a comparatively uncomplicated procedure, as it is regulated by insulin (Khan et al., 2020). This phenomenon can be exemplified through dyes, albeit it does not qualify as a staining methodology. Cancer is typically characterized by an uncontrolled and accelerated proliferation of cells, which may manifest in regions of the body where such cells are not typically found (e.g., secondary melanomas in the brain) or beyond the confines of normal growth, such as in cases where a tumor obstructs the esophagus.

Histological examination of a tissue biopsy using appropriate staining techniques, such as hematoxylineosin, may reveal such anomalies (Tratnyek et al., 2001). However, applying staining directly to the suspected tumor site is often more effective in facilitating prompt remedial measures, such as surgical debulking or laser ablation. Pre-cancerous conditions, such as Barrett's esophagus, can be accurately identified using dyes like methylene blue to demarcate the abnormal tissue before surgical or laser intervention.

The utilization of dye affinity can be employed to trace melanoma microsatellites (metastases) in the bloodstream, which is challenging. The strong attraction between phenothiazinium dyes, specifically methylene blue, and the biopolymer melanin has resulted in using radio-labeled dye as a highly effective tracer (Sprick, 1956). This is achieved by incorporating either radioactive sulfur or an iodine 125 substituent in the ring.

Did You Know?

In the early 1900s, radiologists began experimenting with various dyes to enhance the visibility of internal structures during X-ray imaging.

7.1.2. Tracer Dyes

To date, the biological stains that have been referenced have been utilized to illustrate tissues, specifically in the process of differentiation. Another categorization pertains to using stains to outline fluid pathways commonly observed in the bloodstream or lymphatic system.

Methylene blue (phenothiazinium) and patent blue (triphenylmethane) are used for in vivo tumor tracing and staging (Rodrigues et al., 2009). The efficacy and safety of methylene blue as a tracer for sentinel lymph node biopsy has been demonstrated. This technique has been utilized for monitoring possible axillary spread in breast cancer and is as effective as radio tracing methods. This is especially advantageous considering the non-universal availability of nuclear medicine in hospitals.

Dye-tracing methodologies of comparable nature are currently employed for various neoplastic ailments, encompassing gynecological, colorectal, thyroid, and pulmonary conditions. Conjugating dye molecules with peptides or other biomolecules that target tumors enables selective imaging, with the ultimate objective of utilizing such indicators during intraoperative procedures (Rodrigues et al., 2007). To achieve this objective, several rhodamine derivatives have been synthesized as probes for the somatostatin receptor, which is highly expressed in various tumors. Fluorescent spermidine acridine conjugates have been utilized to target upregulated levels of polyamine transporters in tumors. The utilization of monoclonal antibodies specific to tumors enables the possibility of fluorescent labeling, which can be achieved by applying cyanine dyes.

7.2. DYES AS COLORANTS IN MEDICINES

As the introductory segment states, pharmaceuticals may be intentionally pigmented to enhance their appeal. For instance, the amoxicillin pediatric suspension referenced is purportedly rendered more agreeable to children through banana flavoring and yellow dye amalgamation. The utilization of dichromatic plastic capsules may prove to be advantageous in expediting drug identification (Mahanta & Tiwari, 2005). However, this method cannot be deemed infallible. The rationale behind incorporating coloring agents in tablet coatings for aesthetic purposes may not be understandable, especially when minors could misconstrue them as confectioneries.

The limited availability of pharmaceutical dyes can be attributed to their potential toxicity. The diminutive is expected to diminish in magnitude solely due to the abovementioned rationale. The prolonged administration of a therapeutic agent with a consistent hue may pose a risk of toxicity (Korankye, 2010). Therefore, it is crucial to comprehend the metabolic pathways of ingested dyes, akin to the investigation of conventional drug toxicity. The study encompasses an extensive spectrum of hues, including monochromatic tones, derived from twelve manufactured and four organic colorants, in conjunction with titania and diverse iron oxide inorganic pigments. Approximately 50% of synthetic dyes belong to the azoic classification.

7.3. DYES FOR THERAPY

7.3.1. Conventional Chemotherapy

The prevailing perspective in Western medicine regarding the use of colored therapeutics is to minimize their utilization. The rationale behind this viewpoint, which appears to be shared by both the medical and pharmaceutical communities, typically consists of two primary factors. Firstly, there is concern that patients may experience adverse reactions to alterations in the skin or other bodily pigmentation resulting from therapeutic interventions. Secondly, there is a widely held belief that dyes possess toxic properties (Park et al., 2007).

The initial criterion may hold veracity or lack thereof. It is not feasible to assert that the pigmentation of a patient is invariably affected by administering medications with colorants. The introduction of any substance into the human body has the potential to cause metabolic changes or degradation of said substance. The administration of the substance may result in a temporary alteration of the patient's skin pigmentation (Li et al., 2022). Nonetheless, a trade-off must be considered if the desired outcome is achieved concerning the patient's well-being. This assertion is invalid if there exists a therapeutic alternative that is not color-coded. In a recent clinical trial conducted in Burkina Faso, methylene blue was administered to children suffering from falciparum malaria. It is noteworthy that the administration of this drug resulted in discoloration of the subjects' urine and nappies. However, the affected individuals exhibited widespread acceptance and adherence to the treatment regimen due to effective education and awareness programs.

The generalization of the toxic effects of dyes is a subject that is also subject to scrutiny but is regulated by a comparable argument. The toxicity of structural types in human patients is contingent upon their ADME profile, which is a function of their physicochemical properties and administration, distribution, metabolism, and elimination characteristics. This is due to the diverse range of structural types available. It is verifiable that specific dyes, such as those obtained from benzidine, have been found to elicit detrimental impacts on human health (Shen et al., 2015). Furthermore, it is common for dyestuffs to be produced from hazardous chemicals, which may be present as impurities in the final product. Acting is imperative in addressing both aspects to guarantee the safety of patients. This involves eliminating structural characteristics such as benzidine and ensuring the purity of the product.

There exists a significant association between contemporary pharmaceutical drugs and dyes. Numerous therapeutic agents utilized in contemporary clinical practice originate in dyes manufactured over a century ago or longer. Methylene blue, first documented to possess curative properties against malaria in 1891, was later employed as a primary chemical entity for exploring anti-malarial drugs during the 1920s and 1930s in both Germany and the USSR. The development of acridine and quinoline antimalarials, specifically mepacrine and chloroquine, can be attributed to this particular cause (Yukruk et al., 2005). The fundamental side chain utilized in synthesizing methylene blue derivatives with anti-malarial properties present in chloroquine and mepacrine has also been identified in pharmaceuticals prescribed for managing depression and schizophrenia. Ehrlich's discoveries have contributed to the development of various chemotherapeutic agents. For instance, during World War One, aminoacridines such as proflavine and acriflavine were utilized as synthetic wound antibacterials.

7.3.1.1. Crystal/Gentian Violet

The Gram stain played a pivotal role in the nascent stages of aniline dye development as a means of biological staining. Using gentian violet, also known as crystal violet, as a bacterial dye precipitated its subsequent application as an antibacterial agent (Wheeland & Applebaum, 1990). During World War One, gentian violet and brilliant green were utilized in treating wound sepsis in conjunction with proflavine and acriflavine, as previously discussed. Gentian violet has demonstrated antifungal properties and was utilized in treating cutaneous infections, including ringworm, until recently. In a related domain, South American organizations incorporate dye into donated blood to neutralize the causative agent of Chagas' disease, namely Trypanosoma cruzi. Crystal violet has demonstrated efficacy in treating patients infected with methicillin-resistant *Staphylococcus aureus* (MRSA), a promising development for those engaged in anti-infective research.

7.3.1.2. Methylene Blue

As a result of Ehrlich's endeavors in the latter part of the 19th century, methylene blue underwent extensive experimentation as a plausible remedy for various contagious ailments. Despite exhibiting limited efficacy against bacteria and fungi, this substance demonstrates a notable absence of human toxicity. As previously noted, it is suitable for employment in vital staining and surgical procedures. The successful treatment of two individuals with malaria by Guttmann and Ehrlich in 1890 has been a significant point of reference in anti-malarial research and has informed recent clinical trials conducted in Burkina Faso (Strauss & Resnick, 1993). Methylene blue is a preferred initial therapy for methemoglobinemia and cyanide poisoning due to its advantageous redox potential and low toxicity.

7.3.1.3. Acridines

Before the 'anti-biotic era' emerged, therapeutic dyes such as proflavine and acriflavine were at the forefront of medical treatment. However, these drugs became obsolete with the widespread availability of potent agents like penicillins and tetracyclines. Proflavine and acriflavine are still utilized in certain medical settings, albeit in limited capacities. Proflavine is commonly employed for localized antisepsis, particularly in burns units (Hansen et al., 2003). On the other hand, acriflavine has been utilized as a viral replication inhibitor in individuals with AIDS. Ethacridine, a compound closely associated with the subject matter, is sometimes utilized in shigellosis. The acridine chromophore's DNA-binding ability has been the foundation for numerous contemporary anti-cancer medications.

7.3.1.4. Phenazines

The substitution of the carbon atom located at position 9 in the acridine molecule with a nitrogen atom results in the formation of the phenazine core.

This structural motif is associated with various biological dyes, including but not limited to neutral red and safranin. Despite examining various phenazines for their antimicrobial properties in the period following World War One, it appears that none demonstrated significant efficacy. Since the 1950s, a phenazine molecule has been utilized in the antibacterial arsenal. Barry initially synthesized Clofazimine to discover drugs that could combat leprosy. This is illustrated in Figure 7.2 (Volwiler, 1926).

Clofazimine is a compound that features the phenazine chromophore as a quinoneimine, resulting in its distinct purple/damson hue. Despite its initial target being Hansen's bacillus (Mycobacterium leprae), it demonstrated efficacy against Mycobacterium avium, among the opportunistic pathogens linked to individuals with HIV-positive status (Li et al., 2019).

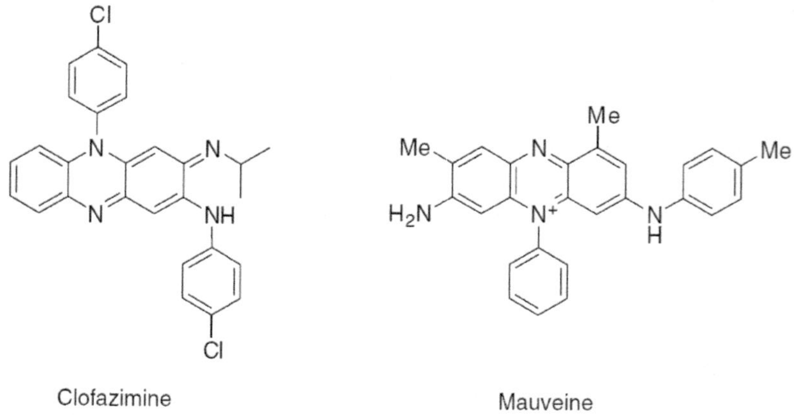

Figure 7.2. *Illustration of clofazimine (quinoneimine) and mauveine.*

Source: M. Wain Wright, Creative Commons License.

Clofazimine has been utilized to treat Crohn's disease and drug-resistant tuberculosis. It is noteworthy to compare the structures of Clofazimine and Perkin's mauve (Figure 7.2), especially considering that Perkin's original intention was to synthesize quinine, a therapeutic molecule for malaria treatment (Lanigan, 1996).

7.3.1.5. Phenanthridines

The extensive research conducted on drugs based on acridine has led to the discovery of several noteworthy compounds derived from the isomeric phenanthridine ring system, which is not unexpected. Phenanthridines that have been appropriately replaced demonstrate comparable characteristics to the acridines that were previously discussed. Amino derivatives, in which the amino groups combine with the ring nitrogen, such as ethidium bromide, are typically characterized by a highly cationic nature and demonstrate strong affinities for nucleic acids (Huilgol et al., 1996).

Phenanthridines were intentionally synthesized for application in tropical medicine, specifically for treating African human trypanosomiasis (also known as sleeping sickness) in the aftermath of World War One. Ethidium bromide and its subsequent derivatives, including isometamidium chloride (depicted in Figure 7.3), have been recognized as effective trypanocidal agents (Swamy et al., 2020). While their use is primarily limited to veterinary applications, these compounds have been shown to target the trypanosomal parasite's DNA specifically.

7.3.1.6. Azoic Compounds

The revelation that the established sulphanilamide, rather than the Bayer dye Prontosil, exhibited antibacterial activity prompted a surge in synthetic production among chemical and pharmaceutical companies during the 1940s (Donelan et al., 2008). This resulted in the creation of numerous analogous compounds within the sulphonamide category.

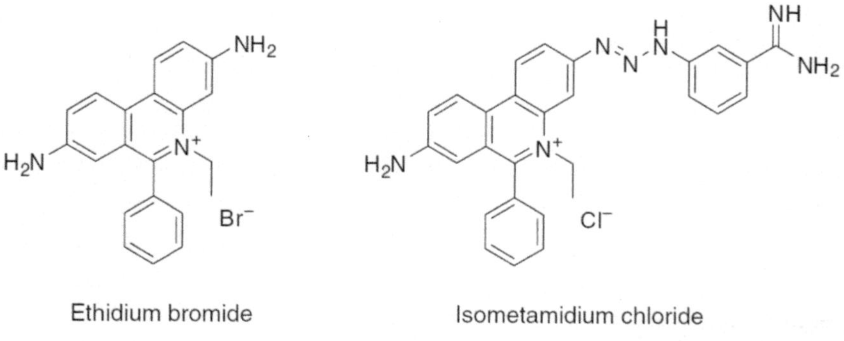

Figure 7.3. *Illustration of phenanthridine-based trypanocidal agents.*

Source: Matthew Clark, Creative Commons License.

It is reasonable to infer that the numerous drug candidates that followed Prontosil and its analogs were sulphanilamide derivatives rather than azoic dyes, as the former requires metabolic activation, specifically the release of the sulphonamide moiety through the reduction in the gut of the azo linkage (as depicted in Figure 7.4), for their effectiveness (Reid et al., 1995). The in vivo reduction of azo has been utilized for clinical benefits due to its potential to deliver drugs to the bowel. The drug sulfasalazine undergoes breakdown in the colon, forming sulfapyridine and 4-aminosalicylic acid (as depicted in Figure 7.4) (Veitch et al., 2017). Both constituents possess anti-inflammatory properties and are utilized in managing ulcerative colitis.

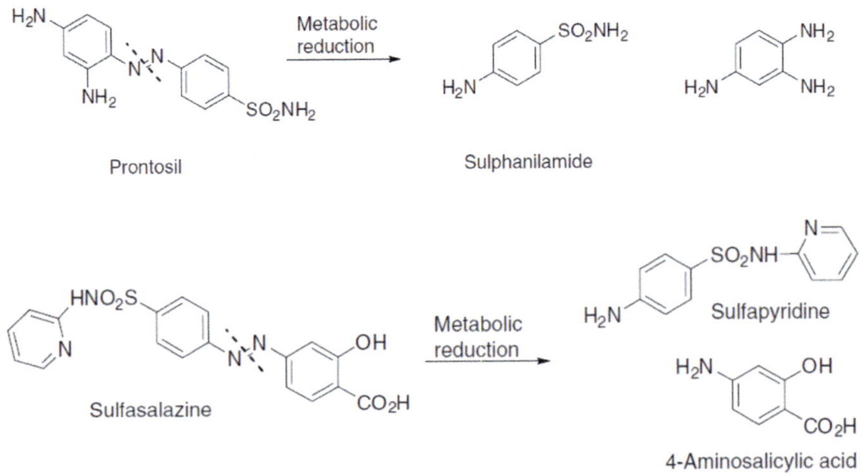

Figure 7.4. *Illustration of drugs based on the reduction of azoic dyes.*

Source: M. Wain Wrights, Creative Commons License.

7.3.2. Antimicrobial Dyes for Medical Textiles

Fungal growth poses a significant challenge to the durability of fabrics, particularly during periods of storage. This issue can be resolved through the utilization of biocidal agents. However, the description mentioned above provides evidence for implementing biocidal textile dyes. Hence, the dyeing process would impact both the coloration and antimicrobial properties.

Did You Know?

Indocyanine green (ICG), a fluorescent dye widely used in medical imaging, was first synthesized in the 1950s.

The abovementioned approach appears to be a highly effective strategy for safeguarding fabrics. However, it has the potential to be expanded to safeguard patients in the ongoing battle against healthcare-associated infections, as a significant number of infections are transmitted through environmental factors such as soft furnishings, bedding, and hospital attire.

Various biocidal agents are utilized in healthcare disinfectants, exhibiting efficacy against MRSA and other significant hospital pathogens (Kenton-Smith & Tan, 1999). Dyes exhibiting analogous structural motifs have been synthesized, such as derivatives of quaternary ammonium salts (depicted in Figure 7.5). The synthesis of biocidal analogs is a relatively straightforward process due to amino functionality in most textile dyes. This has been reported for simple azoic and anthraquinone derivatives, as depicted in Figure 7.5. Cationic cyanine dyes with antibacterial properties have been documented (Figure 7.5) (Wainwright, 2008). However, due to the delocalization of the positive charge throughout the chromophore, the molecular interactions with bacteria may differ from those of the quaternary ammonium salt azoic and anthraquinones mentioned. These latter compounds will likely interfere with bacterial membrane dynamics by presenting point charges.

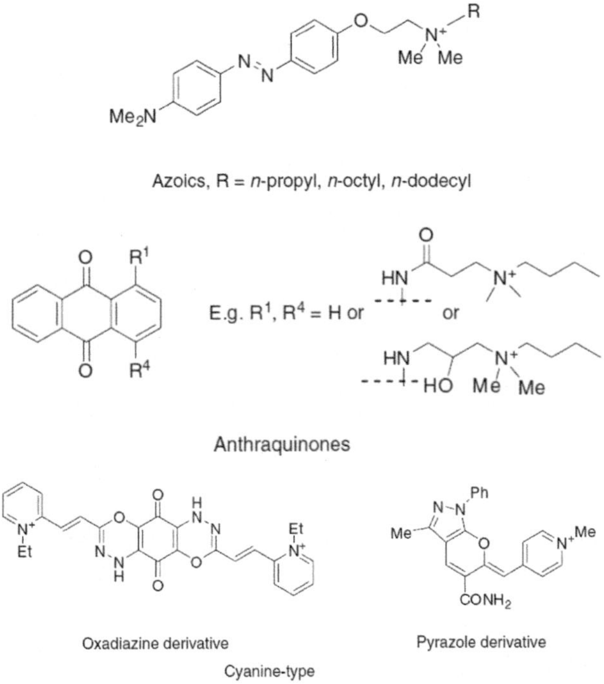

Figure 7.5. *Representation of antimicrobial textile dyes.*

Source: D. M. Lewis, Creative Commons License.

As illustrated in Figure 7.6, pyrazole-based bis azo dyes are effective against bacteria and fungi (Kamkaew et al., 2013). They sometimes outperform fluconazole. Innovative textile agents and drug discovery leads have been found.

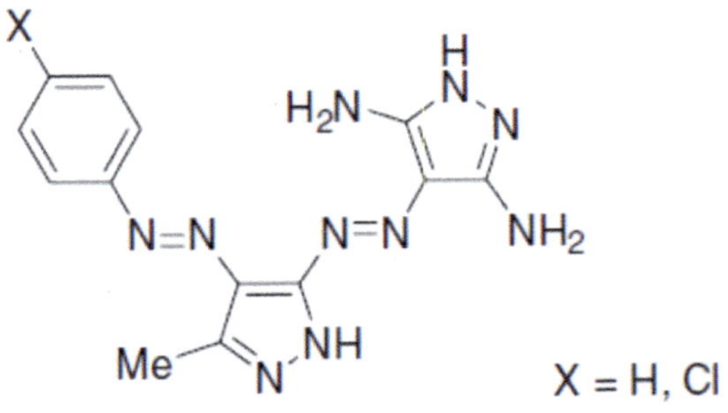

Figure 7.6. *Representation of antibacterial/antifungal bis(azoic) dyes.*

Source: D. M. Lewis, Creative Commons License.

Therapeutic dyes include various, as mentioned. However, medicinal products use antimicrobials sparingly. Brilliant green and chlorhexidine were used as an antibacterial treatment for latex and nitrile examination gloves on one occasion.

Antimicrobial textile dyes use biocidal agents like chlorhexidine digluconate and benzalkonium chloride at higher concentrations than conventional antibacterial drugs (van de Winckel et al., 2018). In textile applications, such performance may be difficult. The material's color is another consideration. Figure 7.6's brilliant orange antifungal bis(azoic) dye is notable. Is this color for operating room scrubs?

7.4. MEDICAL PHOTOSENSITISERS

Photosensitization is when a substance undergoes activation or modification upon exposure to light energy, as the term implies. Photosensitizers are the substances responsible for facilitating this process (Wainwright et al., 2011). The present chapter focuses on the medical application of photosensitizers; however, these compounds are also found in other contexts, such as plant photosynthesis and industrial oxidation processes. Photosensitization is a common cause of reactions in the skin to chemicals, particularly pharmaceuticals.

Using photosensitization to eliminate or render inactive undesired, or unprofitable cells predates the existence of Homo sapiens. Certain plants and fungi employ natural product photosensitizers for this purpose. The plant species known as Sanguinaria canadensis, commonly called bloodroot, secretes a fluid containing sanguinarine, a photosensitizer (Grossweiner et al., 1998). This substance adheres to the exoskeletons of herbivorous insects and induces photodamage when exposed to sunlight. On the other hand, the parasitic fungus Hypocrella bambusae employs compacted aromatic pigments, commonly referred to as hypocrellins, to perform photosensitization and disintegrate the cell walls of plants.

The utilization of photosensitizers by humans has primarily been directed towards two avenues, namely, anti-cancer and antimicrobial applications. Both of these phenomena were exhibited in the initial years of the 20th century; however, they were previously deemed to be of minimal significance regarding clinical implementation. Raab (1900) demonstrated the photo antimicrobial action by reporting the inactivation of paramecia using acridine and eosin. Similarly, Tappeiner & Jesionek (1903) employed similar dyes against neoplasms of the skin.

7.4.1. Photodynamic Therapy (PDT)

The clinical utilization of photosensitizers in treating various types of tumors, such as those affecting the skin, lungs, breast, brain, and others, has predominantly relied on porphyrin-based compounds (Sibani et al., 2008). This is largely due to initial discoveries regarding the fluorescence of natural porphyrin derivatives in cancerous tissues during surgical procedures. Notwithstanding, initial generation photosensitizers, such as hematoporphyrin derivative (HpD), exhibited limitations in impurities, inadequate photo properties, and cutaneous photosensitivity. Successive artificial contenders have been derived from phthalocyanine, phenothiazinium, and cyanine chromophores.

7.4.1.1. Phthalocyanines

Due to the structural resemblance to the porphyrin nucleus, phthalocyanines have been extensively studied by numerous research groups globally as potential substitutes for HpD and other porphyrin derivatives. This is particularly noteworthy due to their significantly higher light absorption properties in the red and near-infrared regions of the electromagnetic spectrum (Jensen et al., 2008). Notwithstanding, even though the dye chemistry linked with phthalocyanines has been firmly established, it is primarily founded on copper phthalocyanine, an exceedingly stable blue pigment. Regrettably, owing to its paramagnetic characteristics, copper(II) exhibits swift deactivation of the excited singlet state, rendering it a highly inadequate photosensitizer. As a result, researchers pursued diamagnetic derivatives, which incorporated a central ion of either zinc or aluminum. The phthalocyanine nucleus is frequently subjected to sulphonation or derived from carboxylic acid-bearing starting materials to generate soluble dyes in water (Craig et al., 2015). Empirically, these methodologies yield amalgamations of products and isomers that, while tolerable in dye production, are unsuitable for pharmaceutical advancement. Undoubtedly, a commonly cited argument opposing the ongoing utilization of HpD pertains to its deficient chemical purity. Despite this, it should be noted that Photosens, a mixture containing sulfonated chloroaluminium phthalocyanine, has obtained official authorization for its utilization in the former Soviet Union (Figure 7.7).

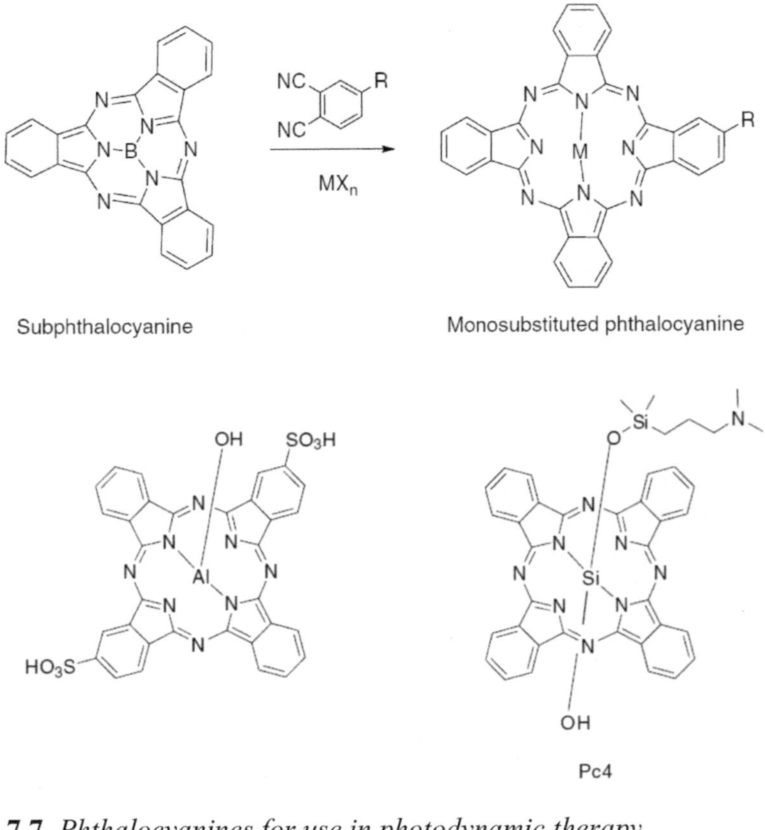

Figure 7.7. *Phthalocyanines for use in photodynamic therapy.*

Source: N. A. Ibrahim, Creative Commons License.

To address the challenge posed by the mixture/isomer issue, scholars have documented the utilization of monosubstitution techniques derived from phthalonitrile derivatives and subphthalocyanine, as well as functionalization of the central metal/metalloid element (Wilson & Pratten, 1995). Recent advancements in the field have shown that silicon has been effectively utilized for this purpose, primarily by incorporating polar, basic/cationic chains. The resultant derivatives exhibit desirable characteristics and efficacy for anti-cancer and antimicrobial applications.

7.4.1.2. Phenothiaziniums

Considering the prevalence of methylene blue in the current chapter, it is unsurprising that it plays a role in creating photodynamic therapy (PDT) drugs. Due to its extended history of use, minimal human toxicity, and specific attraction to tumor tissue, as previously mentioned, in addition to its photosensitizing properties, this compound has become a prominent candidate for developing antitumor and antimicrobial treatments. An analogous circumstance exists regarding toluidine blue, as depicted in Figure 7.7 (Phillips, 2010).

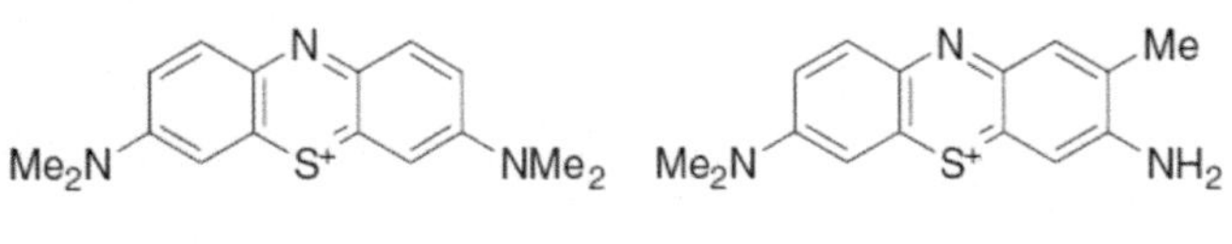

Methylene blue Toluidine blue

The preliminary experiments conducted on these lead-based substances for PDT demonstrated their hydrophilic properties, which resulted in their prompt elimination upon systemic administration. The efficacy of methylene blue as a photosensitizer in treating superficial bladder cancer has been investigated. The auxochromic groups of methylene blue have been modified symmetrically or asymmetrically to alter their shape and hydrophilic lipophilic balance, increasing hydrocarbon content (Wainwright & Giddens, 2003). This includes a homologous series ranging from methylene blue to the bis(dihexylamino) derivative, as reported by previous studies. The study mentioned above demonstrated that the observed variation impacts the uptake and localization of mammalian cells. The derivatives were found to not localize in the nucleus, which negates a significant portion of the reason for potential mutagenic activity in human subjects. The symmetrical and asymmetrical derivatives, which are differentiated based on their auxochrome properties, can be conveniently synthesized using bromine or iodine through oxidation/amination methods. These methods are less detrimental to the molecular structure than the previously employed acid/chromium(VI) salts.

7.4.1.3. Cyanines

Cyanine dyes comprise two heteroaromatic nuclei joined by a chain of linked methine units with an odd number. This structure allows the charge to be delocalized from one end to the other, resulting in a longer highest wavelength of absorption for the chromophore. The formation of cyanine dye analogs is a relatively uncomplicated process, as it only requires adherence to basic structural specifications such as heteroatom variation N,N-alkylation, and polymethine chain length. This results in the generation of a significant quantity of compounds. However, it is noteworthy that there exists a scarcity of photosensitizing drugs derived from this category, with only merocyanine 540 (MC540) and indocyanine green being the notable exceptions (Baldea et al., 2015).

MC540 is a photosensitizer that is anionic and lipid soluble (Millson et al., 1996). These characteristics enable it to aggregate in the outer layer of cellular membranes, characterized by low levels of anionic head. The utilization of MC540 has resulted in its application in the identification and eradication of leukemic cells in comparison to hematopoietic stem cells. Currently, the primary application of MC540 pertains to eliminating leukemic cells from fractions necessary for autologous bone marrow transplantations.

7.4.1.4. Indocyanine Green

It can be observed from Figure 7.7 that the π system of indocyanine green is more extensive in comparison to that of MC540 (Kömerik et al., 2002). As previously mentioned, the green dye has been observed to be relatively non-toxic to humans, which has resulted in its utilization for various staining purposes during surgical procedures. Recently, this technique has been expanded to photodynamic domains, such as in the angiography of individuals undergoing choroidal neovascularization in macular degeneration due to age.

Despite the potential for indocyanine green to function as a photosensitizer, the effects observed are probably attributable to photothermal mechanisms, as opposed to the conventional Type I / Type II photosensitization pathways discussed previously. The extended wavelength range of indocyanine green, which is focused on the maximum wavelength of 780 nm, enables it to be stimulated by an 808 nm diode laser (Josefsen & Boyle, 2008). As a result, the stimulated cyanine transmits its excitation energy to its environment, leading to cellular thermal damage. The impact remains discriminatory due to the necessary amalgamation of chromophore localization and light orientation, albeit the technique can be utilized more broadly to eliminate tissue (photothermal ablation). Indocyanine green is a valuable byproduct in tissue soldering, serving as a substitute for suturing. The heat generated during illumination accelerates the reaction rate of supplementary constituents, including fibrin and albumin.

7.5. POTENTIAL ADVERSE EFFECTS OF DYES

As shown above, dyes have several medicinal uses. The kind and length of exposure to various dyes define their health risks. Compared to a quick topical application of the same dye to heal a surface lesion, continuous food ingestion of a dye may increase the chance of unwanted consequences. Wearing clothing dyed with the same substance poses the least risk.

The complexity of Homo sapiens makes understanding chemical toxicity difficult. To start comprehending the issue, let's examine these human body questions: Please explain the topic. This occurs where? What frequency triggers toxicity? Does everyone experience this? Is the chemical pure? Does metabolism change it? These essential issues include many variations, suggesting a complicated situation.

Unfavorable responses vary. Cytotoxicity causes necrosis or apoptosis. Protein precipitation, membrane breakdown, and other gross responses occur during necrosis (To et al., 2021). However, apoptosis is caused by the chemical's main impact. This cascade turns off vital cell machinery. Mutagenicity is cellular DNA damage or disruption that affects DNA structure chemically or physically. This change erroneously copies base pair templates, creating "mutant" daughter cells. Mutagenicity may cause uncontrolled cell proliferation or carcinogenicity. Mutagenesis does not always cause cancer. Thus, a dye's toxicity may be ambiguous in the literature. The reader should also be wary of single-cell line toxicity research. Dye interaction with epithelial cells does not account for metabolic processes or plasma protein binding and concentration. In vivo, dye binding to DNA or lipoproteins may differ from isolated studies. A mutagenic dye must reach the target organ, be internalized by a cell, find the nucleus, cross the nuclear membrane, and interact with the nucleic acid (Leme et al., 2015). While the dye is unmetabolized and biomolecule-free, several processes must occur.

7.5.1. Toxicology

Saying aniline is dangerous is inaccurate. Aniline catabolism produces several hazardous chemicals (Haritoglou et al., 2008). Anilines are used in color manufacturing; hence their catabolism will increase aniline levels in the body. This may be true in certain cases, but it applies to all artificial colorants. Modern metabolic pathways allow modification of aniline derivative chemistry to avoid these difficulties. Liver reductase enzymes manufacture metabolites based on the dye molecule. P-nitroaniline may be converted to p-phenylenediamine, and its N-methylated derivatives are mutagenic. Water-solubilizing groups diminish mutagenicity. Metabolism increases xenobiotic hydrophilicity, facilitating removal. Dye compounds with hydrophilicity are unlikely to stay in the body.

Due to compelling reasons, Dyes derived from a basic p-phenylenediamine or the benzidine motif are deemed unsuitable for human use. Despite the 1,4-diaminobenzene unit within a more intricate composition, numerous dyes do not exhibit mutagenic properties. This particular unit is commonly observed in a significant proportion of azine dyes. The absence of detectable mutagenicity can be attributed to the failure of metabolism to recognize the distinct generation of p-phenylenediamine, typically caused by its sustained linkage with another component of the azine ring system (Wani et al., 2019). The relevance of hydrophilicity argument concerning swift elimination is reiterated in this context.

7.5.2. Mutagenicity

Dyes that exhibit a propensity to interact with DNA are a clear source of apprehension in situations where human utilization is envisaged. The acridine class serves as a prominent illustration of this phenomenon. The DNA-binding properties of aminocridines and isomeric amino phenanthridines, such as ethidium bromide, have been demonstrated to result in mutagenesis in bacterial and yeast cultures (Peters & Freeman, 1991). The transferability of these effects to humans remains uncertain. Nonetheless, the argument against acridines and phenanthridines is probably founded on experimental data obtained from non-human subjects, and even then, only for a limited selection of compounds. Although it is infeasible to conduct human trials to assess the side effects of these compounds, a significant number of individuals, particularly those in the military, were administered amino acridines during the first half of the 1900s, either as an antiseptic or as an anti-malarial agent. Given the high potential for tumor expression in said compounds, what evidence exists to support an elevated cancer incidence among individuals exposed to them?

The induction of mutagenicity through an intercalative mechanism necessitates a substantial degree of interaction between the intercalator and the DNA base pairs. The notion that intercalation mandates a cationic, planar molecule of roughly acridine dimensions is not necessarily accurate (Elbanna et al., 2017). Research has demonstrated that possessing basic cationic properties is inadequate, as hydrogen bonding is also necessary.

Unlike proflavine, which intercalates acridine in vitro and has an acridine structure with R and R' being hydrogen atoms, N-methyl acridine orange generates molecules that cannot intercalate.

Dye effects on humans are complicated and depend on several aspects. Dye for patient administration must be made like medicines, emphasizing ultrahigh purity. The impurities in ordinary textile dyes, made from basic aniline-type precursors, may enhance the risk of harmful side effects. Dye metabolism should help reduce patient risks.

7.6. SUMMARY

Dyes play a significant role in the medical industry particularly in diagnostic and therapeutic applications. Dyes are often utilized in medical imaging techniques to enhance visualization and aid in the diagnosis of various conditions. Fluorescent dyes are frequently employed in molecular and cellular imaging. Surgeons often use dyes to mark specific areas or structures during surgical procedures. Dyes can be incorporated into pharmaceutical formulations to assist in drug development and delivery systems. These dyes are carefully developed and tested to ensure they meet safety and regulatory requirements.

REFERENCES

1. Assaf, K. I., Suckova, O., Al Danaf, N., von Glasenapp, V., Gabel, D., & Nau, W. M., (2016). Dodecaborate-functionalized anchor dyes for cyclodextrin-based indicator displacement applications. *Organic Letters, 18*(5), 932–935.
2. Chen, B., Pogue, B. W., & Hasan, T., (2005). Liposomal delivery of photosensitising agents. *Expert Opinion on Drug Delivery, 2*(3), 477–487.
3. Chen, Y., Li, L., Chen, W., Chen, H., & Yin, J., (2019). Near-infrared small molecular fluorescent dyes for photothermal therapy. *Chinese Chemical Letters, 30*(7), 1353–1360.
4. Craig, R. A., McCoy, C. P., Gorman, S. P., & Jones, D. S., (2015). Photosensitisers–the progression from photodynamic therapy to anti-infective surfaces. *Expert Opinion on Drug Delivery, 12*(1), 85–101.
5. Descalzo, A. B., & Rurack, K., (2009). On the signaling pathways and CuII□mediated anion indication ofN□meso□substituted heptamethine cyanine dyes. *Chemistry–A European Journal, 15*(13), 3173–3185.
6. Donelan, M. B., Parrett, B. M., & Sheridan, R. L., (2008). Pulsed dye laser therapy and z-plasty for facial burn scars: the alternative to excision. *Annals of Plastic Surgery, 60*(5), 480–486.
7. Elbanna, K., Sarhan, O. M., Khider, M., Elmogy, M., Abulreesh, H. H., & Shaaban, M. R., (2017). Microbiological, histological, and biochemical evidence for the adverse effects of food azo dyes on rats. *Journal of Food and Drug Analysis, 25*(3), 667–680.
8. Ezati, P., Tajik, H., Moradi, M., & Molaei, R., (2019). Intelligent pH-sensitive indicator based on starch-cellulose and alizarin dye to track freshness of rainbow trout fillet. *International Journal of Biological Macromolecules, 132*(1), 157–165.
9. Garden, J. M., Bakus, A. D., & Paller, A. S., (1992). Treatment of cutaneous hemangiomas by the flashlamp-pumped pulsed dye laser: Prospective analysis. *The Journal of Pediatrics, 120*(4), 555–560.
10. Gee, K. R., Brown, K. A., Chen, W. U., Bishop-Stewart, J., Gray, D., & Johnson, I., (2000). Chemical and physiological characterization of fluo-4 Ca^{2+}-indicator dyes. *Cell Calcium, 27*(2), 97–106.
11. Grossweiner, L. I., Fernandez, J. M., & Bilgin, M. D., (1998). Photosensitisation of red blood cell hemolysis by photodynamic agents. *Lasers in Medical Science, 13*(1), 42–54.

12. Hansen, K., Kreiter, C. D., Rosenbaum, M., Whitaker, D. C., & Arpey, C. J., (2003). Long□term psychological impact and perceived efficacy of pulsed□dye laser therapy for patients with port□wine stains. *Dermatologic Surgery, 29*(1), 49–55.
13. Haritoglou, C., Schüttauf, F., Gandorfer, A., & Thaler, S., (2008). An experimental approach towards novel dyes for intraocular surgery. *Vital Dyes in Vitreoretinal Surgery, 42*(1), 141–152.
14. Huilgol, S. C., Barlow, R. J., & Markey, A. C., (1996). Failure of pulsed dye laser therapy for resistant verrucae. *Clinical and Experimental Dermatology, 21*(2), 93–95.
15. Hulse, J. H., (2004). Biotechnologies: past history, present state and future prospects. *Trends in Food Science & Technology, 15*(1), 3–18.
16. Jarman, W. M., & Ballschmiter, K., (2012). From coal to DDT: the history of the development of the pesticide DDT from synthetic dyes till silent spring. *Endeavour, 36*(4), 131–142.
17. Jori, G., & Brown, S. B., (2004). Photosensitized inactivation of microorganisms. *Photochemical & Photobiological Sciences, 3*(5), 403–405.
18. Josefsen, L. B., & Boyle, R. W., (2008). Photodynamic therapy and the development of metal-based photosensitisers. *Metal-Based Drugs, 1*, pp. 6–10.
19. Kamkaew, A., Lim, S. H., Lee, H. B., Kiew, L. V., Chung, L. Y., & Burgess, K., (2013). BODIPY dyes in photodynamic therapy. *Chemical Society Reviews, 42*(1), 77–88.
20. Kenton-Smith, J., & Tan, S. T., (1999). Pulsed dye laser therapy for viral warts. *British Journal of Plastic Surgery, 52*(7), 554–558.
21. Khan, I. S., Ali, M. N., Hamid, R., & Ganie, S. A., (2020). Genotoxic effect of two commonly used food dyes metanil yellow and carmoisine using Allium cepa L. as indicator. *Toxicology Reports, 7*(1), 370–375.
22. Korankye, O., (2010). *Extraction and Application of Plant Dyes to Serve as Colorants for Food and Textiles* (Vol. 1, pp. 6–9). Doctoral dissertation.
23. Lanigan, S. W., (1996). Port wine stains on the lower limb: response to pulsed dye laser therapy. *Clinical and Experimental Dermatology, 21*(2), 88–92.
24. Leme, D. M., Oliveira, G. A. R. D., Meireles, G., Brito, L. B., Rodrigues, L. D. B., & Palma de Oliveira, D., (2015). Eco-and genotoxicological

assessments of two reactive textile dyes. *Journal of Toxicology and Environmental Health, Part A, 78*(5), 287–300.

25. Li, L., Chen, Y., Chen, W., Tan, Y., Chen, H., & Yin, J., (2019). Photodynamic therapy based on organic small molecular fluorescent dyes. *Chinese Chemical Letters, 30*(10), 1689–1703.
26. Li, Y., Gao, J., Wang, S., Li, S., Hou, X., Pan, Y., & Xiao, Y., (2022). Organic NIR-II dyes with ultralong circulation persistence for image-guided delivery and therapy. *Journal of Controlled Release, 342*(1), 157–169.
27. Mahanta, D., & Tiwari, S. C., (2005). Natural dye-yielding plants and indigenous knowledge on dye preparation in Arunachal Pradesh, northeast India. *Current Science, 1*, 1474–1480.
28. Maurya, N. S., Mittal, A. K., Cornel, P., & Rother, E., (2006). Biosorption of dyes using dead macro fungi: effect of dye structure, ionic strength and pH. *Bioresource Technology, 97*(3), 512–521.
29. Michaels, D., (1988). Waiting for the body count: corporate decision making and bladder cancer in the US dye industry. *Medical Anthropology Quarterly, 2*(3), 215–232.
30. Mihm, F. G., Feeley, T. W., & Jamieson, S. W., (1987). Thermal dye double indicator dilution measurement of lung water in man: comparison with gravimetric measurements. *Thorax, 42*(1), 72–76.
31. Mohammad, A. W., Teow, Y. H., Ang, W. L., Chung, Y. T., Oatley-Radcliffe, D. L., & Hilal, N., (2015). Nanofiltration membranes review: Recent advances and future prospects. *Desalination, 356*(1), 226–254.
32. Moreno, G., Pottier, R. H., & Truscott, T. G., (2013). *Photosensitization: Molecular, Cellular and Medical Aspects* (Vol. 15, pp. 5–7). Springer Science & Business Media.
33. Park, H. S., Kim, J. W., Jang, S. J., & Choi, J. C., (2007). Pulsed dye laser therapy for pediatric warts. *Pediatric Dermatology, 24*(2), 177–181.
34. Parker, S., (2013). The use of diffuse laser photonic energy and indocyanine green photosensitizer as an adjunct to periodontal therapy. *British Dental Journal, 215*(4), 167–171.
35. Peters, A. T., & Freeman, H. S., (1991). *Color Chemistry: The Design and Synthesis of Organic Dyes and Pigments* (Vol. 1, pp. 193). London, New York: Elsevier applied science.
36. Phillips, D., (2010). Light relief: photochemistry and medicine.

Photochemical & Photobiological Sciences, 9(1), 1589–1596.

37. Reid, R., Omoto, K. H., Precop, S. L., Berman, N. R., Rutledge, L. H., Dean, S. M., & Pleatment, M., (1995). Flashlamp-excited dye laser therapy of idiopathic vulvodynia is safe and efficacious. *American Journal of Obstetrics and Gynecology, 172*(6), 1684–1701.
38. Rodrigues, E. B., Costa, E. F., Penha, F. M., Melo, G. B., Bottós, J., Dib, E., & Farah, M. E., (2009). The use of vital dyes in ocular surgery. *Survey of Ophthalmology, 54*(5), 576–617.
39. Rodrigues, E. B., Maia, M., Meyer, C. H., Penha, F. M., Dib, E., & Farah, M. E., (2007). Vital dyes for chromovitrectomy. *Current Opinion in Ophthalmology, 18*(3), 179–187.
40. Sabnis, R. W., (2010). *Handbook of Biological Dyes and Stains: Synthesis and Industrial Applications* (Vol. 1, pp. 5–7). John Wiley & Sons.
41. Shen, L., Zhou, G., Zhao, J., Li, P., Xu, Q., Dong, Y., & Zhang, Z., (2015). Pulsed dye laser therapy for infantile hemangiomas: a systemic review and meta-analysis. *QJM: An International Journal of Medicine, 108*(6), 473–480.
42. Sprick, M. G., (1956). Phagocytosis of M. tuberculosis and M. smegmatis stained with indicator dyes. *American Review of Tuberculosis and Pulmonary Diseases, 74*(4), 552–565.
43. Stasiak, N., Kukula-Koch, W., & Glowniak, K., (2014). Modern industrial and pharmacological applications of indigo dye and its derivatives – a review. *Acta Pol Pharm., 71*(71), 215–221.
44. Strauss, R. P., & Resnick, S. D., (1993). Pulsed dye laser therapy for port-wine stains in children: psychosocial and ethical issues. *The Journal of Pediatrics, 122*(4), 505–510.
45. To, K. T., St. Mary, L., Wooley, A. H., Wilbanks, M. S., Bednar, A. J., Perkins, E. J., & Garcia-Reyero, N., (2021). Morphological and behavioral effects in zebrafish embryos after exposure to smoke dyes. *Toxics, 9*(1), 2–9.
46. Tratnyek, P. G., Reilkoff, T. E., Lemon, A. W., Scherer, M. M., Balko, B. A., Feik, L. M., & Henegar, B. D., (2001). Visualizing redox chemistry: probing environmental oxidation–reduction reactions with indicator dyes. *The Chemical Educator, 6*(1), 172–179.
47. van de Winckel, E., Mascaraque, M., Zamarrón, A., Juarranz de la Fuente, Á., Torres, T., & de la Escosura, A., (2018). Dual role of

subphthalocyanine dyes for optical imaging and therapy of cancer. *Advanced Functional Materials, 28*(24), 1705938.

48. Van der Schueren, L., & De Clerck, K., (2010). The use of pH-indicator dyes for pH-sensitive textile materials. *Textile Research Journal, 80*(7), 590–603.
49. Veitch, D., Kravvas, G., & Al-Niaimi, F., (2017). Pulsed dye laser therapy in the treatment of warts: a review of the literature. *Dermatologic Surgery, 43*(4), 485–493.
50. Volwiler, E. H., (1926). Medicinals and dyes. *Industrial & Engineering Chemistry, 18*(12), 1336–1337.
51. Wainwright, M., & Giddens, R. M., (2003). Phenothiazinium photosensitizers: choices in synthesis and application. *Dyes and Pigments, 57*(3), 245–257.
52. Wainwright, M., (2008). Dyes in the development of drugs and pharmaceuticals. *Dyes and Pigments, 76*(3), 582–589.
53. Wainwright, M., (2011). Dyes for the medical industry. In: *Handbook of Textile and Industrial Dyeing* (Vol. 1, pp. 204–230). Woodhead Publishing.
54. Wainwright, M., (2021). Anti-infective dyes in the time of COVID. *Dyes and Pigments, 196*(1), 109–813.
55. Wainwright, M., Smalley, H., & Flint, C., (2011). The use of photosensitizers in acne treatment. *Journal of Photochemistry and Photobiology B: Biology, 105*(1), 1–5.
56. Wani, K. A., Jangid, N. K., & Bhat, A. R. (Eds.)., (2019). Impact of textile dyes on public health and the environment. *IGI Global, 1*, p. 4–6.
57. Wilson, M., & Pratten, J., (1995). Lethal photosensitization of *Staphylococcus aureus* in vitro: Effect of growth phase, serum, and pre-irradiation time. *Lasers in Surgery and Medicine, 16*(3), 272–276.
58. Yukruk, F., Dogan, A. L., Canpinar, H., Guc, D., & Akkaya, E. U., (2005). Water-soluble green perylenediimide (PDI) dyes as potential sensitizers for photodynamic therapy. *Organic Letters, 7*(14), 2885–2887.

CHAPTER 8

Non-Textile Applications of Dyes

Contents

Unit Introduction 262
8.1. Dyes for Displays 263
8.2. Laser Dyes 268
8.3. Dyes For Optical Data Storage 272
8.4. Organic Photoconductors (OPCS) 277
8.5. Nonlinear Optics 282
8.6. Summary 285
References 286

UNIT INTRODUCTION

In the realm of textile dyes, it can be argued that the majority of research efforts in dyestuffs during this century have been directed towards enhancing the properties of chromogens that were previously discovered in the last century. However, there have been a few noteworthy exceptions to this trend, such as the development of phthalocyanine, triphendioxazines, and benzodifuranones (BDFs). A significant portion of the research endeavors have been focused on altering the chromogens to enable their application on both natural and synthetic fibers, which have been prevalent since the beginning of the current century (Freeman & Peters, 2000). The effectiveness of the dying procedure and the caliber of the dyes themselves have both been improved with a lot of labor.

The utilization of dyestuffs is not limited to the textile sector, but also encompasses the photographic business, which is a significant consumer of dyes, as well as several other minor markets such as foodstuffs, drugs, and histological stains (Zhu et al., 2022). Several aspects of these aforementioned applications have been previously emphasized and will consequently be disregarded in this section. This study will examine the utilization of dyes in emerging technological sectors, specifically focusing on select instances of their implementation within the electronic and electrophotographic industries.

The electronics industry has experienced significant growth due to the rapid advancements in research and development, resulting in a multibillion-dollar industry with excellent returns (Sayyed et al., 2020). This has caused envy among many established industries that are slower-growing and older. It is a reality that a limited number of individuals in developed countries are exempt from the impact of the expanding assortment of electronic component-based products. Chemistry has played a significant role in expediting numerous swift advancements, including but not limited to liquid crystals, integrated circuits, printed circuit boards, and semiconductors. Chemistry has been a crucial contributor to the rapid expansion of the electrophotography industry, particularly in the production of photoconductive drums and developing materials (Khan & Husain, 2007). Over the recent years, there has been a noticeable surge in the significance of dyestuffs, a crucial branch of chemistry, in the emerging technology sectors. This chapter will highlight a handful of instances to demonstrate this escalating prominence. The subsequent sections are not meant to be exhaustive, but rather to provide the reader with an impression of the research methodology and the characteristics of the dyes under investigation.

8.1. DYES FOR DISPLAYS

In recent decades, there has been a notable effort to create novel display technologies that can effectively interact with the growing multitude of electronic devices. The function of the display is to present the output generated by a device in a comprehensible format for human interpretation. Its application is diverse, as evidenced by the numerous instruments that necessitate its incorporation, including but not limited to computers, calculators, timepieces, public signage, fuel dispensers, and various others (Christie, 2011). The current market need for displays is significantly high, resulting in the displays industry becoming a multibillion-dollar revenue generator. This industry is experiencing rapid growth and holds a crucial position within the larger electronics industry.

Despite the continued prevalence of cathode ray tube (CRT) displays, several emerging display technologies have demonstrated a discernible influence. The liquid crystal display stands out as a highly noteworthy achievement, particularly in applications where a compact and energy-efficient display is of utmost importance, such as in calculators, watches, and clocks. Significant research endeavors are currently focused on developing a liquid crystal alternative to the CRT utilized in television displays (Gordon, 1990). The properties of liquid crystal materials and their corresponding cell configurations have undergone continuous modifications aimed at enhancing their contrast, viewing angle, multiplexing capabilities, and cost efficiency. Liquid crystal dyes have recently gained significance in the developmental process due to their ability to enhance specific properties. One of the most significant contributions of this technology is its capacity to function in displays with reduced polarizers compared to its predecessors. This feature has resulted in a cost-effective display device and has paved the way for the development of larger area displays. The schematic representation of the functioning of a dye-containing liquid crystal cell is depicted in Figure 8.1 (Abu-Melha, 2019). The liquid crystal cell is comprised of a blend of liquid crystal material, referred to as the host, and a compatible dye, known as the guest. Typically, the liquid crystal substance comprises a specifically formulated blend of multiple individual liquid crystalline materials, which facilitates rapid switching of the cell at low fields across a broad temperature spectrum. The thin rod-shaped cyanobiphenyls (R = alkyl) constitute a significant category of liquid crystals and were one of the initial practical materials in this field (Faustino et al., 2008). At a temperature slightly above their melting point, these molecules exhibit a highly aligned state within their liquid phase, which persists over a temperature range that is contingent upon the specific R characteristics. Subsequently, they transition to an isotropic liquid state at an elevated temperature.

The molecules in this state of high order, frequently referred to as "liquid crystalline," demonstrate a number of significant characteristics (Pishgar et al., 2022). These include their capacity to orient themselves in accordance with surfaces that have been specially prepared, as well as electric fields, which is a result of their dielectric anisotropy. Therefore, a basic cellular structure would consist of a pair of glass plates that are spaced apart to enclose the liquid crystalline substance. The glass plates are subjected to a treatment that induces alignment of the liquid crystalline molecules that are in a single direction, referred to as homeotropic alignment in Figure 8.1 (Husain, 2006). Additionally, the plates are equipped with electrodes that facilitate the switching of the molecules into an alternative alignment, known as homogeneous alignment, upon the application of an electric field (as depicted in Figure 8.1) (Zollinger, 2003).

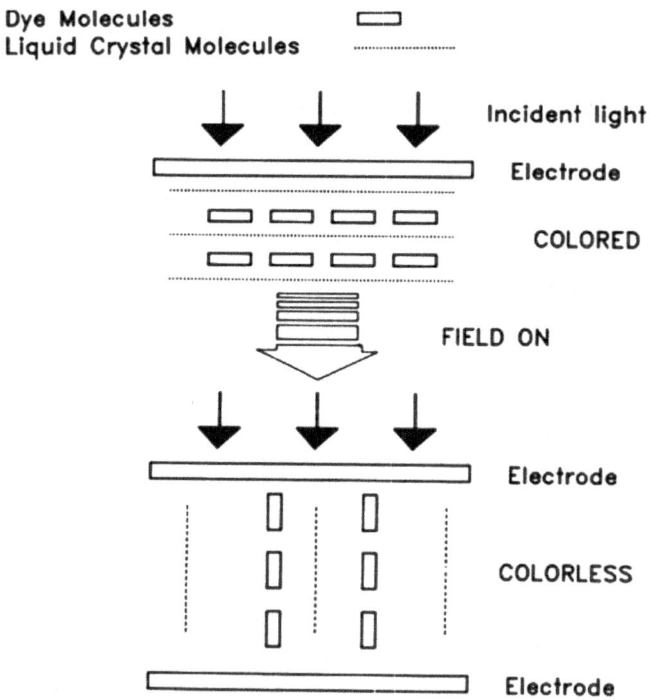

Figure 8.1. *Representation of a dyed liquid crystal display.*

Source: David R. Waring, Creative Commons License.

The arrangement of these cells serves as the fundamental structure for a device that enables the utilization of dyes. The functionality of the device is contingent upon the facile alignment of dye molecules with the liquid crystal molecules, as depicted in Figure 8.1. Assuming that the dipole moment of the transition of the dye is oriented along its longitudinal axis, it can be inferred that the transition from state (a) to state (b) depicted in Figure 8.1 would result in a transformation from a chromatic display to an achromatic one (Towns, 1999). This phenomenon occurs due to the fact that the electric vector of the incoming light is perpendicular to the direction of the propagating wave and parallel to the molecules of the dye within the cell. Consequently, the absorption of light would occur due to the alignment of its electric vector with the transitional dipole moment of the dye molecule. Upon undergoing a switch, the dye molecules experience a compulsion to orient themselves in a parallel manner with respect to the applied electric field. Consequently, the transition dipole moment of the dye becomes perpendicular to the electric vector that is associated with the light wave. In this scenario, the dye would undergo a state of desaturation, resulting in a diminished capacity to absorb light. Consequently, the light would traverse the cell, leading to a visual manifestation of colorlessness.

In actuality, guest-host displays that are commercially viable operate based on the aforementioned principles, albeit with a significantly more intricate cell design than what has been presented. Typically, dyes intended for use in these kinds of displays must exhibit various characteristics which are (Talebian et al., 2021):

- Nonionic character-to avoid turbulence, etc.;
- Solubility in, and compatibility with, the liquid crystal host;
- High purity;
- A high-order parameter (Jaiswal et al., 2011).

A number of the above-mentioned features are readily apparent, particularly the necessity for a parameter of high order. The order parameter is a metric that quantifies the degree of alignment between the dye and the liquid crystal. A recorded value of 0 is observed in the absence of alignment, whereas a value of 1 is recorded in cases of perfect alignment. Dyes exhibiting order parameters approaching unity are of significant interest, as they provide a genuinely high-contrast transition from colored to colorless upon switching (Vanstan, 1961). Conversely, an order parameter less than unity yields a display that consistently exhibits certain hues in the off state, thereby resulting in reduced contrast upon switching.

Considering the molecular structure of liquid crystal material, which comprises elongated and slender rod-shaped molecules, it is reasonable to expect that azo compounds would be the initial dyes tested in dyed liquid crystal blends, given that they can also be tailored to possess a similar rod-like configuration. A number of azo colorants have been efficaciously developed for this particular application. The dyes depicted below are representative of this category (Jang et al., 2022) (Figure 8.2).

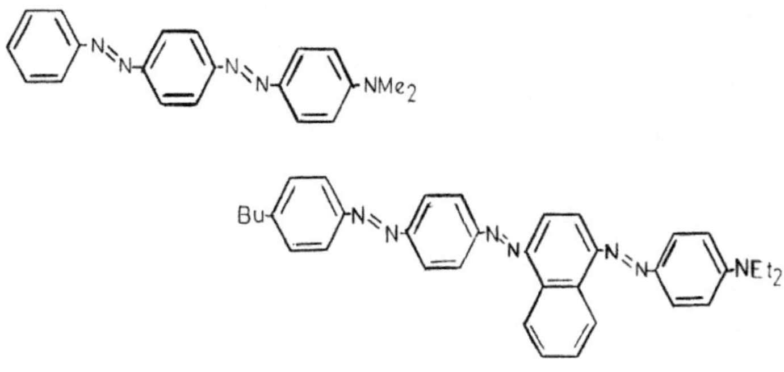

Figure 8.2. *Illustration of azo dyes tried in liquid crystal mixtures.*

Source: Geoffrey Hallas, Creative Commons License.

A noteworthy and contemporary revelation pertains to the utilization of suitably adapted anthraquinones in dyed liquid crystal displays, despite their limited classification as thin rodlike molecules. A variety of anthraquinone dyes have been formulated specifically for application in liquid crystal displays. These dyes include yellow, red, blue (n = 1–12), and green (n = 1–12) variants (Son et al., 2007). The amalgamation of all four constituents culminates in the production of a black pigment. The inclination observed in the textile sector towards the creation of a diverse spectrum of chromatic pigments for the purpose of color amalgamation is analogous to the development of liquid crystals. Anthraquinone dyes have been recognized for their high lightfastness on fibers, and this property is also observed in their high the photostability in liquid crystal media. This similarity is a second parallel that can be drawn between the two. Undoubtedly, the insights acquired from the textile sector have been utilized to the advantage of the electronics sector. The incorporation of n-alkyl chains results in the augmentation of higher-order parameters of the dyes and an increase in their solubility within the LC mixture (El-Apasery et al., 2013). Remarkably, it has been observed that the sulfur derivatives 5 and 6 exhibit high-order parameters even in the absence of long alkyl chains. Notwithstanding the enhanced photostability exhibited by anthraquinones when compared to azo dyes, a significant drawback of the former is their inherent weakness, which necessitates their usage at elevated concentrations in the liquid crystal medium. Hence, it appears probable that the utilization of blends comprising both categories of dye will be prevalent in pragmatic apparatus.

Whilst azo and anthraquinone dyes have been prevalent in guest-host liquid crystal displays, fluorescent pigments, including perylenes (n=1–16 and 11, n=1–16), have emerged as a relatively new addition to the field of dyed liquid crystal displays. The fluorescent nature of these dyes results in vividly colored displays that possess an aesthetically pleasing quality. It would be intriguing to monitor their advancements in the years to come (Figure 8.3).

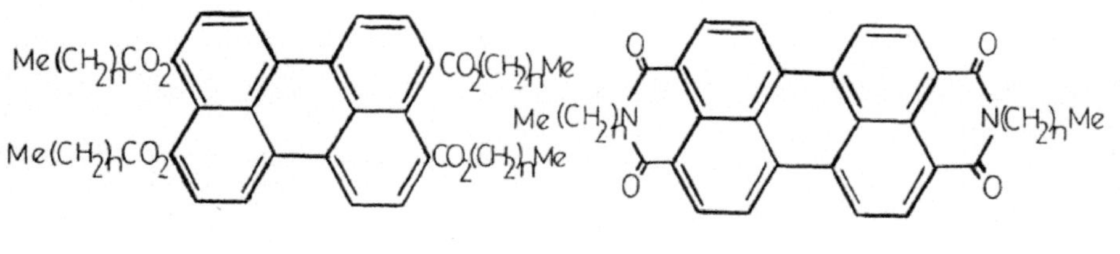

Figure 8.3. *Representation of fluorescent dyes.*

Source: S. M. Burkinshaw, Creative Commons License.

It appears that the dyes created for liquid crystal displays are currently only marginally altered textile dyes, as evidenced by the fact that numerous patent holders in this field are recognized dyestuff corporations. Electrochromic dyes, which lack equivalents in the textile industry, present an instance of dye application in the electronics sector (Sabnis, 2017). This highlights the potential for innovative research in the field of dyes. An electrochromic dye is required to exhibit a minimum of two distinct "colored" states that can be converted through electrochemical processes, as stated in basic terms.

Despite the limited impact of electrochromic displays thus far, the viologen system utilizing bispyridinium displays potential and will be utilized to briefly elucidate the underlying principles. The following procedure illustrates the mechanism of color alteration, which entails an electrochemical decrease of the bispyridinium salt lacking color to the radical cation with a vivid hue. The reaction exhibits complete reversibility as the cation can undergo electrochemical oxidation to revert to its colorless form. In addition to their fully reversible color alteration, viologens possess an additional benefit whereby the radical cation's color can be modified by modifying the substituent at the quaternary centers (Christie, 2013). For instance, the introduction of alkyl groups leads to a purple hue, whereas p-cyanophenyl replaced radical cations exhibit a green coloration. One significant benefit of electrochromic displays is their capacity to display a complete range of gray shades through the manipulation of the electrochemical reaction's magnitude (Figure 8.4).

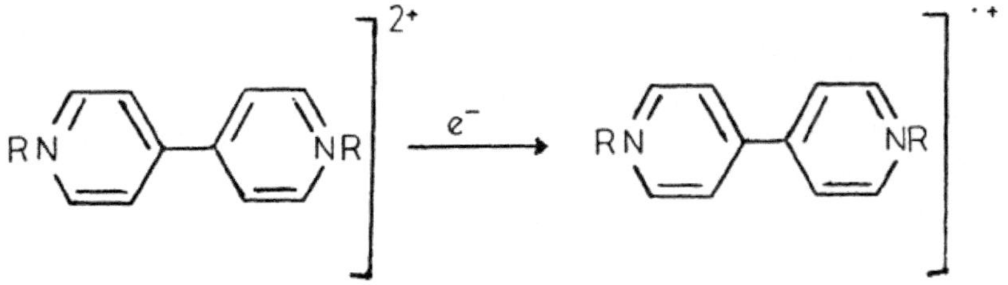

Figure 8.4. *Representation of electrochemical reduction of the colorless bispyridinium salt to the highly colored radical cation.*

Source: P. F. Gordon, Creative Commons License.

8.2. LASER DYES

The majority of individuals residing in developed nations possess knowledge of lasers and possess a rudimentary understanding of their applications, albeit lacking comprehension of their operational principles. The term "laser" is derived from the abbreviation of "Light Amplification by Stimulated Emission of Radiation" (Pavlopoulos, 2002). The inception of laser technology dates back to approximately 1960. Nanotechnologies have progressively permeated various facets of human existence, ranging from industrial applications such as the welding process and hole boring, to healthcare procedures such as treatment for cancer, and even extending to recreational pursuits. An instance of the latter pertains to the utilization of lasers in optical disks, as well as more lately in compact disks for the purpose of audio reproduction. Hence, it is probable that a considerable portion of the population will possess a laser in the upcoming years.

Table 8.1 displays a range of lasers that are presently accessible, emitting light across a spectrum from ultraviolet (UV) to infrared (Fletcher & Bliss, 1978). Upon careful examination of the table, it is evident that the emission spectrum of these lasers is highly limited (±1 nm), thereby rendering a significant amount of the UV-visible and infrared spectrum devoid of laser coverage. The judicious approach would be to employ a limited number of adjustable lasers, each capable of spanning a broad range of wavelengths, as opposed to utilizing numerous lasers, each with a restricted bandwidth. Laser dyes are crucial in facilitating the tuning of fixed lasers over a significantly wider range (Schäfer, 2005).

Table 8.1. Emission of Common Lasers

Laser	Peak Emission (nm)
Helium-cadmium	325
Argon	4,88,514
Helium-neon	632
Semiconductor	780–850 (approx.)
Yttrium-aluminum-garnet (YAG)	1,060
Carbon dioxide	10,600

Source: P. F. Gordon, Creative Commons License.

The operational principles underlying the functionality of laser dyes are comprehensively comprehended, thereby facilitating the selection of suitable dyes. In order for a dye to exhibit laser action, it is necessary for its constituent molecules to possess the ability to be stimulated or pumped into one of their higher energy states. After achieving a substantial population inversion, wherein a significant number of dye molecules exist in an excited state, the dye is required to release the energy it has stored from its excited state to the ground state through a radiative mechanism. The efficacy of the laser dye is significantly reliant on its efficient radiative decay, making the condition highly crucial (Pavlopoulos & Hammond, 1974). In the presence of competing nonradiative decay processes such as internal conversion, the utility of the dye is considerably restricted. From a practical standpoint, it can be inferred that laser dyes that exhibit high fluorescence efficiency are typically considered to be of good quality. However, it should be noted that the converse is not necessarily applicable. Another pragmatic aspect to consider is the necessity for the dye to assimilate at the output wavelength of a suitable laser, as the laser is employed to stimulate the dye into its excited state. The diagram presented in Figure 8.4 depicts a standard correlation among dye absorption, laser output, and dye laser emission (Deligeorgiev et al., 2010). The figure illustrates that the laser dye's output, represented by the dotted line, exhibits longer wavelengths compared to both its absorption, represented by the full line, and the output of the pump lasers. Additionally, the frequency range of the dye laser's emitted laser frequencies is significantly broader than that of the pump laser. The tunable nature of the dye laser allows for effective conversion of the basically monochromatic emission of the pump laser, as it emits over a significant range. Through the use of a set of filters, any desired wavelength of laser light within the envelope of the dye laser's emission can be selected (Figure 8.5).

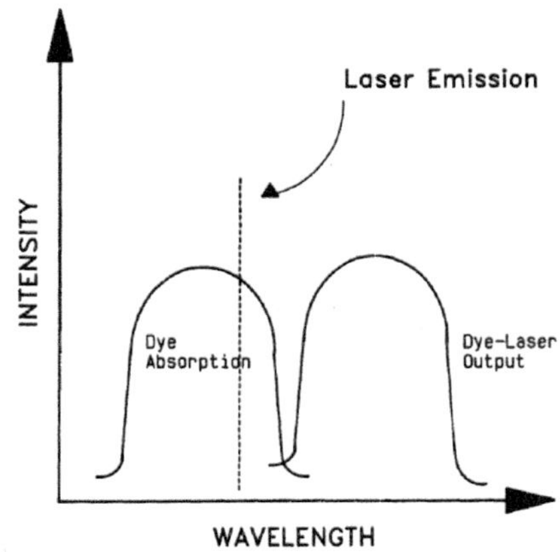

Figure 8.5. *Graph of relative positions for dye absorption and lasing wavelength.*

Source: David R. Waring, Creative Commons License.

Ideally, a singular dye would be utilized to generate a wide range of laser wavelengths with similar effectiveness, utilizing only one laser source. Unfortunately, attaining this objective has been impracticable. Similar to the textile sector, which employs several dyes to produce a diverse range of colors, a multitude of laser dyes are required to generate a "color gamut" of laser wavelengths. After conducting a thorough examination of the product listings from multiple vendors, it has been noted that a diverse array of dyes, amounting to several tens, are accessible and capable of adequately covering the spectrum from less than 400 nanometers to greater than 1,000 nanometers (Basting et al., 1976). The choice of a specific dye is dependent on various factors, including the ideal laser wavelength, properties of the pump laser, cost implications, dye stability, and efficiency, among other considerations. The customary practice for dye application entails their deployment at a diminished concentration of 10–4 M in solvents like alcohols and DMF to avert overheating. Ensuring high purity of dyes is crucial in order to avoid premature quenching of the excited state. Consequently, their worth is substantial, surpassing that of textile dyes by several orders of magnitude, despite their requirement in minuscule quantities compared to the millions of tons of textile dyes sold each year. It may seem paradoxical that many of the earliest formulated dye categories from the previous century are among the most highly efficient laser dyes, despite the relatively recent use of laser dyes. The following paragraphs outline some of the relatively noteworthy ones. The spectral range associated with blue hues is predominantly regulated by dyes classified under the carbostyril and coumarin classes, which are two notable exemplars of this group. Carbostyril displays absorption at a wavelength of 360 nm and exhibits its maximum lasing effectiveness at a wavelength of 425 nm, indicating a positive shift of 65 nm (Rinke et al., 1986). In contrast, the compound with a numerical identifier of 16 demonstrates absorption at a comparatively longer wavelength of 366 nm and displays its maximum lasing wavelength at a significantly greater wavelength of 446 nm, indicating an 80 nm shift (Winters et al., 1974).

The application of coumarin-type dyes in the green region of the spectrum is demonstrated by the absorption of the coumarin dye at 436 nm and its subsequent lasing at 523 nm, which indicates a shift of 87 nm. The category of dyes that offers the greatest benefits is based on the generalized structure 18. This particular dye exhibits the ability to effectively absorb wavelengths within both the green and red regions of the electromagnetic spectrum. Fluorescein, rhodamine 6G, cresyl violet, and Nile blue exhibit absorption capabilities at wavelengths of 501 nm, 530 nm, 601 nm, and 628 nm, respectively. In addition, it should be noted that these dyes demonstrate lasing maxima at wavelengths of 560, 590, 650, and 710 nm, correspondingly (Maslov, 2006). The region of near infrared within the laser spectrum is of considerable significance. Nevertheless, within the realm of textile dyes, this particular area is frequently disregarded, save for its prospective utilization in the realm of camouflage applications. Cyanine dyes exhibit a wide range of applicability as laser dyes within the near-infrared spectral region. Furthermore, cyanine dyes are utilized as a prototype to showcase the systematic implementation of structural alterations that result in predictable variations in the lasing wavelength. The incorporation of supplementary double bonds leads to a displacement of the lasing wavelength from 740 nm to 850 nm (Mula et al., 2013). Furthermore, the introduction of supplementary heterocycles into the rings and substituents on the conjugated chain results in a noteworthy displacement to 980 nm. By means of structural manipulation, it is possible to obtain dyes that demonstrate lasing characteristics at wavelengths greater than one micron (Figure 8.6).

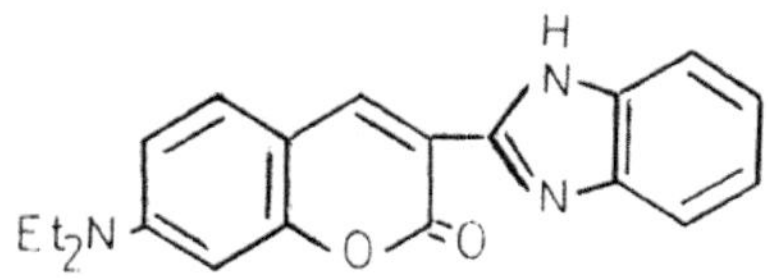

Figure 8.6. *Structure of coumarin dye.*

Source: F. Walker, Creative Commons License.

8.3. DYES FOR OPTICAL DATA STORAGE

Throughout history, various societies have engaged in the practice of data storage for millennia. In the beginning, oral tradition was the primary means of transmitting knowledge from one generation to the next, relying solely on human memory. Over time, data storage media evolved from stone inscriptions and papyrus leaves to paper. Throughout centuries, ink and paper have been the preferred means of data storage. However, the emergence of computers has introduced magnetic data storage media, which have significantly influenced the field. These media offer benefits such as high storage density and quick access to and retrieval of information (Mustroph et al., 2006). In recent times, optical data storage has emerged as a technology that provides remarkably elevated levels of data storage density. Although photographic-type processes remain capable of providing a considerable level of storage density, it is the optical disk coupled with laser optics that has facilitated the realization of ultrahigh densities achievable in optical disks. Figure 8.7 depicts a schematic illustration of a basic optical disk (Åstrand et al., 2000). Commonly, the substrate material employed is a plastic, specifically polycarbonate, onto which a recording layer of minimal thickness is applied. Subsequent coatings are applied onto the recording layer in order to safeguard it against particulate matter, smudges, and other unwanted impurities. The process of modifying particular regions of the recording layer to exhibit distinct light interaction from the unmodified sections facilitates the storage of information. This is conventionally achieved by ensuring a substantial contrast in reflectivity among those two regions of the disk. The technology employed by the compact audio disk (CD), which is currently garnering significant attention, is precisely this. The recording layer in practical applications is typically composed of a metallic film that has been perforated with holes (Waring & Hallas, 2013). This allows a laser to scan the disk's surface and either reflect off the metal or pass through the holes. Subsequently, a detector captures signals that are eventually processed and utilized for information retrieval.

Figure 8.7. *Configuration of a simple optical disk.*

Source: Geoffrey Hallas, Creative Commons License.

The utilization of laser technology is the key factor contributing to the remarkably high storage density, as it has the capability to be concentrated to spot sizes of approximately one micron. The utilization of a semiconductor laser with a wavelength of around 800 nm has become widely accepted in the realm of optical data storage, as evidenced by its use in the compact disk (Emmelius et al., 1989).

The aforementioned methodology of utilizing a pre-stamped metallized disk is deemed suitable for the compact disk configuration; however, it is deemed unsuitable for optical disks. As previously stated, the compact disc (CD) is pre-encoded with all the necessary information, similar to a traditional record. Nevertheless, in numerous data storage applications, it is more desirable to furnish a blank disc that can be subsequently documented on by the end-user.

The production of masters and pressings, as commonly employed for CDs, is deemed excessively expensive and impractical for the creation of a limited quantity of necessary data (Gourevich et al., 2004). Dyes have had a significant impact due to their ability to enable the user to directly encode information through the laser. The process of written communication and comprehension is fundamentally uncomplicated. The disk substrate is coated with a dye in order to serve as the recording layer. During the writing process, the laser operates at nearly full power and selectively creates perforations in the extremely thin dye layer of the disk in accordance with the digitally encoded data. To access the encoded signal, the laser's power is reduced to prevent harm to the recording layer. The surface of the disk is then scanned with a detector to play the information.

The aforementioned delineation of the writing/reading process promptly implies certain characteristics that the dye must possess. The efficacy of hole formation mechanism is contingent upon the absorption of laser radiation by the dye, which subsequently undergoes heating and migration from the hot region during the hole formation stage. Therefore, it is imperative that the dye absorbs the wavelength of the laser output. In actuality, a proficient dye must satisfy multiple criteria, encompassing:

- Absorb at laser output;
- Possess good reflectivity;
- Possess good stability (Lott et al., 2011).

The initial point has incited novel research on dyes, as aforementioned, due to the application of solid-state lasers that emit near infrared radiation at roughly 800 nm. This spectral region does not necessitate conventional textile dyes, which lack strong absorption properties. Numerous dyes have been suggested for this particular application, although they are primarily derived from traditional dye structures that have been appropriately altered. The techniques employed to shift the recognized chromogens towards the proximity of infrared radiation are fundamentally akin to those utilized by subsequent generations of dyestuff chemists to procure dyes of varying hues (Heilemann et al., 2005). Consequently, near-infrared dyes can be regarded as a logical extension of the visible spectrum.

The types of dyes currently receiving most attention can be divided into four classes:

- Cyanines and related dyes;
- Triarylmethanes and related dyes;
- Quinone-based dyes; and
- Metal-complex dyes (Yanez et al., 2009).

The ensuing discourse delineates several customary frameworks for each classification. Cyanine dyes that exhibit absorption in the near-infrared region have been recognized for a considerable duration, primarily due to their application as sensitizers in photography (Belfield & Schafer, 2002). One significant benefit of cyanine dyes is their ability to establish a consistent correlation between color and structure. This allows for effective customization of the dyes to absorb at the wavelength of the emitting laser through manipulation of the conjugated chain, end groups, and substituents in the chain. As an illustration, the cyanine molecule with a zero degree of polymerization exhibits an absorption peak at 224 nm, whereas the cyanine molecule with a degree of polymerization of five, characterized by an increased number of conjugated double bonds, displays an absorption peak at 735 nm (Shiozaki et al., 1987). Empirically, it has been observed that each supplementary double bond induces a bathochromic shift of 100 nm as a general principle. Due to their instability, it is improbable that dyes categorized as type 26 would be utilized in optical data storage. Cyanine dye serves as an example of a more stable dye that can be achieved, with an absorption wavelength of 755 nm. A dye that is closely associated exhibits absorption at a longer wavelength of 780 nm due to the additional conjugation that arises from the two benzene rings that are annelated to the indoline rings (Zhu et al., 2006). This dye has been the focus of numerous patents." A comparable wavelength can be achieved through the utilization of cyanine, whereby the central double bond is constrained within a ring structure.

Numerous modifications of traditional cyanine dyes have garnered significant interest from corporations invested in optical data storage, as evidenced by the recent surge in patents and publications. The Thiopyrylium dyes exhibit absorption in the near infrared region (λmax = 755 nm) despite having a limited number of conjugated double bonds in the chain that links the two heterocyclic rings (Hunger, 2007). When a squarylium nucleus is introduced at the center of the conjugated chain, it results in the production of near-infrared dyes that are potentially useful in video disks. These dyes display interesting spectral properties, such as λmax = 800 nm and λmax = 745 nm. Significant research efforts have been directed towards squarylium-substituted dyes and their analogues, which exhibit absorption properties that align with the emission wavelengths of semiconductor lasers.

The triarylmethane dyes, which were one of the earliest synthetic dyes, bear some resemblance to the cyanines. Some dyes in the violet to green range are still utilized in textile production today, albeit to a limited extent, as indicated by the preceding sections. The planarity enforced in the rings can shift absorbance to the highest of these dyes into the near infrared. The magnitude of the displacement achievable can be observed through a comparison between the optical properties of crystal violet (n= 0 λ_{max} = 589 nm) and the infrared dye (λ_{max}=850 nm) (Spiridon et al., 2015). In a similar manner to cyanines, it can be observed that an increase in conjugation leads to a bathochromic shift. This is exemplified by the absorption of the dye (n = 1) at 770 nm, which is 200 nm more bathochromic compared to n = 0. Moreover, substitution of the central atom, specifically the replacement of carbon with nitrogen, induces a significant bathochromic shift, causing the aza compound to exhibit absorption at 920 nm.

The quest for appropriate infrared absorbers for optical disks has extended beyond cyanines and triarylmethanes, as investigations have been conducted on conventional-looking textile dye categories such as quinones, azo compounds, and phthalocyanines. In the field of optical disks, a number of patents have emerged that pertain to the utilization of substituted anthraquinones. These patents place significant reliance on the use of (donor) substituted anthraquinones in order to achieve the desired shift towards the near infrared.

Anthraquinone exhibits absorption at a wavelength of 750 nm, while the heterocyclic derivative displays absorption at a longer wavelength of 770 nm (Corredor et al., 2006). Analogous methodologies applied to benzoquinones with similar structural characteristics have yielded positive results, as evidenced by the characterization of various substituted benzoquinones that exhibit absorption in the near-infrared spectrum. These compounds are expected to possess the requisite properties for employment in optical data storage applications. The heterocyclic derivative exhibits an absorption peak at a wavelength of 735 nm, while the dicyano compound displays an absorption peak at 770 nm, which is in close proximity to the emission wavelength of semiconductor lasers (Pudavar et al., 1999).

Phthalocyanines are widely recognized for their exceptional durability and resistance to light degradation. Consequently, there has been a concerted effort in research to enhance their light absorption properties in the near-infrared spectrum. The pigment known as copper phthalocyanine exhibits a characteristic blue hue and possesses an absorption capacity of 678 nm (Cui et al., 2016). Similarly, the dihydrogen compound of this pigment demonstrates an absorption capacity of 698 nm. The process of appending supplementary benzene rings to the phthalocyanine molecule leads to a significant shift towards longer wavelengths, known as bathochromic shift, which is observed in naphthalocyanine with a maximum absorption wavelength of 780 nm in the near-infrared region (Park et al., 2002). The electronic absorption is influenced by the selection of the central metal atom, and it appears that metals such as vanadium and lead possess the necessary characteristics for optical data storage. The introduction of electron donors at the outermost region of the phthalocyanine ring structure is a viable strategy for causing a red shift in the absorption spectrum. It is a well-established fact that altering the crystal morphology of phthalocyanines can result in the production of near-infrared absorbing phthalocyanines. This is due to the varying wavelengths at which different crystalline forms of phthalocyanines absorb (Seto et al., 1996). The manipulation of e and Q forms of phthalocyanines has been a longstanding practice in the dyestuffs industry. The aforementioned phenomenon has been leveraged to generate diverse metastable crystalline configurations of phthalocyanines that exhibit absorption at 780 and 830 nm, such as the X and T forms.

8.4. ORGANIC PHOTOCONDUCTORS (OPCS)

Despite the deliberate omission of photography in this chapter, it is noteworthy to include electrophotography as a significant addition to the category of new applications of dyes in high-technology industries. The electrophotographic procedure is a recent technological advancement that surfaced within the past few decades. The photocopier industry has experienced remarkable growth since its inception, evolving into a multibillion-dollar business. Photocopiers are ubiquitous in various settings such as offices, schools, administration buildings, and libraries. Furthermore, the advent of laser printers that can be connected directly to computers has expanded the potential market for photocopiers significantly (Figure 8.8).

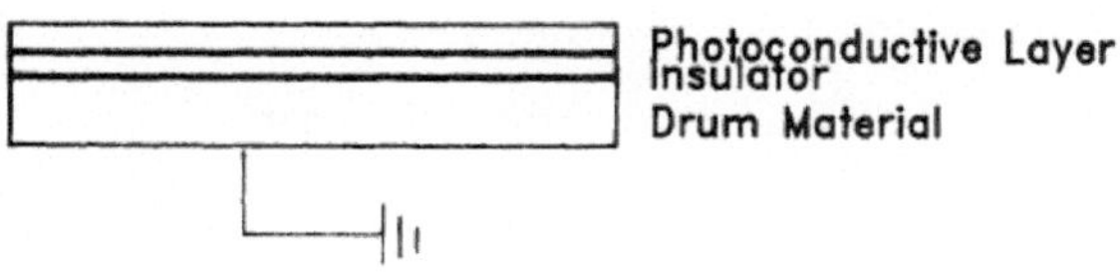

Figure 8.8. *Illustration of photoconductive drum.*

Source: M. A. Weaver, Creative Commons License.

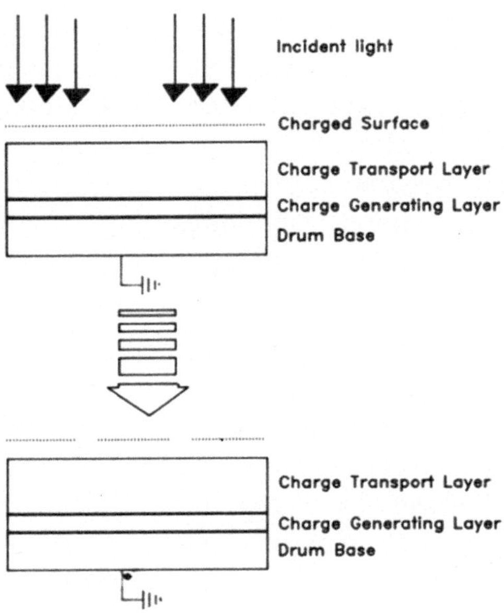

Carlson pioneered the transfer xerographic process, which is widely recognized as the most familiar electrophotographic process. The process entails three main steps: (a) the creation of an electrostatic latent image; (b) the development of said image; and (c) the transfer of the developed image onto a substrate. The process can be conveniently divided into five distinct steps.

- The xerographic plate undergoes ionization through a process such as corona discharge in order to attain a charge (Weiss & Abkowitz, 2010).
- The subject is presented to the plate, resulting in the formation of an electrostatic latent image.
- The development process of the latent image takes place.
- The image that has been created is subsequently transferred onto a substrate, such as paper or plastic.
- The image is rendered immobile through a process such as thermal treatment (Warta et al., 1985).

In its most basic configuration, the xerographic plate comprises a photoconductive layer and a base electrode that are separated by an insulator, as depicted in Figure 8.8 (Bässler, 1993). During the initial phase, the xerographic plate, commonly in the shape of a cylindrical drum or an uninterrupted belt, is uniformly charged. During the second phase, the source document intended for duplication is subjected to light exposure, and the resultant reflected image is directed onto the plate with precision. In regions characterized by high light intensity, the charge undergoes dissipation, whereas in areas devoid of light exposure, the charge persists (Schaffert, 1971). Consequently, a replica of the original document is produced through the creation of an electrostatic image on the xerographic plate, where regions of low or negligible charge density correspond to the lighter sections of the original, while regions of high charge density correspond to the darker sections of the original.

Selenium has been the preferred photoconductor for a considerable duration (Pai, 1983). Nonetheless, in recent years, the organic photoconductor (OPC) market has experienced a significant influx. OPCs present several benefits over selenium. They are cost-effective, non-hazardous, and can be customized to offer sensitivity in any spectral range, including the near-infrared. Additionally, they can be manufactured into diverse configurations. As such, they are optimally suited for employment in personal and compact copying apparatuses, and it is within this particular market segment that they are currently exerting the most pronounced influence. A significant proportion of organic photovoltaic cells (OPCs) are constructed using a dual-layer configuration, wherein the charge generation layer (CGL) is situated in closer proximity to the electrode, and a transparent charge transfer layer is applied on the top. During the stage of latent image formation, the xerographic plate is exposed to light which penetrates the transparent charge transfer layer and reaches the charge generating layers (Borsenberger et al., 1978). As a result, an electron or a hole is injected into the charge transfer layer, depending on the surface charge. Subsequently, the CTL facilitates the transportation of the hole or electron to the surface, thereby neutralizing the charge that was formerly deposited as a result of corona discharge (refer to Figure 8.9) (Weiss & Abkowitz, 2017). In regions devoid of illumination, the aforementioned mechanism fails to function, resulting in the retention of charge on the surface. In the event that the surface charge dissipates to the ground through an unknown mechanism in the absence of light exposure, commonly referred to as dark decay, the operational effectiveness of the optical proximity correction (OPC) is significantly compromised (Yarmchuk & Keefe, 1989).

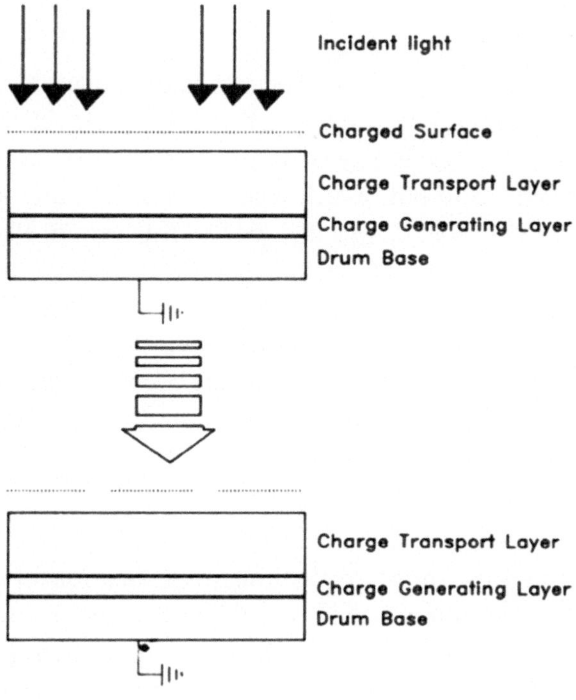

Figure 8.9. *Representation of latent image formation.*

Source: J. Griffiths, Creative Commons License.

The majority of the operational process conditions (OPCs) that are currently in use or have been proposed consist of modified dyestuffs. The OPC's sensitivity to light in the apparent or near infrared regions is a predictable outcome of utilizing a light source of this nature during the exposure stage (Weiss & Abkowitz, 2007). It is evident that dyestuffs are employed for purposes beyond their function as mere colorants. It is important to note that not all dyestuffs exhibit the complete set of desirable properties necessary to function as an effective OPC. Various categories of dyes have been discovered to be highly advantageous, including but not limited to azos, perylenes, phthalocyanines, vat types, and cyanine dyes. Thiopyrylium dyes have been purported to possess utility as OPCs (Waltman et al., 1993). Azo dyes, particularly in their pigmentary state, have been extensively patented and utilized by various companies as photoconductors. The majority of azo compounds exhibit a symmetrical arrangement, consisting of either disazo or trisazo compounds (Schlebusch et al., 1996). Additional dye varieties that exhibit utility include perylenes and vat dyes, such as dibromoanthanthrone. Typically, the earlier mentioned OPCs are paired with a charge transport layer that comprises electron-rich compounds like pyrazolines or hydrazones.

A contemporary advancement in electrophotography pertains to the emergence of near-infrared OPCs. These photoconductors can be paired with semiconductor laser diodes and subsequently employed in laser printers (Karl et al., 1991). As previously mentioned, the emission wavelength of these lasers is around 800 nm, which enables various materials to function as OPCs. Polyazo compounds are included in this group. Phthalocyanines have been adapted to function as optical power limiters, similar to their application in optical data storage, which is noteworthy. Various crystal structures of dihydrogen phthalocyanines have demonstrated remarkable photoconductivity, and a specific machine has been developed utilizing this material. Chloroaluminum phthalocyanine, along with several other metal-substituted phthalocyanines, exhibits potential for utilization as an organic photoconductive material.

The preceding discourse and succinct portrayal of OPC substances evince that dyes are assuming a novel and stimulating function in xerography as OPCs. Dyestuffs, in addition to their role in the photoconductor application, also serve a function in the development stage of the xerographic process (Popovic & Menzel, 1979). During the development stage, the photoconductor drum undergoes a transformation from an imperceptible electrostatic image to a discernible image through the use of a developer. Typically, the latter component comprises a carrier substance, which accounts for 95% of the mixture, and a toner substance, which accounts for the remaining 5%. The carrier material comprises of iron particles that have been milled to a size range of 50–500 μ (Cais et al., 1992). The toner component constitutes 59 cents of the developer composition and is responsible for providing the chromaticity to the resultant duplicate. The toner particles, which measure between 10 and 15 micrometers in size, are applied as a coating onto the comparatively larger iron carrier particles (Borsenberger & Weiss, 2018). Consequently, as the charged xerographic plate draws near, it elicits an opposing charge in the toner, which subsequently adheres to the electrostatic picture on the plate upon application of the developer. The process for a selenium drum with a positive charge is depicted schematically in Figure 8.10 (Kalinowski et al., 1996).

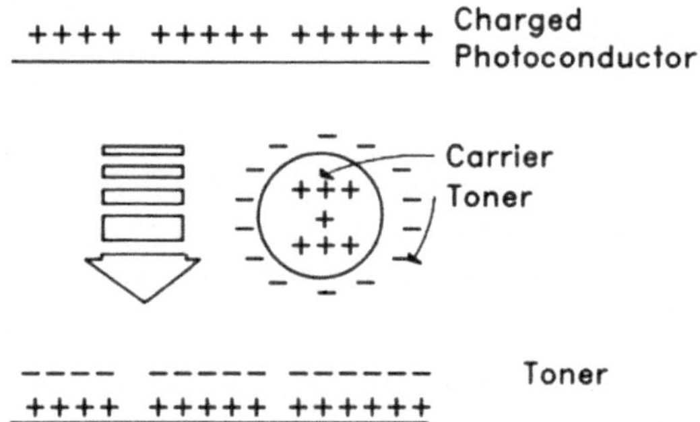

Figure 8.10. *Development of latent image.*

Source: F. Walker, Creative Commons License.

The toner is composed of three constituents, namely a thermoplastic heat fusible resin accounting for 90% of the composition, a pigment constituting 8%, and a charge control agent (CCA) comprising 2% (Lardon et al., 1967). Remarkably, the dyes are employed not for their coloring properties, as carbon black is typically favored for this purpose, instead serving as the CCA. The CCA appears to function as a buffer in regulating the charge of toner particles, resulting in a uniform, high-contrast, and densely printed final document. The exact mode of action of these substances remains unclear (Bässler, 1994). Nevertheless, complexed azo dyes appear to be well-suited for the purpose of positively charging xerographic plates, such as those made of selenium metal. As a standard practice, the dyes commonly employed are chromium complexes of azo compounds, which can be illustrated by the compound in question. In contrast, nigrosines have been found to be beneficial for photoconductors with a negative charge.

8.5. NONLINEAR OPTICS

The initial segment of the preceding quartet of sections has demonstrated the utilization of dyestuffs solely for their capacity to function as pigments. The subsequent three segments employed dye molecules for alternative purposes, wherein their hue is inconsequential, yet their capacity to assimilate light holds significance. In such instances, subsequent to the absorption of light, there is a reemission of the light at a longer wavelength. This phenomenon results in either the induction of local heating that leads to the formation of pits or a modification in the medium's conductivity (Ray, 2010). The utilization of dyestuffs in nonlinear optics is a noteworthy application, wherein the relevance of their color is negligible. It is noteworthy that the absorption of radiation is not a prerequisite for dyestuffs to be employed in nonlinear optics.

Nonlinear optics is a field of study that focuses on the interaction between light waves and dielectric media, resulting in the modification of certain properties of light, which include phase or frequency. The incident light's oscillating electric vector elicits an oscillating polarization in the dielectric medium, which subsequently emits radiation at the same frequency as its oscillation. Hence, it is evident that the characteristics of the reirradiated light are contingent upon the interplay between the incoming light and the medium. In a linear optical medium, the polarization (P) exhibits a linear relationship with the polarizing radiation (E), as expressed in equation (1). This relationship results in the manifestation of observable effects, which include refractive index (Marder et al., 1991).

$$P = \varepsilon XE \tag{1}$$

In a medium exhibiting nonlinear optical property, the relationship between polarization and the polarizing field is no longer linear, as demonstrated by equation (2) (Prasad & Reinhardt, 1990).

$$P = \varepsilon X(1)E + \varepsilon X(2)E^2 + \varepsilon X(3)E^3 + \cdots \tag{2}$$

The contrast between equations (1) and (2) is conspicuous as the latter introduces supplementary terms that generate further beneficial outcomes, as illustrated in Table 8.1 (Lacroix et al., 2016).

The utility of dye studs in unconventional applications can be discerned by examining equations (2) and (3) in relation to the second term.

$$p = \alpha E + \beta E^2 + \cdots \tag{3}$$

Equation (2) shows a relationship between the macroscopic polarization and the incident electric field; a rather similar relationship holds at the molecular level as shown in equation (3) (Burland, 1994). The hyperpolarizability term, β, can be calculated using molecular orbital (MO) programs and it appears that molecular features which give rise to a high β are the presence of electron donors, electron acceptors, and a conjugated pathway—the very features possessed by many colored molecules (Muhammad et al., 2013). Unfortunately, designing efficient materials is not quite as simple as making compounds with high β since, for X^2 to be large, a symmetry requirement must be satisfied. The symmetry rule requires that the molecules be assembled in a noncentrosymmetric arrangement with all the molecules pointing in approximately the same direction, otherwise cancellation occurs to give a zero X^2, whether or not the individual molecules possess a high β. Hence, an ideal material would have the following attributes:

- High β;
- Noncentric structure (Zhang et al., 2009).

The second equation demonstrates a correlation among macroscopic polarization and the electric field that is introduced, while a comparable correlation is evident at the molecular level, as depicted in the third equation. The calculation of the hyperpolarizability parameter, β, can be achieved through the utilization of MO software. It has been observed that molecular characteristics that contribute to a heightened β value include the existence of electron-donating and electron-accepting groups, as well as a conjugated pathway. These attributes are commonly found in various pigmented molecules (Asher et al., 1998). Regrettably, the process of creating effective materials is not as straightforward as synthesizing compounds with elevated β, as the fulfillment of a symmetry prerequisite is necessary for X2 to attain significant values. The principle of symmetry dictates that the molecules must be arranged in a noncentrosymmetric configuration with all molecules oriented in a similar direction (Mutailipu et al., 2017). Failure to adhere to this arrangement results in cancellation, yielding an X2 value of zero, regardless of the individual molecules' high β. Therefore, an optimal substance would possess the subsequent characteristics:

- The high β value is indicative of a strong positive correlation between two variables in a statistical analysis.

- The noncentric structure refers to a geometric arrangement (Bredas et al., 1994).

Numerous dyes already fulfill the initial requirement as they possess donors, acceptors, and a conjugated pathway. Examples of such dyes include azo dyes, nitro dyes, and anthraquinones. The outstanding issue pertains to the arrangement of the molecules within a noncentrosymmetric framework. Crystallizing molecules are a highly appealing alternative due to the ability to manipulate crystals through cutting, polishing, and integration into various devices. The primary issue lies in the fact that merely a minority of compounds possess crystal structures that lack centrosymmetry (Würthner & Yao, 2000).

Despite this, a number of colored compounds exhibiting a noncentric configuration and elevated nonlinear coefficient have been discovered, surpassing those of presently utilized inorganic substances. Nitroanilines have been found to be effective nonlinear optic materials, with similarities to nitro dyes. Regrettably, 4-nitroaniline exhibits limited utility due to its tendency to form a centrosymmetric crystal structure (Luo et al., 2007). Nevertheless, the introduction of a methyl group at the 2-position (located ortho to the amine) results in a highly advantageous substance. Additional techniques have been employed to ensure the absence of centric structures, including the integration of chiral moieties or the utilization of hydrogen bonding phenomena.

Other chromophores can also serve as nonlinear optical materials, besides nitroanilines. Various other established categories of dyes are present, including stilbenes, azo compounds, and hemicyanine dyes, which have been documented to produce significant second-order effects. Moreover, it seems that their molecular efficiency surpasses that of the aforementioned nitroanilines. The crystallization of the stilbazolium salt occurs in a space group that lacks a center of symmetry and exhibits a significantly greater effect compared to previously described molecules (Lin et al., 2020). Both the stilbene azo molecule and the hemicyanine exhibit beneficial properties; however, neither has been employed as a straightforward crystal due to the implementation of alternative techniques to achieve a noncentrosymmetric configuration. An example of this is the incorporation of stilbene into a liquid crystal blend, which is subsequently oriented into a noncentrosymmetric configuration using an electric field. Through the integration of hydrophobic and hydrophilic moieties, the azo compound and hemicyanine have been effectively altered to enable their organization into multiple layers. The Langmuir-Blodgett method was employed to deposit one molecular layer at a time, resulting in the development of nonlinear optical systems that exhibit exceptional efficiency (Yu et al., 2016).

8.6. SUMMARY

As evidenced by the preceding five sections, research on dyestuffs is currently thriving and continues to evolve, in part due to its involvement in various emerging technological industries. The chapter's coverage of the topics is cursory, and the enumeration of novel dye applications is not exhaustive. Nonetheless, it is evident that numerous innovative uses necessitate the dyestuff chemist's expertise in devising dyes that possess absorption properties at precise wavelengths. In addition to the aforementioned condition, the dyes must satisfy supplementary standards such as lasing, pit creation for optical disks, or photoconductivity. However, the chemist specializing in dyestuds possesses the capability to fulfill these demands, as numerous of these occurrences utilize characteristics that are familiar and researched by the chemist. In the field of nonlinear optics, where light absorption is not a prerequisite, the expertise of dyestuffs chemists appears to be valuable. This is because the techniques employed by these chemists to shift dyes towards longer wavelengths, such as enhancing conjugation and electron donor/acceptor capabilities, have been observed to yield materials with greater molecular nonlinearities.

It is probable that novel applications for chromophores will emerge in the future, either due to their distinctive electronic characteristics or purely based on their chromatic responsiveness. Furthermore, it is possible that the molecules that have been engineered for the purpose of optical data storage, laser dyes, and OPCs exhibit a spin-od value in alternative domains. It can be inferred that the field of dyestuffs research holds promising prospects, and the utilization of dyes in non-textile applications is expected to witness a surge in the foreseeable future. It is possible that in a future century, non-textile applications of dyes will become more prevalent, relegating textile uses to a secondary, albeit significant, role for these substances.

REFERENCES

1. Abu-Melha, S., (2019). Synthesis of new diazobenzene dyes clubbed with sulphonamide moiety and their biological applications. *Pigment & Resin Technology, 48*(5), 397–403.
2. Asher, S. A., Holtz, J., Weissman, J., & Pan, G., (1998). Mesoscopically periodic photonic-crystal materials for linear and nonlinear optics and chemical sensing. *Mrs Bulletin, 23*(10), 44–50.
3. Åstrand, P. O., Ramanujam, P. S., Hvilsted, S., Bak, K. L., & Sauer, S. P., (2000). Ab initio calculation of the electronic spectrum of azobenzene dyes and its impact on the design of optical data storage materials. *Journal of the American Chemical Society, 122*(14), 3482–3487.
4. Banchero, M., (2020). Recent advances in supercritical fluid dyeing. *Coloration Technology, 136*(4), 317–335.
5. Bässler, H., (1993). Charge transport in disordered organic photoconductors. A Monte Carlo simulation study. *Physica Status Solidi B (Basic Research); (Germany), 175*(1), 3–6.
6. Bässler, H., (1994). Non-dispersive and dispersive transport in random organic photoconductors. *Molecular Crystals and Liquid Crystals Science and Technology. Section A. Molecular Crystals and Liquid Crystals, 252*(1), 11–21.
7. Basting, D., Ouw, D., & Schäfer, F. P., (1976). The phenoxazones: a new class of laser dyes. *Optics Communications, 18*(3), 260–262.
8. Belfield, K. D., & Schafer, K. J., (2002). A new photosensitive polymeric material for WORM optical data storage using multichannel two-photon fluorescence readout. *Chemistry of Materials, 14*(9), 3656–3662.
9. Borsenberger, P. M., & Weiss, D. S., (2018). Photoreceptors: Organic photoconductors. In: *Handbook of Imaging Materials* (Vol. 1, pp. 387–442). CRC Press.
10. Borsenberger, P. M., Chowdry, A., Hoesterey, D. C., & Mey, W., (1978). An aggregate organic photoconductor. II. Photoconduction properties. *Journal of Applied Physics, 49*(11), 5555–5564.
11. Bredas, J. L., Adant, C., Tackx, P., Persoons, A., & Pierce, B. M., (1994). Third-order nonlinear optical response in organic materials: Theoretical and experimental aspects. *Chemical Reviews, 94*(1), 243–278.

12. Burland, D., (1994). Optical nonlinearities in chemistry: introduction. *Chemical Reviews, 94*(1), 1–2.
13. Cais, R. E., Nozomi, M., Kawai, M., & Miyake, A., (1992). Antiplasticization and abrasion resistance of polycarbonates in the charge-transport layer of an organic photoconductor. *Macromolecules, 25*(18), 4588–4596.
14. Christie, R. M., (2011). Fluorescent dyes. *Handbook of Textile and Industrial Dyeing, 1*, 562–587.
15. Christie, R. M., (2013). Chromic materials for technical textile applications. In: *Advances in the Dyeing and Finishing of Technical Textiles* (Vol. 1, pp. 3–36). Woodhead Publishing.
16. Corredor, C. C., Huang, Z. L., & Belfield, K. D., (2006). Two-photon 3D optical data storage via fluorescence modulation of an efficient fluorene dye by a photochromic diarylethene. *Advanced Materials, 18*(21), 2910–2914.
17. Cui, Y., Li, B., He, H., Zhou, W., Chen, B., & Qian, G., (2016). Metal–organic frameworks as platforms for functional materials. *Accounts of Chemical Research, 49*(3), 483–493.
18. Deligeorgiev, T., Vasilev, A., Kaloyanova, S., & Vaquero, J. J., (2010). Styryl dyes–synthesis and applications during the last 15 years. *Coloration Technology, 126*(2), 55–80.
19. Drexhage, K. H., (1976). Fluorescence efficiency of laser dyes. *Journal of research of the National Bureau of Standards. Section A, Physics and chemistry, 80*(3), 421.
20. Dulmage, W. J., Light, W. A., Marino, S. J., Salzberg, C. D., Smith, D. L., & Staudenmayer, W. J., (1978). An aggregate organic photoconductor. I. Chemical composition, preparation, physical structure, and optical properties. *Journal of Applied Physics, 49*(11), 5543–5554.
21. Emmelius, M., Pawlowski, G., & Vollmann, H. W., (1989). Materials for optical data storage. *Angewandte Chemie International Edition in English, 28*(11), 1445–1471.
22. Fabian, J., Nakazumi, H., & Matsuoka, M., (1992). Near-infrared absorbing dyes. *Chemical Reviews, 92*(6), 1197–1226.
23. Faustino, H., El-Shishtawy, R. M., Reis, L. V., Santos, P. F., & Almeida, P., (2008). 2-Nitrosobenzothiazoles: useful synthons for new azobenzothiazole dyes. *Tetrahedron Letters, 49*(48), 6907–6909.

24. Fletcher, A. N., & Bliss, D. E., (1978). Laser dye stability. Part 5: Effect of chemical substituents of bicyclic dyes upon photodegradation parameters. *Applied Physics, 16*(1), 289–295.
25. Freeman, H. S., & Peters, A. T. (Eds.)., (2000). *Colorants for Non-Textile Applications* (Vol. 1, pp. 5–8). Elsevier.
26. Gordon, P. F., (1990). Nontextile applications of dyes. *The Chemistry and Application of Dyes, 1*, 381–406.
27. Gourevich, I., Pham, H., Jonkman, J. E., & Kumacheva, E., (2004). Multidye nanostructured material for optical data storage and security labeling. *Chemistry of Materials, 16*(8), 1472–1479.
28. Heilemann, M., Margeat, E., Kasper, R., Sauer, M., & Tinnefeld, P., (2005). Carbocyanine dyes as efficient reversible single-molecule optical switch. *Journal of the American Chemical Society, 127*(11), 3801–3806.
29. Hu, L., Duan, Y., Xu, Z., Yuan, J., Dong, Y., & Han, T., (2016). Stimuli-responsive fluorophores with aggregation-induced emission: implication for dual-channel optical data storage. *Journal of Materials Chemistry C, 4*(23), 5334–5341.
30. Hunger, K. (Ed.)., (2007). *Industrial Dyes: Chemistry, Properties, Applications* (Vol. 1, pp. 3–7). John Wiley & Sons.
31. Jaiswal, A., Kapoor, A. K., & Kapoor, S., (2011). Health conditions of the textiles workers and their association with breathing condition. *The Asian Man-An International Journal, 5*(1), 28–33.
32. Jang, S. H., Lee, G., Lee, S. Y., Kim, S. H., Lee, W., Jung, J. W., & Choi, J., (2022). Synthesis and characterisation of triphenylmethine dyes for color conversion layer of the virtual and augmented reality display. *Dyes and Pigments, 204*(1), 110–419.
33. Kalinowski, J., Stampor, W., & Di Marco, P., (1996). Electromodulation of luminescence in organic photoconductors. *Journal of the Electrochemical Society, 143*(1), 314–315.
34. Karl, N., Marktanner, J., Stehle, R., & Warta, W., (1991). High-field saturation of charge carrier drift velocities in ultrapurified organic photoconductors. *Synthetic Metals, 42*(3), 2473–2481.
35. Khan, A. A., & Husain, Q., (2007). Decolorization and removal of textile and non-textile dyes from polluted wastewater and dyeing effluent by using potato (Solanum tuberosum) soluble and immobilized polyphenol oxidase. *Bioresource Technology, 98*(5), 1012–1019.

36. Lacroix, P. G., Malfant, I., & Lepetit, C., (2016). Second-order nonlinear optics in coordination chemistry: An open door towards multi-functional materials and molecular switches. *Coordination Chemistry Reviews, 308*(1), 381–394.
37. Lardon, M., Lell-döller, E., & Weigl, J. W., (1967). Charge transfer sensitization of some organic photoconductors based on carbazole. *Molecular Crystals and Liquid Crystals, 2*(3), 241–266.
38. Lin, H., Wei, W. B., Chen, H., Wu, X. T., & Zhu, Q. L., (2020). Rational design of infrared nonlinear optical chalcogenides by chemical substitution. *Coordination Chemistry Reviews, 406*(1), 213150–21355.
39. Lott, J., Ryan, C., Valle, B., Johnson III, J. R., Schiraldi, D. A., Shan, J., & Weder, C., (2011). Two-photon 3D optical data storage via aggregate switching of excimer-forming dyes. *Advanced Materials, 23*(21), 2425–2429.
40. Luo, J., Huang, S., Cheng, Y. J., Kim, T. D., Shi, Z., Zhou, X. H., & Jen, A. K. Y., (2007). Phenyltetraene-based nonlinear optical chromophores with enhanced chemical stability and electrooptic activity. *Organic Letters, 9*(22), 4471–4474.
41. Marder, S. R., Sohn, J. E., & Stucky, G. D., (1991). *Materials for Nonlinear Optics Chemical Perspectives* (Vol. 1, pp. 5–10). American Chemical Society Washington DC.
42. Maslov, V. V., (2006). Spectral and fluorescent characteristics of laser dyes for 650–800 nm range. *Functional Materials, 1*, pp. 5–10.
43. Muhammad, S., Xu, H. L., Zhong, R. L., Su, Z. M., Al-Sehemi, A. G., & Irfan, A., (2013). Quantum chemical design of nonlinear optical materials by sp^2-hybridized carbon nanomaterials: issues and opportunities. *Journal of Materials Chemistry C, 1*(35), 5439–5449.
44. Mula, S., Hablot, D., Jagtap, K. K., Heyer, E., & Ziessel, R., (2013). Design and synthesis of sulfobetainic diketopyrrolopyrrole (DPP) laser dyes. *New Journal of Chemistry, 37*(2), 303–308.
45. Mustroph, H., Stollenwerk, M., & Bressau, V., (2006). Current developments in optical data storage with organic dyes. *Angewandte Chemie International Edition, 45*(13), 2016–2035.
46. Pai, D. M., (1983). Organic photoconductors in electrophotography. *Journal of Non-Crystalline Solids, 59*(1), 1255–1263.
47. Park, H., Kim, E. R., Kim, D. J., & Lee, H., (2002). Synthesis of metal-azo dyes and their optical and thermal properties as recording materials

for DVD-R. *Bulletin of the Chemical Society of Japan, 75*(9), 2067–2070.

48. Pavlopoulos, T. G., & Hammond, P. R., (1974). Spectroscopic studies of some laser dyes. *Journal of the American Chemical Society, 96*(21), 6568–6579.
49. Pavlopoulos, T. G., (2002). Scaling of dye lasers with improved laser dyes. *Progress in Quantum Electronics, 26*(4–5), 193–224.
50. Pishgar, M., Gharanjig, K., Yazdanshenas, M. E., Farizadeh, K., & Rashidi, A., (2022). Synthesis and characterization of novel fluorescent reactive dyes for dyeing of cotton fabrics. *The Journal of The Textile Institute, 113*(4), 595–605.
51. Popovic, Z. D., & Menzel, E. R., (1979). Electric field-induced fluorescence quenching in organic photoconductors. *The Journal of Chemical Physics, 71*(12), 5090–5096.
52. Prasad, P. N., & Reinhardt, B. A., (1990). Is there a role for organic materials chemistry in nonlinear optics and photonics? *Chemistry of Materials, 2*(6), 660–669.
53. Pudavar, H. E., Joshi, M. P., Prasad, P. N., & Reinhardt, B. A., (1999). High-density three-dimensional optical data storage in a stacked compact disk format with two-photon writing and single photon readout. *Applied Physics Letters, 74*(9), 1338–1340.
54. Rinke, M., Guesten, H., & Ache, H. J., (1986). Photophysical properties and laser performance of photostable UV laser dyes. Part 1. Substituted p-quaterphenyls. *The Journal of Physical Chemistry, 90*(12), 2661–2665.
55. Sabnis, R. W., (2017). Manufacture of dye intermediates, dyes, and their industrial applications. *Handbook of Industrial Chemistry and Biotechnology, 1*, 581–676.
56. Sayyed, R. Z., Bhamare, H. M., Sapna, Marraiki, N., Elgorban, A. M., Syed, A., & Dailin, D. J., (2020). Tree bark scrape fungus: a potential source of laccase for application in bioremediation of non-textile dyes. *PLOS One, 15*(6), e0229968.
57. Schäfer, F. P., (2005). Dye lasers and laser dyes in physical chemistry. *Dye Lasers: 25 Years, 1*, pp. 19–36.
58. Schaffert, R. M., (1971). A new high-sensitivity organic photoconductor for electrophotography. *IBM Journal of Research and Development, 15*(1), 75–89.

59. Schlebusch, C., Kessler, B., Cramm, S., & Eberhardt, W., (1996). Organic photoconductors and C60. *Synthetic Metals, 77*(1–3), 151–154.

60. Seto, J. E., Tamura, S. I., Asai, N., Kishii, N., Kijima, Y., & Matsuzawa, N., (1996). Macrocyclic functional dyes: Applications to optical disk media, photochemical hole burning and non-linear optics. *Pure and Applied Chemistry, 68*(7), 1429–1434.

61. Son, Y. A., Park, Y. M., Park, S. Y., Shin, C. J., & Kim, S. H., (2007). Exhaustion studies of spiroxazine dye having reactive anchor on polyamide fibers and its photochromic properties. *Dyes and Pigments, 73*(1), 76–80.

62. Spiridon, M. C., Iliopoulos, K., Jerca, F. A., Jerca, V. V., Vuluga, D. M., Vasilescu, D. S., & Sahraoui, B., (2015). Novel pendant azobenzene/polymer systems for second harmonic generation and optical data storage. *Dyes and Pigments, 114*(1), 24–32.

63. Talebian, A., Habibi, S., & Neshat, P., (2021). Green dyeing of weld on Corona discharge treated wool fabric. *The Journal of the Textile Institute, 112*(1), 144–151.

64. Towns, A. D., (1999). Developments in azo disperse dyes derived from heterocyclic diazo components. *Dyes and Pigments, 42*(1), 3–28.

65. Vanstan, I., (1961). Ancient Peruvian textile arts: Patchwork and tie-dye from Pachacamac. *Expedition, 3*(4), pp. 33–34.

66. Waring, D. R., & Hallas, G. (Eds.)., (2013). *The Chemistry and Application of Dyes* (Vol. 1, pp. 6–10). Springer Science & Business Media.

67. Warta, W., Stehle, R., & Karl, N., (1985). Ultrapure, high mobility organic photoconductors. *Applied Physics A, 36*(1), 163–170.

68. Weiss, D. S., & Abkowitz, M., (2010). Advances in organic photoconductor technology. *Chemical Reviews, 110*(1), 479–526.

69. Weiss, D. S., & Abkowitz, M., (2017). Organic photoconductors. *Springer Handbook of Electronic and Photonic Materials, 2*(1), 1–1.

70. Weiss, D., & Abkowitz, M., (2007). Organic photoconductors. *Springer Handbook of Electronic and Photonic Materials, 1*, pp. 953.

71. Winters, B. H., Mandelberg, H. I., & Mohr, W. B., (1974). Photochemical products in coumarin laser dyes. *Applied Physics Letters, 25*(12), 723–725.

72. Würthner, F., & Yao, S., (2000). Dipolar dye aggregates: a problem for nonlinear optics, but a chance for supramolecular chemistry. *Angewandte Chemie, 112*(11), 2054–2057.

73. Yanez, C. O., Andrade, C. D., Yao, S., Luchita, G., Bondar, M. V., & Belfield, K. D., (2009). Photosensitive polymeric materials for two-photon 3D WORM optical data storage systems. *ACS Applied Materials & Interfaces, 1*(10), 2219–2229.

74. Yarmchuk, E. J., & Keefe, G. E., (1989). High-resolution surface charge measurements on an organic photoconductor. *Journal of Applied Physics, 66*(11), 5435–5439.

75. Yu, H., Young, J., Wu, H., Zhang, W., Rondinelli, J. M., & Halasyamani, P. S., (2016). Electronic, crystal chemistry, and nonlinear optical property relationships in the dugganite A3B3CD2O14 family. *Journal of the American Chemical Society, 138*(14), 4984–4989.

76. Zhang, Q., Chung, I., Jang, J. I., Ketterson, J. B., & Kanatzidis, M. G., (2009). Chalcogenide chemistry in ionic liquids: Nonlinear optical wave-mixing properties of the double-cubane compound [Sb7S8Br2] (AlCl4) 3. *Journal of the American Chemical Society, 131*(29), 9896–9897.

77. Zhao, C. F., Gvishi, R., Narang, U., Ruland, G., & Prasad, P. N., (1996). Structures, spectra, and lasing properties of new (aminostyryl) pyridinium laser dyes. *The Journal of Physical Chemistry, 100*(11), 4526–4532.

78. Zhu, L., Lin, J., Pei, L., Luo, Y., Li, D., & Huang, Z., (2022). Recent advances in environmentally friendly and green degumming processes of silk for textile and non-textile applications. *Polymers, 14*(4), 658–659.

79. Zhu, M. Q., Zhu, L., Han, J. J., Wu, W., Hurst, J. K., & Li, A. D., (2006). Spiropyran-based photochromic polymer nanoparticles with optically switchable luminescence. *Journal of the American Chemical Society, 128*(13), 4303–4309.

80. Zollinger, H., (2003). *Color Chemistry: Syntheses, Properties, and Applications of Organic Dyes and Pigments* (Vol. 1, pp. 6–11). John Wiley & Sons.

INDEX

Symbols

2-chloro-4-nitroaniline 112
π electronic transitions 149

A

Absorption process 19
Acid dyes 8, 14
Acridine 87
Acridine Orange 134
Acriflavine 242, 243
Acrylic fibers 106, 107, 108, 109, 110, 115, 119, 120, 121, 123, 124, 125, 127
Affinity 8
Alizarin 3, 4, 9, 10, 11
Alkali fusion 9, 10, 13, 14
All-trans chromophore 150
Alum 3
Amalgamation 20
American Cyanamid 133
Amines 174
Ammonia 4, 16
Amoxicillin 236
Analytical research 179
Aniline 5, 6, 7, 10
Aniline Purple 6, 7, 9
Anthranilic acid 13
Anthraquinone 9, 10, 14, 15, 19, 42, 44
Anthraquinone dyes 30, 40, 42, 43, 44, 50, 58, 59
Anthraquinone red 30
Anthraquinones 17
Aromatic amines 34, 53
Arylamines 14
Astrazon Red FG 124
Auramine 120
Auramine dyes 85
Authentic equilibria 20

Azacarbocyanines 125
Aza cyanine 108
Azines 122
Azoacetoacetanilides 37
Azo dye Bismarck Brown 8
Azoic dyeing 2, 11
Azonaphthol 36
Azophenol dyes 36
Azopyrazolone 41
Azopyridones 37
Azopyridone yellow dyes 41
Azosol Fast Brilliant Red BA 64
Azo yellow 30

B

Bacterial infection 236
Badische Anilin 10
Basic Blue 4 123
Bayer dye Prontosil 245
Becquerel effect 174, 197
Bed linens 237
Benzene 5, 6, 9, 19
Benzodifuranone (BDF) dyes 44
Benzo Orange S 64
Berberine 62
Binderwald 35
Biological stains 238, 240
Biopolymers 238
Birefringent dichroic filters (BDFs) 44
Blending proportion 208
Blending textile fibers 206
Blotchy dyeing 212
Blue alcohol-soluble dyes 74
Blue disperse styryl dyes 31
Blue reactive triphen-dioxazine dyes 31
British Standard (BS) 177
Bromamine acid 14
Bromination 9

C

Carbinol 63, 70, 71, 72, 73, 75, 81, 83
Carbocyanines 125
Carbocyclic azo dyes 39
Carbostyril 271
Cathode ray tube (CRT) 263
Cationic chromophores 62
Cationic dyes 106, 107, 108, 109, 110, 113, 115, 119, 124, 127, 132, 134, 135, 137, 138, 139, 140, 141, 142, 143, 144
Cellular entity 238
Cellulosic fabrics 3
Characteristic Fading (CF) Curves 184
Characteristic Fastness-Grade (CFG) Curves 182
Charge generation layer (CGL) 279
Chemical-based finishing 212
Chemical composition 30, 55
Chemical transformation 153
Chemistry 2, 6, 7, 8, 9, 13, 15, 17, 18, 19, 20, 22, 23, 25, 27, 28
Chlorophyll 147, 155, 165, 200
Chloropyrimidines 17
Chloroquinoxalines 17
Chromogen 44, 46, 50, 53
Chromophore 14, 15, 16
Chromophoric systems 17
Chromoxane Brown 5R (LIII) 78
Chronological manner 2
Chrysoidine 11, 134
CI 10305 53
CI Acid Yellow 1 53
Ciba 123, 134

CI Direct Blue 86 51
CI Sulfur Black 1 53
Class I dyes 162, 163
Clean methods 222
Clofazimine 244
Collodion 180
Color and constitution theory 2, 18
Colorants 2, 3, 14
Color Index 31, 48
Color Index classification 31
Coloring agent 3
Commercial dyes 106, 124, 125, 129
Commercialization 4, 9, 17
Compact audio disk (CD) 272
Comparison Method 185
Concoction 71, 92
Concomitant anion 107
Concurrent advancement 2
Condensation 118, 123, 125, 130, 134
Congo Red 12
Conjugated polyene systems 20
Conjugation 19, 20
Contemporary standards 3
Copper phthalocyanine 16
Correlation 33, 39
Cotton 4, 11, 14, 17, 25, 30, 50, 55
Cotton dyeing 4
Coupling 33, 34, 35, 41
Covalent bonds 214
CO π-bond 171
Crystal Violet 66, 68, 71, 74, 79, 81, 82, 83, 93, 94
Cuprous cyanide 16
Cyanation 48, 49
Cyanine-type chromogens 30
Cyclization 13
Cyclohexadiene residue 19
Cyclohexatriene 9

D

Decomposition 63, 70
de Dierbach 16
Delocalized or distributed dyes 119
Dendridge 16
Dependability 3
Deuterium 153
Diabetes 239
Dialkylaminoalkylamine 115, 117
Diazahemicyanine 129, 131
Diazahemicyanine dyes 47
Diazahemicyanines 127
Diazonium ions 8, 10, 12
Diazonium salts 15
Diazotization 110
Diazotization process 11
Dibromobenzene 16
Dihydro isoalloxazine moiety 152
Dimerization 160
Diminution 153
Disazo dyes 38
Distillation 5
Donor-acceptor chromogens 30
Du Pont 116
Durability 42, 50
Duranols 15
Dyebath 15
Dyebath substances 206
Dye chemistry 2, 3, 4, 5, 8, 9, 12, 13, 14, 15, 18, 24, 25, 27
Dye chromophores 2
Dye fixation 214, 219, 220
Dyehouse scientist 30
Dyeing 2, 3, 11, 12, 13, 15, 17, 22, 25
Dye isolation procedures 6
Dyestuffs 262, 276, 280, 282, 285

Dye toxicity 239
Dye-tracing methodologies 240

E

Economic constraints 3
Economic viability 3
Economy 206
eErythrosine 158
Ehrlich 236, 238, 242, 243
Electric field 264, 265, 283, 284
Electric vector 265, 282
Electromagnetic radiation 19
Electron 18, 19, 21
Electron donation 154
Electron-pair bond 18
Electron-withdrawing (-M) groups 19
Electrooptics 18
Elucidation 51
Eosin 147, 158, 159, 161, 162
E radical 32
European nation 9

F

Fade-Ohmmeter hour 177
Fade-Ometer 177
Fading-rate arcs 179
Faraday 5
Ferrocenyl group 76
Fiber 3, 15, 17
Fiber blend constituents 208
Fiber blending 207, 222
First-order kinetic law 149
First synthetic colorant 4
Flash photolysis method 147
Flavin mononucleotide 153
Flax 206, 222
Fluorene analogs 82
Fluorescein 85, 147, 159, 202
Fluorescence 153, 165, 169, 175, 195, 200, 202
Fluorescent methine dyes (XXXVI) 118
Fluoropyrimidines 17
Formazan dyes 54
Free radicals 149, 155, 160
French innovations 9
Fuchsin (XL) 119
Fungi 243, 247, 248, 258

G

Genacryl Red 4B 124
Gentian violet 243
Germany 9, 10
Germ Theory 238
Giles 177, 182, 183, 185, 188, 189, 192, 193, 194, 196
Glycoside 3
Gram 236, 238, 242
Gram staining technique 236, 238
Greenford Green plant 6
Griess 8, 10, 11
Guttmann 243

H

Helvetia Blue 63
Hemicyanines 47
Hemophilus influenzae 236
Heterocyclic azos 31
Hofmann 5, 7, 8
Hückel molecular orbital (HMO) theory 21
Humidity 187
Hydrogen abstraction 168, 169
Hydrogen bonds 214
Hydrolytic stability 119, 124
Hydroxyazo dyes 36
Hydroxy groups 9

Hypocrella bambusae 249
Hypsochromic spectral shift 90

I

Illumination 92
Indanthrone 14
Indian dye 4
Indigo 3, 4, 5, 11, 12, 13, 14, 15
Indigofera tinctoria 3
Indocyanine Green 252
Indoles 41
Indoxyl 3, 4, 13
Infection control 238
Insulin 239
Integrated circuits 262
Intermolecular effects 167
Iodopsin 150
Ion exchange 214
Ionic bonding 214
Ionization 34
Irradiation 149, 152, 156, 171, 177, 193, 198, 204
Irreversible deterioration 13
Isoalloxazine nucleus 152
Isolation 6, 9
Isopropanol 73

J

Jute 206

K

Katanol 67
Katanol O 67
Kauffmann's theory 20
Kekule's theory 18
Ketones 147, 153, 168, 169, 171
Knitting 207, 209, 211, 212
Koch 236, 238, 259
Koizumi 159, 160

L

Lactoflavin 152
Lasers 18, 23
Laser therapy 222
Leaf fermentation 3
Leather 65, 96
Leprosy 244
Leuco bases 63
Leucothionine 166
Light Amplification by Stimulated Emission of Radiation 268
Lightfast quinophthalone yellows (XXXVII) 118
Liquid crystal material 263, 266
Liquid crystals 262, 263, 266
Lonamine dyes 15
Luco form 53
Lumi chrome (III) 153
Luminosity 39, 40, 42, 44, 49, 50

M

Madder 3, 9, 10
Magenta 7, 8, 9
Malachite Green 62, 68, 69, 70, 76, 77, 81, 82, 83, 85, 86, 87, 89, 91, 93, 94
Malachite Green (XXXIX) 119
Manifold record systems 94
Mannich base 112
Mannich reaction 112
Marckwald 155
Martius 8, 10
Mauveine 6
Merocyanine colorants 180
Mesomerism 20
Metal-free phthalocyanine 16
Metallization 39, 44
Metallochromic indicators 85
Methemoglobinemia 243

Methicillin-resistant Staphylococcus aureus (MRSA) 243
Methylene blue (CX) 89
Methylene blue (phenothiazinium) 240
Methyl Orange 39
Methyl Red 39
Michler's ketone 74, 78, 81, 83, 120
Mineral sources 3
Molecular orbital (MO) theory 21
Molecular structure 7, 18
Monastral Fast Blue BS 16
Monoazo dyes 32
Mordant 3, 4, 11, 15
Mordant Blue 1 (CI 43830) 178
Mordant dye 4
M-phenylenediamine 8
Mutagenicity 253, 254

N

Naphthalene 40, 59
Naphthol 11
Naphtholate 35
Naphthostyril Dyes 132
Natanson 7
Natural dyes 3, 4, 22, 25, 27
Natural material 13
N-cyclohexyl-1-naphthylamine 74
Negative first discoloration 181
Nicholson 7, 8
Nile Blue 65
Nitration 6
Nitric acid 4
Nitrocellulose lacquers 74, 88
Nitrodiphenylamine dyes (XXXVIII) 118
Nitronium ion (NOT) 34
Nitrosation 123
Nitrous acid 8
N-methylsulfonic acid 15
N-nitroso compound 33
Nomenclature 30, 46
Nonlinear optics 282
Non-textile domain 14
Novel dye products 13
Nullification 6
Nylon 30, 40, 55

O

O-carboxyphenylglycine 13
Oleophilic anions 86
O-nitrocinnamic acid 12
Onium 68
Organic compounds 7
Organic photoconductor (OPC) 278
Organic photovoltaic cells (OPCs) 279
Orlon 17
Orlon Type 42 polyacrylonitrile fiber 106
Oxazine 87, 89
Oxidation 3, 5, 6, 7, 11, 15, 23
Oxidizing agents 158, 166
Oxygen-deprived photobleaching 148

P

Pararosaniline (IV) 63
Partial valence hypothesis 20
Pendant 106, 109, 110, 115, 131, 134, 139, 140, 141
Pendant ammonium dyes 109
Penicillin 236, 237, 238
Perkin's Aniline Purple 5
Perkin's French patent 6
Permeability 189
Permutations 32, 33

Peter Greiss 33
Pfeiffer-Schetty 39
Phenanthridines 244, 245
Phenazines 243
Phenol 4, 11
Phenolic disintegration product 149
Phenolphthalein 85
pH level 84, 92, 93
Phosphomolybdic acid (PMA) 91
Phosphotungstic acid (PTA) 91
Phosphotungstomolybdic acid (PTMA) 91
Photocatalytic processes 166
Photochemical changes 146
Photochemical cis-trans isomeric formation 150
Photochemistry 146, 147, 148, 156, 167, 194, 195, 197, 198, 201
Photodecomposition 149, 154, 157, 164, 165
Photodynamic Therapy (PDT) 249
Photofading 148, 149, 153, 192, 199
Photoflash 147
Photographic sensitizers 14
Photooxidative bleaching 157, 158, 160, 161
Photoprocessing 149, 152
Photoreactions 146, 147, 148, 149, 167
Photoreduction 147, 157, 158, 162, 163, 165, 166, 172
Photosens 250
Photosensitization 248, 258
Photosensitized oxidization 162, 168, 169
Photosensitizers 248
Photostability 124, 128
Phototropic color modifications 154
Phthalic anhydride 15, 16
Phthalocyanine blue 30
Phthalocyanines 46, 50, 51, 57, 60, 249, 250
Picric acid 4
Pigment diffusion 214
Pigmented substances 5
Pigments 237, 241, 249
Plantations 13
Plant photoperiodism 150
Pollutants 212
Polyacrylonitrile 39, 40, 46, 47, 48
Polyacrylonitrile polymers 66
Polyamide 17
Polyazo dyes 32
Polycyclic aromatic carbonyl dyes 45
Polyene chromogens 30
Polyester 30, 40, 41, 44, 48, 49, 52, 54
Polymethine 31, 47, 48, 58, 108
Polymethine dyes 46
Potassium ferricyanide 158
Prelumirhodopsin 150
Prereduced vat dyes 2
Printed circuit boards 262
Printing 211, 212, 231
Proflavine 242, 243, 255
Protein fibers 8
Prout 4
Pseudo-first-order principle 180
P-toluenesulfonic acid 73
Pyrazolones 41, 111, 115
Pyridones 41
Pyronine B 87

Q

Quantum mechanics 20, 21
Quantum yield 149, 155, 162, 168
Quaternised dialkyl aniline couplers

113
Quaternization 115, 116, 117, 127, 129, 130, 131, 133
Quinones 14
Quinonoid structures 19
Quinonoid system 45

R

Reaction 4, 7, 8, 10, 12, 14, 16
Red disperse benzodifuranone (BDF) dyes 31
Red disperse tricyanovinyl styryl dyes 31
Red dyes 3
Reiterating 31
Re-oxidation 152
Resistone 67
Resonance transmission 20
Rhodamine 3B 124
Rhodamine 6G 64, 67, 69, 91
Rhodamine B 65, 67, 68, 69, 88
Rhodanile Blue 64, 65
Riboflavin 147, 148, 152, 153, 154, 162, 202
Rosaniline 7
Roussin 11
Royal College of Chemistry 5
Rubia tinctorum 3

S

Salicylic acid 12
Scottish Dyes Corporation 16
Scottish Dyes Ltd 50
Semiconductors 262
Semiquinone 165, 168, 174
Sevron 124
Shading 212
Silk 207, 210, 216, 221, 222, 225, 226
Singlet 149, 153, 165, 166, 167, 169, 171, 187, 193, 196
Sisal 206
Society of Chemical Industry 15
Soda-Fabrik 10
Solubility 8, 15, 62, 63, 64, 85, 86, 88, 94
Spealman 157
Spun yarns 207
SRA Colors 15
Standard fading period (SFP) 177
Structural tautomerism 20
Styryl dyes 48
Substrates 2
Sulfonated N-phenyl derivatives 8
Sulfonation 8, 10
Sulfonphthalein 84, 85
Sulfonphthalein (LXXIX) 84
Sulfur dyes 53, 56
Supplementary heterocycles 271
Synthetic adaptability 35, 43
Synthetic dyes 2, 3, 7, 18, 19, 22, 23, 24, 25, 27

T

Tanninol BM 67
Tautomeric ratios 37
Ternary textile 208
Terylene 17
Tetrazo component 12
The Atom and the Molecule 18
Thermodynamics instability 154
Thiazine 87, 89, 90
Three-dimensional intermediate 148
Thymol Blue 85
Toluidine 7
Tracer Dyes 240

Trial and error 8
Triarylmethanes 120
Triphenylated Pararosaniline (VI) 63
Triphenylmethane dye 7, 20
Turkey Red method 3
Tyrian Purple 46

U

Ultraviolet (UV) light 75
Uncontrolled proliferation 239
Utilization 2, 3, 4, 8, 9, 11, 12, 15, 20

V

Valence bond theory 20, 21
Valence tautomerism 20
Valencies 20
Valuable insights 179
Vanadium pentoxide 15
van der Waals forces 214
Vastness 62
Vat dye spectrum 15
Vegetable 3
Vegetable-based natural blue colorant 3
Verguin 7
Victoria Blue B 74
Vinylsulfones 17
Visible spectrum 44

W

Wegmann 121, 122
William Henry Perkin 5
Woad plant 3
Woulfe 4

X

Xanfhene 87
Xerographic process 278, 280
X-ray data 91
X-ray irradiation 89

Y

Y-methoxy derivative 131

Z

Zincke 35
Zine 87